AF342041

Porous Polymers
Design, Synthesis and Applications

Monographs in Supramolecular Chemistry

Series Editors:
Professor Philip Gale, *University of Southampton, UK*
Professor Jonathan Steed, *Durham University, UK*

Titles in this Series:

How to obtain future titles on publication:
A standing order plan is available for this series. A standing order will bring
delivery of each new volume immediately on publication.

For further information please contact:
Book Sales Department, Royal Society of Chemistry, Thomas Graham House,
Science Park, Milton Road, Cambridge, CB4 0WF, UK
Telephone: +44 (0)1223 420066, Fax: +44 (0)1223 420247
Email: booksales@rsc.org
Visit our website at http://www.rsc.org/Shop/Books/

Porous Polymers
Design, Synthesis and Applications

Shilun Qiu
Jilin University, Changchun, China
Email: sqiu@jlu.edu.cn

Teng Ben
Jilin University, Changchun, China
Email: tben@jlu.edu.cn

THE QUEEN'S AWARDS
FOR ENTERPRISE:
INTERNATIONAL TRADE
2013

Monographs in Supramolecular Chemistry No. 17

Print ISBN: 978-1-84973-932-0
PDF eISBN: 978-1-78262-226-0
ISSN: 1368-8642

A catalogue record for this book is available from the British Library

Published by The Royal Society of Chemistry,
Thomas Graham House, Science Park, Milton Road,
Cambridge CB4 0WF, UK

Registered Charity Number 207890

Visit our website at www.rsc.org/books

Printed in the United Kingdom by CPI Group (UK) Ltd, Croydon, CR0 4YY, UK

Preface

This book is intended as an overview of the recent developments in the synthesis, characterization and application of organic-framework-based nanoporous polymers, which are often considered as ideal host materials for applications including selective separation, gas storage, and carriers of catalysts. As such, we hope this book may help those who are interested in porous polymers to gain a better understanding of some key principles and challenges with regards to porous polymers.

The scope of discussions of porous polymers in this book is limited to polymers synthesized by the widely known concave-like monomers, which lead to high international molecular free volumes. The resulting free volume is interpenetrated and is fully open or accessible to make the polymer act as a host. The typical porous polymers discussed in this book are hypercrosslinked polymers (HCPs), polymers of intrinsic microporosity (PIMs), covalent organic frameworks (COFs), conjugated microporous polymers (CMPs), and porous aromatic frameworks (PAFs).

The book was initially constructed with a historical development sequence of porous polymers combined with illustrations of structure–property correlations. Each chapter provides an example of a particular element of porous polymers. Chapter 1 provides a summary of porous polymers and discusses the relationship between structure and function. In Chapter 2, the design principles of porous polymers are discussed and modification methods are introduced, while Chapter 3 introduces the synthetic routes and reactions used in polymerization. An understanding of these reactions is essential if we are to understand the origin of the ordered or amorphous structure of porous polymers. Chapter 4 describes the first porous polymers, developed in the 1990s and named hypercrosslinked polymers or "Davankov-type" resins. Chapter 5 focuses on the first soluble polymer with intrinsic microporosity that was reported in 2002. Meanwhile, Chapter 6

Monographs in Supramolecular Chemistry No. 17
Porous Polymers: Design, Synthesis and Applications
By Shilun Qiu and Teng Ben
© The Royal Society of Chemistry 2016
Published by the Royal Society of Chemistry, www.rsc.org

demonstrates the crystalline covalent organic frameworks that were first published in 2005. Chapter 7 shows a series of conjugated microporous polymers synthesized in 2007 with various attractive functions. Chapter 8 focuses on a new porous polymer named porous aromatic frameworks, which show ultra-high surface area combined with high physicochemical stability. After introducing some typical porous polymers, in Chapter 9 we describe the functionalization methods, such as pre- or post-synthetic structural modification strategies, that fine tune the properties of porous polymers. In Chapter 10 we summarize the applications of porous polymers, and lastly, Chapter 11 brings to a conclusion the structure design, synthesis methods, structure–function correlations, function modifications, and applications of porous polymers.

We would like to express our sincere appreciation and gratitude to all our colleagues who have made huge contributions toward the completion of this book. We would also like to acknowledge the great contributions of Prof. Bien Tan (Chapter 4), Prof. Dapeng Cao (Chapter 6) and Prof. Abbie Trewin (Chapter 7). We also appreciate Prof. Soumyajit Roy, Dr Cuiying Pei, and Dr Yanqiang Li for their great help in forming this book.

Shilun Qiu and Teng Ben

Contents

Monographs in Supramolecular Chemistry No. 17
Porous Polymers: Design, Synthesis and Applications
By Shilun Qiu and Teng Ben
© The Royal Society of Chemistry 2016
Published by the Royal Society of Chemistry, www.rsc.org

CHAPTER 1

Introduction

The developing field of porous polymers[1,2] has gained significant attention in recent times owing to their huge applicability towards various areas like gas adsorption and storage,[3,4] gas separation and selective permeation,[5,6] adsorption of organic pollutants,[7] catalysis,[8–10] as photoconductors,[11,12] molecular motors[13] and in clean energy storage.[14,15] Such huge applicability is due to the combined properties from the porous materials and polymers. Depending on the pore size, such polymers can be categorized into three classes, *viz.*, macroporous (pore size > 50 nm), mesoporous (pore size = 2–50 nm) and microporous (pore size < 2 nm) polymers.[16] Porous materials having tailor-made functionalities and built from simple molecular synthons are now a major aspect of exploration in materials science. Such materials, especially the microporous polymers, by virtue of the nature of their pore size, provide a large scope of research in order to utilize the inherent huge spaces inside and thus the extraordinarily high surface area. Several reviews have been published exploring the design of synthetic strategies, and properties and applications of different porous polymeric networks.[1,2,17,18] However, this expanding field has been further enhanced with new types of porous materials with improved properties. One such kind of material, porous aromatic frameworks (PAFs), is an important milestone in the field of porous networks, producing exceptionally high surface area with great stability. No book has chronicled the detailed design procedure, and functions and applications of this new class of porous material. There is also a need to correlate the properties of various porous polymers with their basic structures and thus compare and contrast the different classes of porous networks with respect to the framework–function relationship.

In this book we have taken a step forward to address the synthesis design of porous polymers, pre- or post-synthetic modification and functionalization to

Monographs in Supramolecular Chemistry No. 17
Porous Polymers: Design, Synthesis and Applications
By Shilun Qiu and Teng Ben
© The Royal Society of Chemistry 2016
Published by the Royal Society of Chemistry, www.rsc.org

obtain pre-designed structures, connection of functions with the basic framework, the characteristics of individual structural components in the obtained network and their interactions that add up to the sum total properties of the entire framework, and the useful applications of different classes of porous polymers, namely hypercrosslinked polymers (HCPs),[19] polymers of intrinsic microporosity (PIMs),[20] covalent organic frameworks (COFs),[21] conjugated microporous polymers (CMPs)[22] and PAFs.[23] We shall emphasize new and less explored material, PAFs especially to compare them with other classes of porous materials in terms of surface area, pore size, free volume and associated applications such as gas storage, selectivity, gas permeation and separation.

Surface area, pore size and pore geometry are the most important aspects of porous polymers on which depends their applicability. The surface area of such a network is principally governed by pore size and pore surface. The molecular dimension of monomeric synthons and the introduction of functional moieties tune the physical structure of the pores and the pore surface functionalities.[24–28] The strategy of synthesis to have tailor-made functionalities is to build from simple molecular synthons in the framework. The desired constitution of the framework is dependent on the type of reactions, the reaction conditions and the monomers employed. The reactions employed for the synthesis of the networks range from various cross-coupling reactions, like Sonogashira–Hagihara,[29] Yamamoto–Ullmann[30] and Suzuki–Miyaura[31] for CMPs,[41] to Yamamoto-type Ullmann reactions for PAFs[32] and Buchwald reactions for HCP syntheses;[33] homocoupling reactions have been employed for the synthesis of CMPs, while self-condensation reactions have been employed for CMPs and COFs,[2,34] and non-reversible condensation reactions have been used to synthesize PIMs. In addition, cyclotrimerization reactions are used for CMPs, and cross-linking and copolymerization reactions are required for HCP synthesis.[2] Solvothermal (COF), ionothermal (COF, and PAF), electrochemical (CMP) and microwave rapid method (COF) conditions have also been stategized and employed to improve the properties, such as porosity, free volume and surface area, of these frameworks.[2] Obtaining rigidity in the porous framework is an important aspect in designing these networks to retain stability with a huge applied area. Aromatic monomers linked together directly or *via* other groups, like alkenes or alkynes, provide such rigidity and are thus utilized mostly during synthesis.[35,36] Organic porous networks have several advantages of their own, *e.g.*, they can be synthesized from light elements, like C, N, O, H *etc.*, they can be modified to achieve a huge variety employing the enormous range of synthetic procedures available and they are highly stable.

Theoretical studies, like first-principles calculations,[37] grand canonical ensemble Monte Carlo (GCMC) simulations,[37,38] second order Møller–Plesset perturbation theory (MP2) calculations[39] and density functional theory (DFT) calculations,[40] have been utilized to investigate optimal structures and their properties. Combined experimental and theoretical data provide a window to the plan of design of these network structures and lead to a new direction to investigate porous networks.

Pre-designed framework modification of porous polymers and the insertion of functionalization can be facilitated by pre- or post-synthetic manipulation. This can be utilized to serve two purposes: firstly, to impose new functions onto these polymers, *e.g.*, PIMs consisting of phthalocyanine, porphyrin or hexaazatrinaphthylene show high catalytic activity on insertion of certain transition metal ions into their network;[45] secondly, to improve the existing properties, for instance, CMP networks can be tuned to vary the surface area, pore volume and pore size by starting with 1,4- or 1,3,5-substituted monomers of different strut lengths.[46] Pore dimension and surface area can be controlled in a quantized fashion by controlling the strut length. Pore size varies inversely with side-chain length. An increase in the length of the side chain decreases the pore dimensions.

Crystalline networks have uniform pore sizes, whereas amorphous networks have a wider pore size range. Crystalline networks of porous polymers are formed using reversible bond chemistry. In COFs, controlled functionalization of their walls with organic groups is used to design networks with desired higher surface areas.[47] The goal of logical tuning of pre-designed pore structures and inclusion of customized functionalities with sufficient framework stability has been reached with the discovery of PAFs, especially PAF-1. It possesses an ultra-high surface area, which is very unlikely for amorphous polymers. These are the first of their kind to not require long range order for their outstanding gas storage properties owing to their ultra-high surface area (Figure 1.1).

Properties and functions of porous polymers are determined mostly by their framework structures. The properties of the individual constituents of the porous frameworks, *viz.*, monomers, linkers, coordinating ligands, functional groups and their interactions, are summed to give the total feature of these networks. Thus, to build the targeted network, chemical syntheses have been planned in a manner to facilitate the coming together of different components. Other reaction parameters, like concentration, temperature and nature of solvents, are also kept in mind during synthesis, on which the structure of the porous polymers also depends. It has been found that increasing temperature may lead to enhance surface area for many porous polymers, whereas increased strut lengths reduce surface area and increase micropore volume and interpenetration. Optical properties can also be tuned by varying the monomer ratio. In the case of CMPs it has been seen that changing the ratio of thiophene and benzene diboronic acids can tune the absorption and emission wavelengths of CMP networks.[48] The stability of COF networks comes from the pure covalent bonds, like C–C, C–O, B–O and C–B, constituting their framework. Such strong covalent bonds composed of light elements form highly porous networks with high surface areas, making COFs ideal candidates for gas storage. An azine linkage-based formation of COFs makes the resultant COFs highly crystalline, porous and chemically stable.[49] COF-108 has been reported as has the lowest density material ever reported for a crystalline solid $(0.17 \mathrm{~g\,cm^{-3}})$.[21] Reversible condensation reactions may be involved to synthesize crystallize two-dimensional (2D) COFs.[42] Layered 2D COFs are the result of out of plane π interactions facilitating transport of charge carriers and excitons, making them ideal candidates for semiconduction. This also leads to variation in the pore size, shape and microenvironment of the COFs.

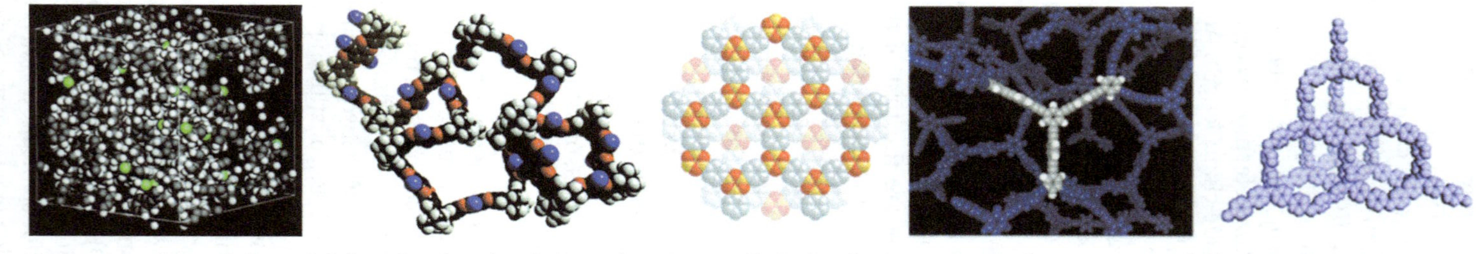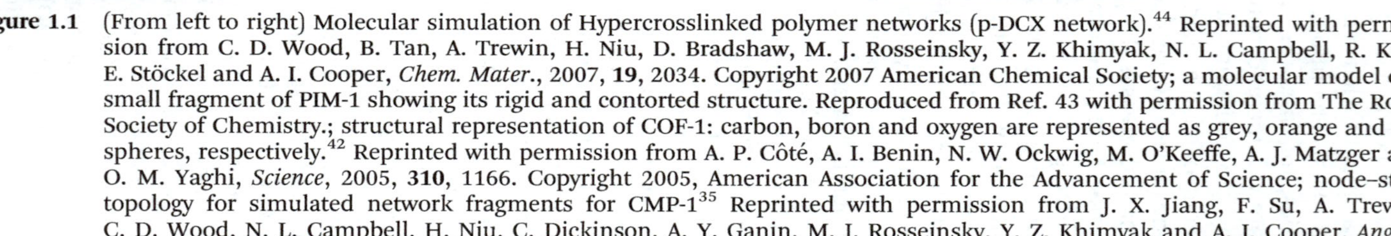

Figure 1.1 (From left to right) Molecular simulation of Hypercrosslinked polymer networks (p-DCX network).[44] Reprinted with permission from C. D. Wood, B. Tan, A. Trewin, H. Niu, D. Bradshaw, M. J. Rosseinsky, Y. Z. Khimyak, N. L. Campbell, R. Kirk, E. Stöckel and A. I. Cooper, *Chem. Mater.*, 2007, **19**, 2034. Copyright 2007 American Chemical Society; a molecular model of a small fragment of PIM-1 showing its rigid and contorted structure. Reproduced from Ref. 43 with permission from The Royal Society of Chemistry.; structural representation of COF-1: carbon, boron and oxygen are represented as grey, orange and red spheres, respectively.[42] Reprinted with permission from A. P. Côté, A. I. Benin, N. W. Ockwig, M. O'Keeffe, A. J. Matzger and O. M. Yaghi, *Science*, 2005, **310**, 1166. Copyright 2005, American Association for the Advancement of Science; node–strut topology for simulated network fragments for CMP-1[35] Reprinted with permission from J. X. Jiang, F. Su, A. Trewin, C. D. Wood, N. L. Campbell, H. Niu, C. Dickinson, A. Y. Ganin, M. J. Rosseinsky, Y. Z. Khimyak and A. I. Cooper, *Angew. Chem., Int. Ed.*, 2007, **46**, 8574. Copyright 2007 WILEY-VCH Verlag GmbH & Co. KGaA, Weinheim; and a structural model of PAF-1. Reproduced from Ref. 23 with permission from The Royal Society of Chemistry.

A ladder-like structure with sites of contortion in PIMs enhances their solubility towards a variety of solvents. Solubility and hindered packing provide high free volume and permeability. Shorter or branched alkyl chains in PIMs provide greater microporosity, whereas longer chains provide lower microporosity.

Permanent microporosity in HCPs is a result of extensive cross-linking preventing the polymer chains from collapsing into a dense, non-porous state. High cross-linking provides high thermal stability. In the case of the 'Davankov-type resin', microporosity is created by converting pendant chloromethyl groups into methylene bridges and several aromatic rings, linked to their neighbors by at least three bonds.[50]

The diamond-like structural framework, consisting of pure covalent bonding with a very small pore size, makes PAFs not only good gas absorbers but also provides selectivity towards different gases. Moreover, these kinds of structures also provide exceptional structural properties and stability, making the PAF an ideal material for gas storage. By replacing C–C bonds in the diamond lattice by a set of organic linkers of variable length, width and number of aromatic rings, the properties can be improved to a great extent, especially the gas adsorption property. This is due to inclusion of an aromatic ring, which exhibits large polarizability and a good interaction with gas molecules *via* London dispersion forces. Long-sized linkers enhance the pore volume and decrease the weight of the network, whereas wide linkers introduce large internal surface area and high heat of adsorption. All of these structural interactions nullify the need for long range order for an ultra-high surface area and make PAFs exceptional gas storage materials.

Gas storage is undoubtedly the most important utility of porous frameworks and is primarily due to their porosity and high surface area structures. Generally, COFs and PAFs have better gas storage capacities for H_2, CO_2, CH_4 and NH_3 due to their remarkable high surface areas. Metal organic frameworks (MOFs) are also equally capable for the same, but since MOFs and COFs suffer from low hydrothermal stability, PAFs provide a better option for storage purposes owing to their better stability. PAFs also exhibit excellent abilities to adsorb organic chemical pollutants, such as benzene, methanol and toluene, at saturated vapor pressures and room temperature, and selectively adsorb greenhouse gases. Membrane gas permeation experiments showed PIM–polyimides to be among the most permeable of all polyimides with selectivity close to the upper bound for several important gas pairs.

In this book we have emphasized the importance of the structure–function relationship in the design of the synthetic strategy. The employed synthetic route paves the way for the formation of the targeted framework having desired properties owing to its individual components. These obtained attributes, such as high surface area, conjugation of π electrons, presence of catalytically active moieties *etc.*, can be exploited in various fields of gas sorption, optoelectronics and catalysis. Based on the same correlation between structure and properties, we will justify how PAFs are comparable or better in some cases than other microporous polymer frameworks. A comparison of different porous polymers with respect to their structural framework, pore size, surface area and major applications has been tabulated (Table 1.1).

Table 1.1 Structural framework, pore size, surface area and major applications of porous polymers.

Porous polymer	Basic structure and connectivity	Pore size (A^0)	BET surface area (m^2/g)	Principle applications
CMPs	Crystalline, conjugated polymers. π–π stacking between aromatic subunits gives rise to conductive properties due to conjugation and electron flow	10 to 50	522 to 1018	Organo-photocatalysis, gas storage, as photoconductors, synthetic metals, organic semiconductors, sensors
COFs	Frameworks composed of purely covalent bonds (C–C, C–O, C–B and B–O)	9 to 34	711 to 4230	Gas storage, heterogeneous catalysis, as photoconductors, light harvesting devices
PIMs	Ladder-like structure with sites of contortion, hindered packing inside lattice	5 to 15	450 to 1760	Membrane gas permeation, gas storage, heterogeneous catalysis
HCPs	Extensive cross-linking in the network	10 to 50	600 to 2090	Gas storage, catalytic supports, hydrocarbon and water absorption
PAFs	Diamond-like structural framework consisting of pure covalent bonding with very small pore size. C–C bonds in the diamond lattice replaced by a set of organic linkers having variable length, width and number of aromatic rings	5 to 50	891 to 6461	Gas storage, adsorption of organic chemical pollutants, selective permeation, molecular rotors

This book will be of interest to chemistry and materials science students at advanced undergraduate and graduate level, as well as to researchers in academia and industry who wish to familiarize themselves with modern porous polymers.

References

1. D. Wu, F. Xu, B. Sun, R. Fu, H. He and K. Matyjaszewski, *Chem. Rev.*, 2012, **112**, 3959.
2. R. Dawson, A. I. Cooper and D. J. Adams, *Prog. Polym. Sci.*, 2012, **37**, 530.
3. T. Ben, C. Pei, D. Zhang, J. Xu, F. Deng, X. Jing and S. Qiu, *Energy Environ. Sci.*, 2011, **4**, 399.
4. T. Ben, H. Ren, S. Ma, D. Cao, J. Lan, X. Jing, W. Wang, J. Xu, F. Deng, J. M. Simmons, S. Qiu and G. Zhu, *Angew. Chem., Int. Ed.*, 2009, **48**, 9457.
5. X. Gao, X. Zou, H. Ma, S. Meng and G. Zhu, *Adv. Mater.*, 2014, **26**, 3644.
6. J. Li, J. Sculley and H. Zhou, *Chem. Rev.*, 2012, **112**, 869.
7. H. Ren, T. Ben, F. Sun, M. Guo, X. Jing, H. Ma, K. Cai, S. Qiu and G. Zhu, *J. Mater. Chem.*, 2011, **21**, 10348.
8. F. Schuth, K. S. W. Sing and J. Weitkamp, *Handbook of Porous Solids*, Wiley-VCH, New York, 2002.
9. V. Valtchev, S. Mintova and M. Tsapatsis, *Ordered Porous Solids: Recent Advances and Prospects*, Elsevier B. V., Oxford, 2009.
10. A. Gomotti, S. Bracco, P. Valsesia, M. Beretta and P. Sozzani, *Angew. Chem., Int. Ed.*, 2010, **49**, 1760.
11. X. Ding, J. Guo, X. Feng, Y. Honsho, J. Guo, S. Seki, P. Maitarad, A. Saeki, S. Nagase and D. Jiang, *Angew. Chem., Int. Ed.*, 2011, **50**, 1289.
12. S. Wan, J. Guo, J. Kim, H. Ihee and D. Jiang, *Angew. Chem., Int. Ed.*, 2009, **48**, 5439.
13. A. Gomotti, S. Bracco, P. Valsesia, M. Beretta and P. Sozzani, *Angew. Chem., Int. Ed.*, 2010, **49**, 1760.
14. M. P. Suh, H. J. Park, T. K. Prasad and D. Lim, *Chem. Rev.*, 2012, **112**, 782.
15. P. Makowski, A. Thomas, P. Kuhn and F. Goettmann, *Energy Environ. Sci.*, 2009, **2**, 480.
16. K. S. W. Sing, D. H. Everett, R. A. W. Haul, L. Moscou, R. A. Pierotti, J. Rouquerol and T. Siemieniewska, *Pure Appl. Chem.*, 1985, **57**, 603.
17. Q. Liu, Z. Tang, B. Ou, L. Liu, Z. Zhou, S. Shen and Y. Duan, *Mater. Chem. Phys.*, 2014, **144**, 213.
18. M. T. Gokmen and F. E. Du Prez, *Prog. Polym. Sci.*, 2012, **37**, 365.
19. M. P. Tsyurupa and V. A. Davankov, *React. Funct. Polym.*, 2002, **53**, 193.
20. N. B. McKeown, B. Gahnem, K. J. Msayib, P. M. Budd, C. E. Tattershall, K. Mahmood, S. Tan, D. Book, H. W. Langmi and A. Walton, *Angew. Chem., Int. Ed.*, 2006, **45**, 1804.
21. H. M. El-Kaderi, J. R. Hunt, J. L. Mendoza-Cortés, A. P. Côté, R. E. Taylor, M. O'Keeffe and O. M. Yaghi, *Science.*, 2007, **316**, 268.

22. A. I. Cooper, *Adv. Mater.*, 2009, **21**, 1291.
23. T. Ben and S. Qiu, *CrystEngComm*, 2013, **15**, 17.
24. R. B. Pernites, E. L. Foster, M. J. L. Felipe, M. Robinson and R. C. Advincula, *Adv. Mater.*, 2011, **23**, 1287.
25. X. Liu and A. Basu, *J. Am. Chem. Soc.*, 2009, **131**, 5718.
26. J. Rzayev and M. A. Hillmyer, *J. Am. Chem. Soc.*, 2005, **127**, 13373.
27. X. Yang, Y. Jin, Y. Zhu, L. Tang and C. Li, *J. Electrochem. Soc.*, 2008, **155**, J23.
28. A. Nagai, Z. Guo, X. Feng, S. Jin, X. Chen, X. Ding and D. Jiang, *Nat. Commun.*, 2011, **2**, 536.
29. Q. Liu, Z. Tang, M. Wu and Z. Zhou, *Polym. Int.*, 2014, **63**, 381.
30. Y. Luo, S. Zhang, Y. Ma, W. Wang and B. Tan, *Polym. Chem.*, 2013, **4**, 1126.
31. R. Cella, A. T. G. Orfaob and H. A. Stefani, *Tetrahedron Lett.*, 2006, **47**, 5075.
32. T. Ben, H. Ren, S. Ma, D. Cao, J. Lan, X. Jing, W. Wang, J. Xu, F. Deng, J. M. Simmons, S. Qiu and G. Zhu, *Angew. Chem., Int. Ed.*, 2009, **121**, 9621.
33. N. Marion, O. Navarro, J. Mei, E. D. Stevens, N. M. Scott and S. P. Nolan, *J. Am. Chem. Soc.*, 2006, **128**, 4101.
34. H. Zhong, H. Lai and Q. Fang, *J. Phys. Chem. C*, 2011, **115**, 2423.
35. J. X. Jiang, F. Su, A. Trewin, C. D. Wood, N. L. Campbell, H. Niu, C. Dickinson, A. Y. Ganin, M. J. Rosseinsky, Y. Z. Khimyak and A. I. Cooper, *Angew. Chem., Int. Ed.*, 2007, **46**, 8574.
36. R. Dawson, F. B. Su, H. J. Niu, C. D. Wood, J. T. A. Jones, Y. Z. Khimyak and A. I. Cooper, *Macromolecules.*, 2008, **41**, 1591.
37. J. Lan, D. Cao, W. Wang, T. Ben and G. Zhu, *J. Phys. Chem. Lett.*, 2010, **1**, 978.
38. L. Wang, Y. Sun and H. Sun, *Faraday Discuss.*, 2011, **151**, 143.
39. Y. Sun, T. Ben, L. Wang, S. Qiu and H. Sun, *J. Phys. Chem. Lett.*, 2010, **1**, 2753.
40. B. Lukose, M. Wahiduzzaman, A. Kuc and T. Heine, *J. Phys. Chem. C.*, 2012, **116**, 22878.
41. G. Cheng, T. Hasell, A. Trewin, D. J. Adams and A. I. Cooper, *Angew. Chem., Int. Ed.*, 2012, **51**, 12727.
42. A. P. Côté, A. I. Benin, N. W. Ockwig, M. O'Keeffe, A. J. Matzger and O. M. Yaghi, *Science*, 2005, **310**, 1166.
43. N. B. McKeown and Peter M. Budd, *Chem. Soc. Rev.*, 2006, **35**, 675.
44. C. D. Wood, B. Tan, A. Trewin, H. Niu, D. Bradshaw, M. J. Rosseinsky, Y. Z. Khimyak, N. L. Campbell, R. Kirk, E. Stöckel and A. I. Cooper, *Chem. Mater.*, 2007, **19**, 2034.
45. N. B. McKeown, P. M. Budd, K. J. Msayib, B. S. Ghanem, H. J. Kingston, C. E. Tattershall, S. Makhseed, K. J. Reynolds and D. Fritsch, *Chem. – Eur. J.*, 2005, **11**, 2610.
46. J. X. Jiang, F. Su, A. Trewin, C. D. Wood, H. Niu, J. T. A. Jones, Y. Z. Khimyak and A. I. Cooper, *J. Am. Chem. Soc.*, 2008, **130**, 7710.

47. Q. Fang, S. Gu, J. Zheng, Z. Zhuang, S. Qiu and Y. Yan, *Angew. Chem., Int. Ed.*, 2014, **53**, 2878.
48. J. Brandt, J. Schmidt, A. Thomas, J. D. Epping and J. Weber, *Polym. Chem.*, 2011, **2**, 1950.
49. S. Dalapati, S. Jin, J. Gao, Y. Xu, A. Nagai and D. Jiang, *J. Am. Chem. Soc.*, 2013, **135**, 17310.
50. S. N. Sidorov, L. M. Bronstein, V. A. Davankov, M. P. Tsyurupa, S. P. Solodovnikov and P. M. Valetsky, *Chem. Mater.*, 1999, **11**, 3210.

Design Principle of Porous Polymers

2.1   Design Principle of Porous Polymers

Tailorable pore size, functionalized framework, sufficient void space and high physicochemical stability often endow porous materials with excellent properties and wide applications in molecular storage, separation and catalysis. Hence, the ultimate goal is the design and synthesis of porous materials with target structures and properties. Execution of the experimental synthetic route aided by computational modeling is an effective approach compared to time-consuming and expensive repeated experimental explorations. The optimal structure with the appropriate functional group, surface area, free volume and heat of adsorption can be obtained from theoretical calculations. Furthermore, the molecule sorption mechanism can be proposed with the help of adsorption process modeling. On the other hand, experimental results can verify the accuracy of the simulations and in turn can furnish data for further modification of the theoretical model.

2.2   Theoretical Simulation

The typical computational methodologies can be divided into quantum mechanical (QM) calculations, grand canonical Monte Carlo (GCMC) simulations and molecular dynamics (MD) simulations. Among them, GCMC simulation is the connecting bridge between the computational result and the macroscopic experimental data. For accurate description of molecule uptake in the sorption process, a multiscale simulation method combined with first principles calculations and a GCMC simulation has been used increasingly in the study of host–guest interactions. Generally, the

Monographs in Supramolecular Chemistry No. 17
Porous Polymers: Design, Synthesis and Applications
By Shilun Qiu and Teng Ben
© The Royal Society of Chemistry 2016
Published by the Royal Society of Chemistry, www.rsc.org

first principles methods contain *ab initio* methods and density functional theory (DFT). For example, MP2/6-311**//B3LYP/6-31g* is one of the commonly used method in first principles calculations. Then, by fitting the calculated interaction energies and optimized interaction distances to potential functions, force field (FF) parameters can be obtained. Finally, based on the atomic and molecular scale of the FF and the interaction in the GCMC simulation, we can evaluate molecule adsorption by porous materials.[1]

2.3 Pore Size Tailoring

Pore size is a key point with regards to the application of porous organic frameworks (POFs). Materials with a fixed pore size can be utilized to filter molecules with different sizes and shapes. A molecule with a bigger dynamic diameter than the pore size of a material is prevented from going through the channel, while the smaller ones are permitted. Even if all the guest molecules are small enough to cross the channel of POFs, the ones with stronger binding affinity could be separated over the others. In addition, POFs with a specific pore shape can be used to separate isomers, especially for chiral resolution.[2] On the other hand, pore size dominates the interaction strength and behavior between guest molecules and the host framework. It mainly shows in two aspects: first, a molecule that has the matching dynamic diameter to the pore size exhibits a strong interaction. Hence, the physical sorption enthalpy of the matching guest molecule is stronger than the others. For example, in hydrogen storage, a porous material with a pore size of two gas diameters is preferred.[3,4] In addition, this pore size dimension is also fitting for methane storage.[3,4] Gogotsi *et al.* reported that pores smaller than 0.8 nm contributed the most to CO_2 uptake at 1 bar, and the effect of 0.5 nm pores exhibited more uptake at 0.1 bar.[5] Second, a molecule with a matching size to the pore shows a slower diffusion speed in the channel of POFs. In this regard, the retention time of these molecules is longer than the other small ones in the separation process. Olefins with smaller diameters can enter the diamondyne channel more easily, thus facilitating selective adsorption of the olefin over paraffin. Furthermore, this diffusion impact exhibits more clearly in the POF's film separation applications.

The strategy to tune the pore size in pre-modification is to choose the building blocks with different length and width in the construction of POFs. As shown in Figure 2.1, the C–C bond in diamond can be replaced by a series of different organic units. Alternated with a different number of phenyl rings between two neighboring tetrahedral bonded carbon atoms obtains porous aromatic framework-301 (PAF-301), PAF-302, PAF-303 and PAF-304. If there is no interpenetration, the surface area and pore volume will increase with the length of the linkers. PAF-304 is expected to be a mesoporous material and exhibits the highest gravimetric hydrogen uptake among with regards to the others. The value could reach 6.53 wt% at 298 K and 100 bar due to the GCMC simulation results. Long linkers generally bring a large pore volume

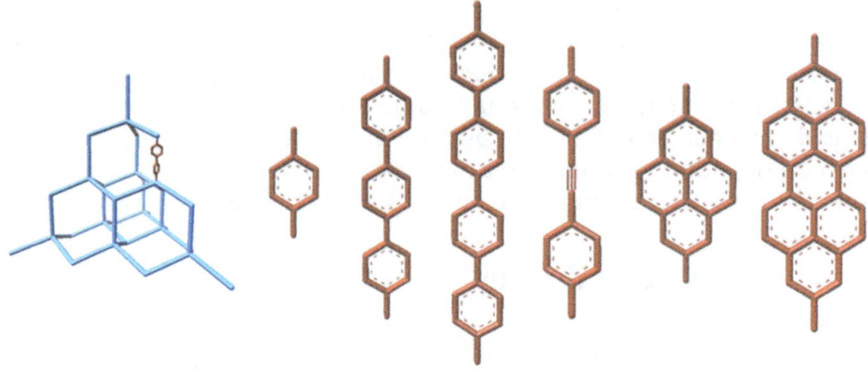

Figure 2.1   Diversity of PAF structural units.

and low bulk density, while wide linkers contribute to the stability and high electron density. The last two cases in Figure 2.1 show wide linkers used in PAF-pyrn and PAF-PTCDA. The structure used here uses polycyclic aromatic hydrocarbons, which enhance the electron density of the framework and will increase the binding affinity to guest molecules. Kuc *et al.* reported research on the mechanical and hydrogen adsorption properties of these materials by the density functional-based tight-binding (DFTB) method and quantized liquid density functional theory (QLDFT).[6–9] The result indicated strong mechanical stability in materials with wide linkers and enhanced hydrogen enthalpy due to the high electron density.

Post-synthetic modification (PSM) also can tune the pore size. Generally, it decreases the pore size in the framework since the pore can be blocked by functional groups or hybrid units. Akin to the pre-modification method, a change in length or width of the organic linker effects porosity and the nature of the framework. However, the influence made by modification is not limited to tuning the pore size of the framework.

2.4   Framework Modification

Any type of modification, either through the introduction of functional groups or hybrid element alternatives, will change the pore size of the framework. Hence, an effect of the pore size on the properties of the materials exists in these cases. However, this section mainly discusses other influences, such as electron density and distribution, on the nature of POFs. Framework modification could be divided into two parts: pre-modification and PSM. Pre-modification means design of a functionalized monomer before the extremely efficient and high yielding cross-coupling reaction. The advantage of the pre-modification method is that it unblocks the pore during the parent material functionalization and effectively utilizes terminal groups to introduce specific units. For example, on tuning the alkyl chain length of the triptycene monomer of hexahydroxytriptycene-based polymers of intrinsic microporosity (PIMs), the surface area of the whole network can be successfully controlled. The shortest alkyl chain (methyl) gives rise to the highest surface area, whereas the longest possible chain (octyl) results in the lowest surface area. PSM is defined as chemically synthesizing and modifying POFs after their formation.[10–12] It means that a chemical bond forms and breaks during the functionalization process. PSM is superior for a wide variety of functional groups and gives a controllable degree of modification. Introduction of polar functional groups and coordinate sites to the framework mainly acts to enhance the binding capacity. These are divided into five categories: covalent PSM, dative PSM, post-synthetic deprotection (PSD), tandem PSM and carbonization. PSM on POFs will be discussed in detail in Chapter 9. Framework modification could enhance the interaction between host adsorbent and guest molecule by increasing the electron density of the framework. In particular, the interaction between an electrostatic ion of the heterocyclic rings and the quadrupole of a CO_2 molecule is stronger than the one between a π-conjugated phenyl and the quadrupole of CO_2, which makes the binding sites around the heterocyclic rings more accessible to the CO_2 molecule. For example, post-modification of PIM-1 has been reported *via* a $[2+3]$ cycloaddition reaction from a tetrazole derivative of PIM-1, providing excellent gas separation properties.[13,14] PIMs consisting of phthalocyanine, porphyrin or hexaazatrinaphthylene show high catalytic activity on insertion of certain transition metal ions into their networks. Such a study has been demonstrated in the case of a metal-containing porphyrin or phthalocyanine network, which showed excellent catalytic properties towards H_2O_2 degradation.[13] The applicability of porous materials relies principally on the porosity and crystallinity of the materials. Tuning of the π-electronic interlayer interaction among the different layers empowers one to hold control over the porosity and crystallinity of two-dimensional (2D) covalent organic frameworks (COFs).[15] Arenes and fluoro-substituted arenes have been incorporated into the edges of imine-linked copper–porphyrin COFs. A tetrafluoro-substituted arene linked copper–porphyrin COF shows the strongest self-complementary electronic interactions. For such π-stacked COFs, the crystalline nature of the COFs is dependent on the improved interlayer π interactions.

2.5   Carbonization

Porous carbon is one of the most popular materials in both industry and daily life. It can be prepared by the direct annealing of porous materials and deposition on the surface of a template. One of the advantages of porous carbon is the high physicochemical stability. The inert carbon element makes porous carbon materials stable to common organic solvents, water and over a large temperature range. Porous polymers are the ideal candidates for precursors or templates. In the carbonization process, chemical bonds break and the ratio depends on the conditions such as temperature, pressure and the atmosphere. Generally, the pore shrinks to match the size of the guest molecules, which leads to an overlap force field, and performances related to an increased sorption enthalpy. Embedding hybrid elements makes porous carbons diversify and expands the application fields. For example, sulfur-doped metal organic framework- (MOF-) derived microporous carbon could be used as a lithium–sulfur battery.[16] Both the direct annealing product of PAF-1 (PAF-1-X) and KOH-activated PAF-1 (K-PAF-1-X, where X is the carbonized temperature) are high efficiency, recyclable environmental superabsorbents.[17] It should be noted that KOH not only adds excess binding sites to the framework but also acts as a template to obtain a unique bimodal microporous structure. As a result, K-PAF-1-X exhibits outstanding CO_2 sorption properties in both the high pressure range and low pressure range. POF template carbon is required to construct nanoporous carbons featuring ordered porous structures and/or narrow pore size distributions. It may be achieved by carbonization of the precursor carbon source or chemical vapour deposition of carbon within the template. Furfuryl alcohol (FA) is a typical precursor. Pyrolysis of the introduced FA in PAF-1 could obtain PAF-1/C-900. As expected, the pore size of PAF-1/C-900 reduces to half that of PAF-1 and the heat of sorption rises up to 27 kJ mol^{-1}.[18] Introduction of FA into the micropores of MOF-5 and carbonization at 1000 °C yields another nanoporous carbon, which shows a BET surface area as high as 2872 m^2 g^{-1}. In general, the MOF structure will collapse during the carbonization due to inferior thermal stability. A reinforced framework strategy, in which the internal and external surfaces of MOFs were functionalized with aminosilanes coordinated to unsaturated chromium sites before impregnation of carbon precursors, could effectively prepare the partially ordered microporous carbons.[19]

Besides, nanoporous carbon materials with high surface areas are potential electrodes for supercapacitors. For instance, Z-*X* (*X* being the annealing temperature), achieved through direct carbonization of zeolitic imidazolate framework-8 (ZIF-8), shows high capacitance and well-developed rectangular capacitance–voltage (*CV*) curves. The volumetric capacitance of Z-900 at 5 mV s^{-1} is estimated to be 200 F g^{-1}. Even after 250 cycles at a scan rate of 50 mV s^{-1}, no loss of capacitance was observed. It shows excellent stability as a capacitor material.[20] The capacitance of K-PAF-1-X was measured both in an aqueous electrolyte and different organic electrolytes. Based on the high microporous volume, which benefits charge storage and the mesoporous volume that allows fast ion diffusion in the pores, K-PAF-1-750 shows capacitance as high as 280 F g^{-1} and 203 F g^{-1} at a current density of 1 A g^{-1} in 6.0 M KOH and 1-ethyl-3-methylimidazolium bis(trifluoromethylsulfonyl)imide (EMImTFSI), respectively.[21] In addition, capacitance of the carbon materials could be tailored by micropore engineering in electrical double-layer capacitors (EDLCs) by diverse carbonization of the porous polymers. Electrochemical measurements of K-PAF-1-750 in different organic electrolytes with varying size demonstrate the effect of geometry packing between the electrolytes ions in the pores. The effect of diffusion and commensurate matching of the electrolyte ions with the pores of carbonized materials significantly influence and control the capacitance of these materials.

Combining these excellent properties with the ease of synthesis, good thermal stability and commercial availability, carbonization of POFs is a valuable strategy to prepare functional materials. The porous carbons obtained through different carbonization methods demonstrate different pore structures, which extend to wide application areas.

2.6   Interpenetration Control and Utilization

POFs with large open frameworks always form the interpenetrative structures to enhance the stability because filling the void space leads to repulsive forces that minimize entropy of the materials. The interpenetration structure reduces the pore size and pore volume. However, it increases the bulk density and binding sites, which effectively enhance the volumetric hydrogen uptake. If the binding sites are not all blocked, the surface area of the POFs could be maximized. The interpenetration structure can be controlled or even avoided by organic building blocks. Wide or large functional building blocks will limit the formation of the second network in the constructed open skeleton. Combined with the stable effect of pendant functional groups, a non-interpenetration framework is obtainable. It should be noted that the interpenetration framework is not a disadvantage of POFs. In contrary, appropriate blocking of the channel decreases the pore size to better match the dynamic diameter of guest molecules, which will enhance the interaction between the host framework and the guest molecules.[22]

Interpenetration of crystalline COFs have already been proved and reported by Yaghi *et al.*[23] A crystalline imine-linked 3D COF-300 was prepared by condensation of 4,4′,4″,4‴-methanetetrayltetraaniline and terephthalaldehyde (Figure 2.2). Powder X-ray diffraction analysis revealed COF-300 shared a five-fold interpenetrated diamond net. Though many COFs have been synthesized to date, few interpenetration structures have been reported because this analysis requires high crystallinity of the porous polymer.

It is difficult to determine the interpenetration in amorphous porous polymers. Haranczyk *et al.*[24] set up some non-interpenetration and interpenetration models of porous polymer networks (PPNs) based on the ideal crystalline model with dia topology. They systematically compared the simulated pore diameter, framework density, simulated pore volume, and experimental pore diameter and pore volume of non-interpenetration and interpenetration models of PAF-1 (PPN-6), PPN-4 and other PPNs (PPN-2, PPN-3, PPN-5). The results indicated that for PPN-4, PPN-5 and PPN-6 (PAF-1), the simulated methane adsorption isotherm of the non-interpenetrated structure compared favourably with the experimental data. This strongly supports that the experimental structure can be well modeled by the non-interpenetrated dia net. On the contrary, the experimental data of the pore

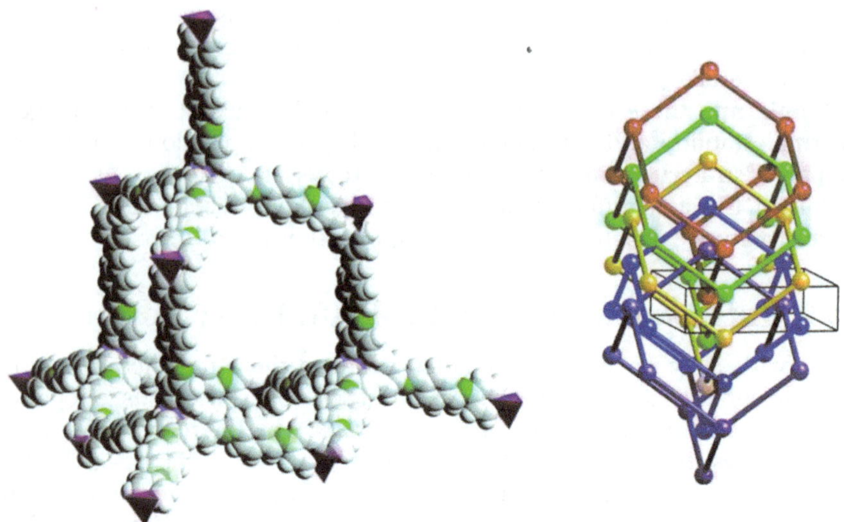

Figure 2.2   Left: single framework (space filling; C grey and pink, N green, H white). Right: representation of the **dia-c5** topology.
Reprinted with permission from F. J. Uribe-Romo, J. R. Hunt, H. Furukawa, C. Klöck, M. O'Keeffe and O. M. Yaghi, *J. Am. Chem. Soc.*, 2009, **131**, 4570. Copyright 2009 American Chemical Society.

volume of PPN-3 was much lower than the simulated pore volume based on the non-interpenetrated model, as well as the difference of the restricting pore diameter. This can be explained by the partial interpenetration of the dia net. Based on the interpenetrated model of PPN-3, it is estimated to be approximately 72% interpenetrated. Though it is difficult to say how accurate the model is with regards to interpenetration in amorphous porous polymers, this study gave a good analysis for further understanding the structure of PAFs and PPNs.

As discussed above, it is very difficult to determine the interpenetration in amorphous polymers. For crystalline COFs, though there are some interpenetrated COFs that have been revealed, we are still far from being able to say how to control it. MOFs are a kind of hybrid crystalline porous coordination polymer that has been well explored for decades. It gives a perfect model for crystal engineering and the control of interpenetration methods has been well developed. Usually, longer organic linkers lead to two-fold interpenetration during the synthesis of IRMOF; however, dilute reaction solutions are required to prepare non-interpenetration MOFs.[25] Moreover, interpenetrated MOFs are always found when synthesis is performed at a higher temperature.[26] A suitable template can introduce non-interpenetrated MOFs with larger pores.[27] Larger side group ligands are anticipated to suppress the framework interpenetration in MOFs.[28] It is believed that all of the above-mentioned methods of interpenetration control can be applied to pure organic porous polymers. Though the interpenetrated structure of the porous polymer is very difficult to be accurately solved, the resulting porous polymer must show different physical properties.

2.7 Structural Order and Disorder

According to the structural order, all condensed materials can be divided into ordered, crystalline materials, and disordered, amorphous materials. In the same way, porous polymers also appear as crystalline and amorphous materials. It seems the crystallinity of porous polymers is determined by the polymerization reactions. Generally speaking, reversible polymerization reactions, such as the formation of boronate- and boroxine-esters and the formation of an imine bond, lead to crystalline COFs.[23,29,30] Reversible nitrile cyclization at high temperature readily synthesizes ordered covalent triazine-based frameworks (CTFs).[31] Irreversible polymerization reactions are anticipated to generate amorphous porous polymers. For example, Yamamoto coupling, Suzuki cross-coupling *etc.* will tend to amorphous porous polymers such as PAFs, hypercrosslinked polymers (HCPs), conjugated microporous polymers (CMPs) *etc.* Ordered porous polymers exhibit aligned functional monomers and a uniform pore or channels. This can be very interesting and may realize the so-called targeted design. It should be noted that the chemical stability of such a linking bond derived from reversible polymerization is usually not very high under acidic, basic or even aqueous solutions, while the amorphous porous polymer prepared by an irreversible reaction always exhibits higher physicochemical stability. Both amorphous and crystalline porous polymers show very high surface area. That means that choosing the reversible or irreversible reaction does not determine the porosity of the porous polymer.

References

1. Z. Xiang, D. Cao, J. Lan, W. Wang and D. P. Broom, *Energy Environ. Sci.*, 2010, **3**, 1469.
2. A. Kewley, A. Stepphenson, L. J. Chen, M. E. Briggs, T. Hasell and A. I. Cooper, *Chem. Mater.*, 2015, **27**, 3207–3210.
3. Z. Xiang and D. Cao, *J. Mater. Chem. A*, 2013, **1**, 2691.
4. Y. Yan, I. Telepeni, S. Yang, X. Lin, W. Kockelmann, A. Dailly, A. J. Blake, W. Lewis, G. S. Walker, D. R. Allan, S. A. Barnett, N. R. Champness and M. Schröder, *J. Am. Chem. Soc.*, 2010, **132** , 4092.
5. V. Presser, J. McDonough, S. Yeon and Y. Gogotsi, *Energy Environ. Sci.*, 2011, **4**, 3059.
6. B. Lukose, M. Wahiduzzaman, A. Kuc and T. Heine, *J. Phys. Chem. C*, 2012, **116**, 22878.
7. A. F. Oliveira, G. Seifert, T. Heine and H. A. Duarte, *J. Brazil. Chem. Soc.*, 2009, **20**, 1193.
8. G. Seifert, D. Porezag and T. Frauenheim, *Int. J. Quantum Chem.*, 1996, **58**, 185.

9. S. Patchkovskii and T. Heine, *Phys. Rev. E: Stat., Nonlinear, Soft Matter Phys.*, 2009, **80**, 031603.

10. Z. Wang and S. M. Cohen, *Chem. Soc. Rev.*, 2009, **38**, 1315.

11. K. K. Tanabe and S. M. Cohen, *Chem. Soc. Rev.*, 2011, **40**, 498.

12. S. M. Cohen, *Chem. Rev.*, 2012, **112**, 970.

13. N. B. McKeown and P. M. Budd, *Chem. Soc. Rev.*, 2006, **35**, 675.

14. B. S. Ghanem, K. J. Msayib, N. B. McKeown, K. D. M. Harris, Z. Pan, P. M. Budd, A. Butler, J. Selbie, D. Book and A. Walton, *Chem. Commun.*, 2007, **1**, 67.

15. S. Kandambeth, A. Mallick, B. Lukose, M. V. Mane, T. Heine and R. Banerjee, *J. Am. Chem. Soc.*, 2012, **134**, 19524.

16. H. B. Wu, S. Wei, L. Zhang, R. Xu, H. H. Hng and X. W. Lou, *Chem. – Eur. J.*, 2013, **19**, 10804.

17. Y. Li, T. Ben, B. Zhang, Y. Fu and S. Qiu, *Sci. Rep.*, 2013, **3**, 2420.

18. V. Presser, J. McDonough, S. Yeon and Y. Gogotsi, *Energy Environ. Sci.*, 2011, **4**, 3059.

19. Y. Meng, G. H. Wang, S. Bernt, N. Stock and A. H. Lu, *Chem. Commun.*, 2011, **47**, 10479.

20. W. Chaikittisilp, M. Hu, H. Wang, H. Huang, T. Fujita, K. C. W. Wu, L. C. Chen, Y. Yamauchi and K. Ariga, *Chem. Commun.*, 2012, **48**, 7259.

21. Y. Li, S. Roy, T. Ben, S. Xu and S. Qiu, *Phys. Chem. Chem. Phys*, 2014, **16**, 12909.

22. H. Ren, T. Ben, E. Wang, X. Jing, M. Xue, B. Liu, Y. Cui, S. Qiu and G. Zhu, *Chem. Commun.*, 2010, **46**, 291.

23. F. J. Uribe-Romo, J. R. Hunt, H. Furukawa, C. Klöck, M. O'Keeffe and O. M. Yaghi, *J. Am. Chem. Soc.*, 2009, **131**, 4570.

24. R. L. Martin, M. N. Shahrak, J. A. Swisher, C. M. Simon, J. P. Sculley, H. C. Zhou, B. Smit and M. Haranczyk, *J. Phys. Chem. C.*, 2013, **117**, 20037.

25. M. Eddaoudi, J. Kim, N. Rosi, D. Vodak, J. Wachter, M. O'Keeffe and O. M. Yaghi, *Science*, 2002, **295**, 469.

26. J. Zhang, L. Wojtas, R. W. Larsen, M. Eddaoudi and M. J. Zaworitko, *J. Am. Chem. Soc.*, 2009, **131**, 17040.

27. (a) D. Sun, S. Ma, Y. Ke, D. J. Collins and H. C. Zhou, *J. Am. Chem. Soc.*, 2006, **128**, 3896; (b) S. Ma, D. Sun, M. Ambrogio, J. A. Fillinger, S. Parkin and H. C. Zhou, , *J. Am. Chem. Soc.*, 2007, **129**, 1858; (c) S. Ma, J. Eckert, P. M. Forster, J. W. Yoon, Y. K. Hwang, J. S. Chang, C. D. Collier, J. B. Parise and H. C. Zhou, *J. Am. Chem. Soc.*, 2008, **130**, 15896.

28. Y. B. Go, X. Wang and A. J. Jacobson, *Inorg. Chem.*, 2007, **46**, 6594.

29. A. P. Côté, A. I. Benin, N. W. Ockwig, M. O'Keeffe, A. J. Matzger and O. M. Yaghi, *Science*, 2005, **310**, 1166.

30. H. M. El-Kaderi, J. R. Hunt, J. L. Mendoza-Cortes, A. P. Côté, R. E. Taylor, M. O'Keeffe and O. M. Yaghi, *Science*, 2007, **316**, 268.

31. P. Kunhn, M. Antonietti and A. Thomas, *Angew. Chem. Int., Ed.*, 2008, **47**, 3450.

CHAPTER 3

Chemical Synthesis of Porous Polymers

3.1 Introduction

The chemical synthetic strategy behind the construction of complex architectures of microporous organic frameworks is to incorporate desired components with targeted applications as building units. It is therefore logical to procure tailor-made functionalities built from simple molecular synthons in such frameworks. The choice of monomers, mode of synthesis and reaction parameters, such as temperature, pressure, solvent *etc.* play a crucial role in 'stitching up' molecular synthons in a particular fashion in order to form the desired framework. Usually, reversible reactions lead to ordered framework structures, while irreversible reactions always produce amorphous porous polymers. A plethora of reactions ranging from cross-coupling and self-condensation to cyclotrimerization reactions have been carried out to synthesize porous networks with targeted structures and properties, which in turn can be capitalized in several fields, for instance gas storage, selective permeation, separation and catalysis. This section shall discuss the chemical synthetic methods of different classes of porous organic polymers, the reaction conditions employed, the logic behind using a particular scheme of reactions and their consequential effect on the structural layout and properties of these frameworks.

3.2 Amorphous Porous Polymers

3.2.1 Insoluble Porous Polymers

3.2.1.1 *Hypercrosslinked Polymers*

The extensively cross-linked hypercrosslinked polymers or HCPs can be broadly categorized into two classes: those derived from non-styrenic

Monographs in Supramolecular Chemistry No. 17
Porous Polymers: Design, Synthesis and Applications
By Shilun Qiu and Teng Ben
© The Royal Society of Chemistry 2016
Published by the Royal Society of Chemistry, www.rsc.org

monomers, for instance carbinol HCPs, and the other based on cross-linked polystyrene, also known as 'Davankov resins'. The former kind is synthesized in such a fashion that, say in case of carbinol HCPs, the biphenyl groups of the starting material are linked together with groups of C–OH, forming a network that is highly microporous. This is accomplished by lithiation of dibromobiphenyl and subsequent addition of dimethyl carbonate.[1,2] The design of Davankov resins, on the other hand, involves preparation of a polymer precursor, which is then further cross-linked by Friedel–Crafts alkylation with a Lewis acid. Rigorous research has been done in the synthesis of the aforementioned polystyrene-based Davankov resins. Various chemical strategies using different kinds of monomers have been employed to design HCPs with a wide range of surface areas. One of the very first strategies adopted was the preparation of a lightly cross-linked co-polymer *via* vinylbenzyl chloride polymerization using divinylbenzene as a cross-linker. This copolymer is then swollen in a suitable solvent and subjected to iron(III) chloride, akin to Lewis-acid-aided Friedel–Crafts alkylation for hypercrosslinking of polymers.[3–5] The basic principle behind HCP synthesis is obtaining a rigid, extended and highly solvated three dimensional network.[6] The inherent rigidity of the constituting monomers and reduced degree of chain entanglement ensures the 'unique ability to swell in both good solvents and non-solvents' by loose chain packing with a high free volume.[4] Each of the steps involved in this reaction aims to achieve this goal. While the polymerization step yields the polymer precursor, the swelling and cross-linking stages avoid microphase separation. The initial mild addition of the cross-linker reduces the mobility of the polymer coil, thus preventing contraction.[4,7] The network remains solvated during the entire course of polymer formation. Chloromethyl methyl ether (CMME) or dimethylformal, bis-chloro methylated benzene and biphenyl, as well as tris-chloromethylated mesethylene (TCMM), which is 1,3,5-tris-chloromethyl-2,4,6-trimethylbenzene, have typically been used as cross-linking agents.[4] These substances alkylate two (or three, in the case of TCMM) polystyrene rings in the presence of Friedel–Crafts catalysts. Apart from their obvious role of 'hypercrosslinking', the cross-linking bridges remain uniformly distributed throughout the reaction system and initiate the cross-linking reaction at several sites in the solution. In this way these cross-linkers fix the already existing intermolecular contacts and thus preserve the favorable, expanded conformation of the so-formed one-phase solvated polystyrene gel.[4,7] The requirement of design of a mildly cross-linked polymer precursor has been eradicated with the use of bis(chloromethyl) aromatic monomers such as dichloroxylene (DCX), bis(chloromethyl)biphenyl (BCMBP) and bis(chloromethyl)anthracene (BCMA)[8] (Figure 3.1). These monomers are used to synthesize hyper-crosslinked porous networks directly by Friedel–Crafts alkylation in the presence of a Lewis acid. Other such direct methods include copolymer-ization of BCMBP along with fluorene, 9,9'-spirobi(fluorene), dibenzofuran and dibenzothiophene.

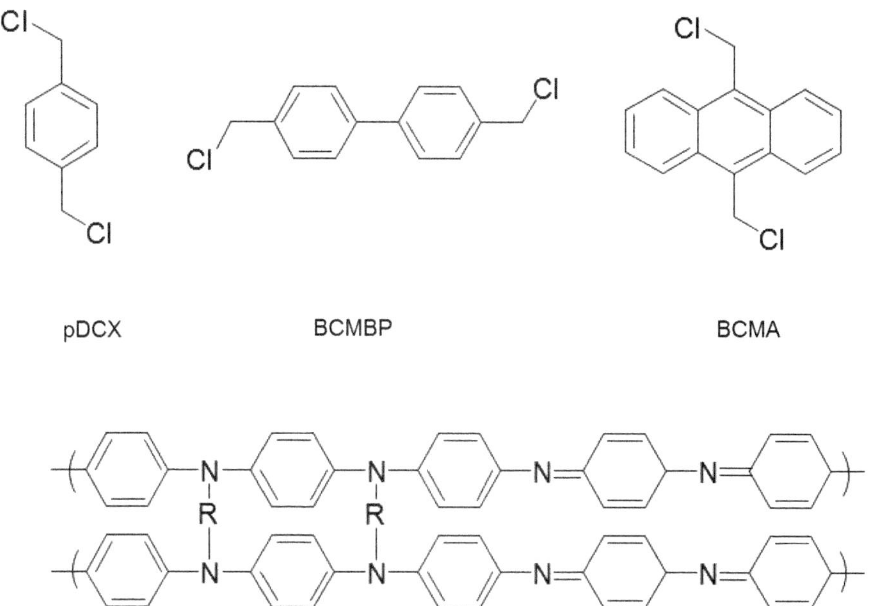

Figure 3.1 Monomers used in the synthesis of hypercrosslinked networks by direct polymerization (top) and the structure of hypercrosslinked polyaniline networks (bottom).[6]
Reprinted with permission from R. Dawson, A. I. Cooper and D. J. Adams, *Prog. Polym. Sci.*, 2012, **37**, 530. Copyright 2011 Elsevier Ltd. All rights reserved.

Tan *et al.*[9] explored the synthesis of monomers that are monofunctional. They have synthesized two hypercrosslinked polymer networks of bishydroxymethyl monomers, *e.g.*, 1,4-benzenedimethanol (BDM), and monohydroxymethyl compounds, *e.g.*, benzyl alcohol (BA) by self-condensation. Precursors of polymers like polystyrene[7] and poly(chloromethylstyrene)[5,10,11] that are swollen, as well as polyfunctional benzyl chlorides,[8,12,13] are synthesized by Friedel–Crafts alkylation in the presence of a Lewis acid. Hydroxymethyl and chloromethyl form a bond with the benzene ring in the presence of an acid catalyst,[14,15] leading to the design of a new series of hypercrosslinked polymers prepared by directly using the building blocks of hydroxymethyl aromatics (Figure 3.2).

In some reports, linear polyanilines and polypyrroles have been used as polymer precursors and diiodomethane/paraformaldehyde has been used as a cross-linker, forming networks.[16–18] However, this is different from the previously mentioned lithiation route as it does not utilize any lithiating agents or Lewis acids; thus, this strategy not only skips the lithiation step but also prevents release of copious amounts of HCl gas produced in other cases[18] (Figure 3.1).

The Ullmann and Buchwald coupling reaction has also been employed for the synthesis of HCP networks by coupling polyanilines and

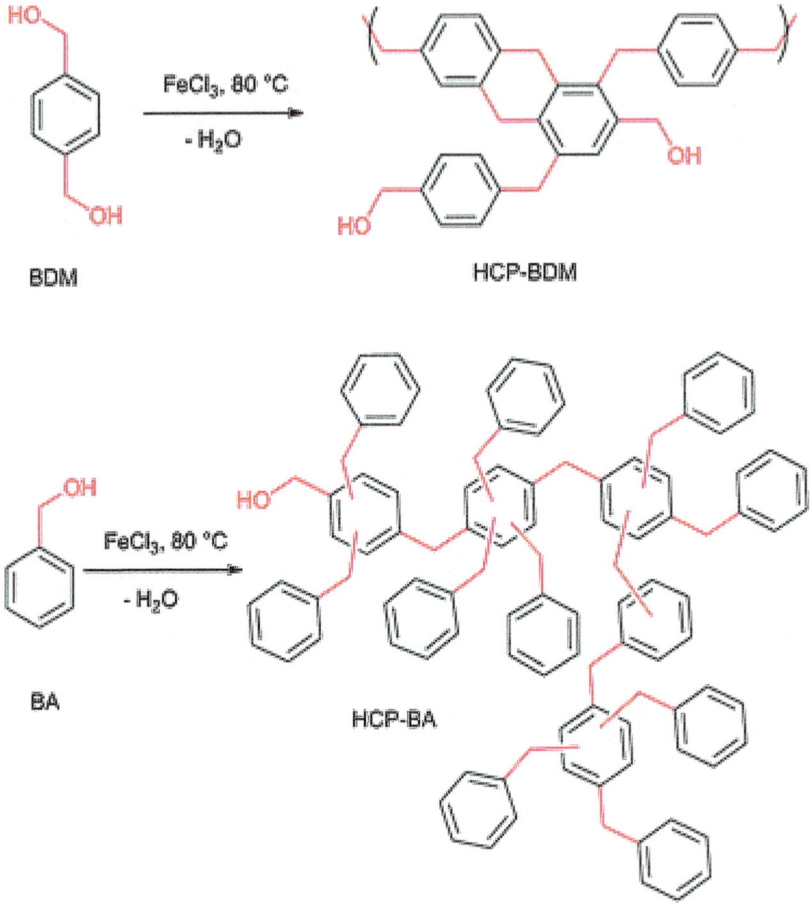

Figure 3.2 Synthesis of HCP–BDM and HCP–BA by Friedel–Crafts catalyzed self-condensation.
Reproduced from ref. 8 with permission from The Royal Society of Chemistry.

diaminobenzene with simple monomers such as diiodobenzene and tribromobenzene.[19] The design strategy in this case is focused on synthesizing networks consisting of aromatic rings connected through linking groups as small as possible. It was demonstrated by Germain *et al.* in 2007 that longer, flexible cross-links show smaller surface areas than rigid cross-links.[17] Additionally, cross-links that are smaller tend to add less non-adsorbing mass to the final porous structure of the framework. Another aspect involved in this synthesis strategy is to avoid substitution of electron-withdrawing groups on the aromatic groups as these have been found to significantly compromise the hydrogen adsorption capacity of the so-formed structure[17,20] (Scheme 3.1). In order to stitch aromatic rings together, the ideal starting materials are aromatic rings substituted with reactive groups, for example, aryl halides and amines. Precursors, such as aryl amines and

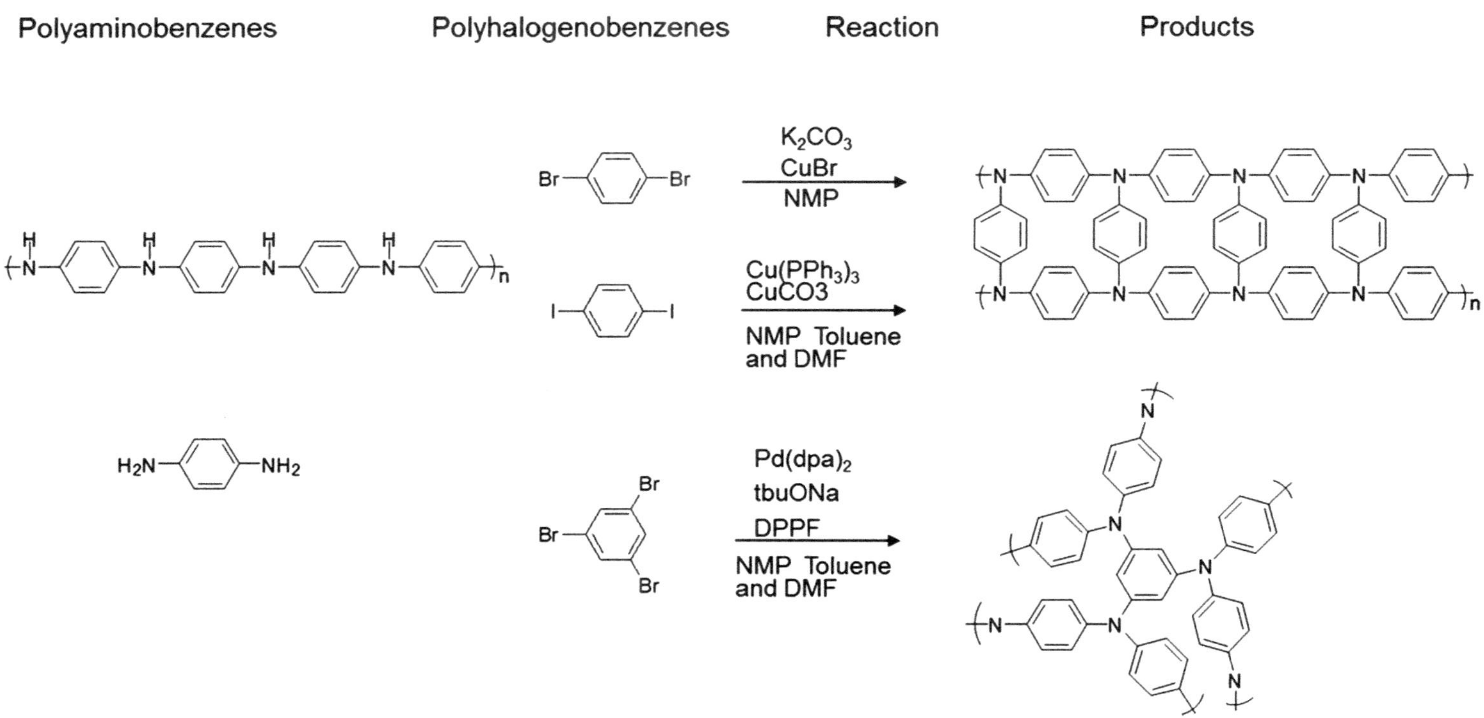

Scheme 3.1 Components and reactions used to prepare nitrogen-linked nanoporous networks of aromatic rings and representative structures thereof: (top) Ullman reaction catalyzed by insoluble CuBr, (middle) Ullman reaction catalyzed by soluble Cu(PPh$_3$)$_3$Br and (bottom) Buchwald reaction catalyzed by soluble Pd(dba)$_2$. Reprinted with permission from J. Germain, F. Svec and J. M. J. Fréchet, *Chem. Mater.*, 2008, **20**, 7069. Copyright 2008, American Chemical Society.

polyhalogenobenzenes, are coupled *via* Ullmann and Buchwald reactions catalyzed by soluble[21] or insoluble[22] copper salts and soluble palladium,[23–25] respectively, to yield amine-linked networks of aromatic rings. In case of cross-linking with diiodobenzene and tribromobenzene di- and tri-functionalized aromatic rings connected *via* amines are obtained, respectively.[19] In the case of polypyrroles, different cross-linkers have been used, namely diiodomethane, triiodomethane and triiodoborane, resulting in CH_2, CH and B as cross-linkers, respectively. Out of the two synthetic routes, the Buchwald approach is preferred as it yields halogen-free materials with higher surface areas, unlike Ullmann coupling where the polymer contains copper and iodine, thus adding unnecessary mass to the porous materials. It was also observed that the solvent used for swelling influences the final hypercrosslinked product's surface area. The solvents with the highest Hildebrand solubility parameters yielded polymers with the highest surface area. For instance, toluene with a Hildebrand solubility parameter of 18.2 $MPa^{1/2}$ affords polymeric nanopores having lower surface areas than those synthesized in *N*-methyl-2-pyrrolidone (NMP; 23.1 $MPa^{1/2}$) or dimethylformamide (DMF; 24.8 $MPa^{1/2}$).[19,26]

The synthesis of a hypercrosslinked network of polymers has been achieved *via* self-condensation of DCX, BCMBP and BCMA, and is outlined by the step growth polycondensation of DCX and other bischloromethyl monomers. This study shows the importance of the choice of monomer to be used. By varying the structure of the precursor one can fabricate networks with a surface area as high as 1900 $m^2\,g^{-1}$ [Brunauer–Emmett–Teller (BET) surface area] or 3000 $m^2\,g^{-1}$ (Langmuir surface area). Other studies include the preparation of analogues of covalent organic frameworks (COFs) and covalent triazine-based frameworks (CTFs) with N_3B_3 rings by reacting different amines with boron trihalides.[27] Cyclotrimerizations of ketones forming truxenes and truxenones have also yielded HCP networks.

3.2.1.2 *Conjugated Microporous Polymers*

The first conjugated microporous polymers (CMPs) were reported in 2007[28] and have employed Sonogashira–Hagihara Pd coupling to connect aromatic halides to aromatic alkynes. It thus forms poly(arylene ethynylene) (PAE) networks.[29,30] Four CMPs have been synthesized with 522–834 $m^2\,g^{-1}$ BET surface areas (Figure 3.3). The CMP-1 network possessed the shortest length of strut with the highest surface area, while CMP-3 had the longest strut length and showed the lowest surface area. Such behavior has been explained by an increase in conformation degrees of freedom and flexibility with increased strut length.[31] It is likely that, with an increase in strut length, the growing network of fragments interpenetrate more, filling the space. This CMP series was expanded by two more networks with a shorter (CMP-0) and longer (CMP-5) strut length. The one with the shortest strut, leads to a higher BET surface area (1018 $m^2\,g^{-1}$).[31] However, it is important to note

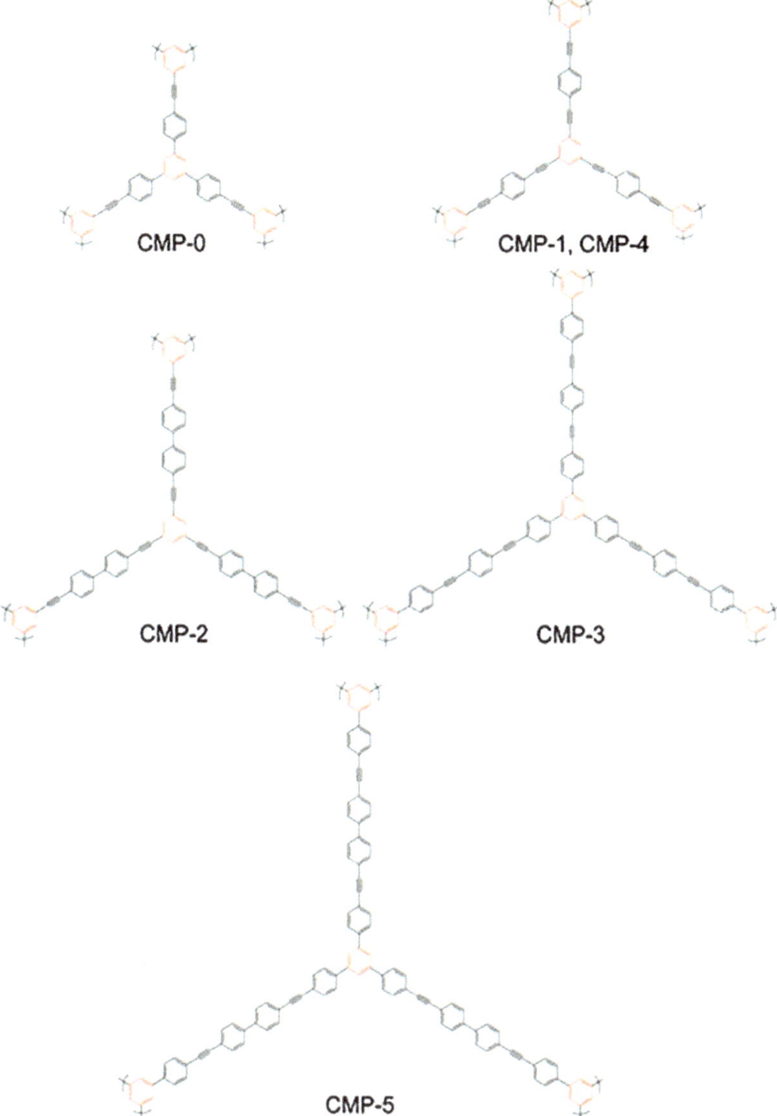

Figure 3.3 Schematic two-dimensional (2D) structures for a series of CMP networks.[6]
Reprinted with permission from R. Dawson, A. I. Cooper and D. J. Adams, *Prog. Polym. Sci.*, 2012, **37**, 530. Copyright 2011 Elsevier Ltd. All rights reserved.

that the strut length influences the micropore volume and the pore size. The smallest struts give rise to small pores with the largest micropore volumes. On the other hand, the largest struts give rise to the lowest micropore volume and the biggest pore size. A series of copolymers have been prepared using 1,3,5-triethynylbenzene and different ratios of halide monomers employed for synthesizing CMP-1 and CMP-2 (CPN-1 to CPN-6).[31] These statistical copolymerizations remarkably give a control over surface area, micropore volume and average pore size. Such control has been considered to be applicable to crystalline compounds like COFs and metal organic frameworks (MOFs).[30] Networks of CMPs of 1,3,5-triethynylbenzne (HCMP-1) and 1,4-diethynylbenzene (HCMP-2) are also synthesized controllably with surface areas of 842 and 827 $m^2 g^{-1}$, respectively.[32] Nitrogen-containing CMP networks have also been made.[33]

Such possibility of nitrogen induction in the structure gives opportunities of functionalization. Thomas *et al.* reported the design of a CMP-type material using spirobifluorene-type monomers[34–36] (Figure 3.4). In this case, 2,2′,7,7′-tetrabromo-9,9′-spirobifluorene and 1,4-diethynylbenzene are coupled by the Sonogashira–Hagihara pathway to create a network with a surface area of 510 $m^2 g^{-1}$ (BET) or 1030 $m^2 g^{-1}$ (small-angle X-ray scattering, SAXS).[35] Using 2,5-thiophene diboronic acid and a spiro monomer, and changing their ratios, the adsorption and emission wavelengths of the networks can be tuned.[37] The Suzuki and Sonogashira–Hagihara reactions are used to form networks using hexakis(4-bromophenyl)benzene, benzene-1,4-diboronic acid and 1,4-diethynylbenzene.[38] The monomer of 2,3,5,6-tetrakis(4-bromophenyl)pyrrolo[3,4-*c*]pyrrole-1,4-(2*H*,5*H*)-dione is used to prepare a series of networks by Yamamoto and Sonogashira–Hagihara Pd-coupling reactions.[39] Coupling of bromo-aromatics by the Yamamoto method has also been reported.[40] Such chemistry has proven important in this area. Homocoupling of 2,2′,7,7′-tetrabromo-9,9′-spirobifluorene produces a network with a BET surface area of 1275 $m^2 g^{-1}$. Copolymerizing 2,2′,7,7′-tetrabromo-9,9′-spirobifluorene and 1,4-dibromobenzene in a ratio of 1:1 gives a network with a surface area of 887 $m^2 g^{-1}$ (Figure 3.4). However, smaller surface areas are produced using 1,3- (5 $m^2 g^{-1}$) and 1,5-substituted dibromobenzene (361 $m^2 g^{-1}$) analogues. When the amount of 1,4-dibromobenzene is increased, networks of lower surface area and reduced micropore volume are formed due to increased strut length, as mentioned previously in the discussion of CMP-type compounds.[29,31] Use of 1,3,5-tribromobenzene alone leads to a homopolymer (surface area = 1255 $m^2 g^{-1}$), whereas its copolymerization with dibromobenzene

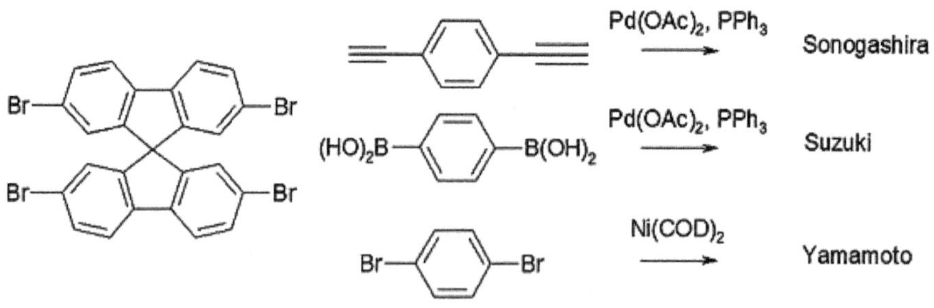

Figure 3.4   Synthesis of CMPs from 2,2′,7,7′-tetrabromo-9,9′-spirobifluorine using the Sonogashira, Suzuki and Yamamoto reactions [Ni(COD)₂, bis(cyclooctadiene)nickel(0)].[35]

leads to increased strut lengths and reduced micropore volumes and surface areas. The need of twisted monomers has been bypassed using Yamomoto coupling. For instance, the homocoupling of 1,3,5-tris(4-bromophenyl)-benzene gives rise to a porous aromatic framework-5 (PAF-5) network having a BET surface area of 1503 m^2 g^{-1}.[41] Cyclotrimerization of 2,2′,7,7′-tetra-ethynyl-9,9′-spirobifluorene (PS4AC2) and tetra(4-ethynylphenyl)methane (PT4AC) with BET surface areas of 1043 and 762 m^2 g^{-1}, respectively, have been reported. Suzuki cross-coupling has been used by Chen *et al.*[42,43] to produced networks. A CMP based on polyphenylene with a light-harvesting capacity prepared from 1,2,4,5-tetrabromobenzene and 1,4-benzene diboronic acid shows a BET surface area of 1083 m^2 g^{-1} and 1.56 nm pores.[42] A heterogeneous catalyst based on a porphyrin-connected network has been prepared and explored.[43] Using a Co$_2$(CO)$_8$ catalyst, a conjugated network based on 1,3,5-triethynylbenzene having a surface area around 1246 m^2 g^{-1} has been produced, exploiting the cyclotrimerization of related di- and tri-alkynes.[44] The cubic polyhedral oligomeric silsesquioxane was used to synthesize the microporous compound *via* Sonogashira–Hagihara and Yamamoto pathways (Figure 3.5). The above monomers, when coupled with alkyne, break the extended conjugation in such materials when subjected to homocoupling using copper chloride.[45–47]

Sonogashira reactions involving 1,4-diethynylbenzene (PSN-1) and 4,4′ diethynylbiphenyl (PSN-2) give rise to BET surface areas of 846 m^2 g^{-1} and 1042 m^2 g^{-1}, respectively.[45] Alkyne-coupled polyhedral oligomeric silses-quioxane (POSS) networks lead to a BET surface area of around 1000 m^2 g^{-1}.[47] Related monomers were used involving iodophenyl-substituted POSS in a thermal decoupling reaction leading to BET surfaces areas of up to 555 m^2 g^{-1}.[48] The adamantane-centered 1,3,5,7-tetrakis(4-iodophenyl)ada-mantane monomer and 1,4-diethynylbenzene *via* Sonogashira–Hagihara cross-coupling produce a material with a BET surface area of 665 m^2 g^{-1}, half of the related carbon-centered analogue.[49]

There are many ways for the production of new CMP compounds. For example, some strategies are quite applicable for the synthesis of microporous polymers.[50–53] One of the important developments is to produce CMPs that are solution-processable. This makes thin film devices, for example, 'conjugated PIMs'.[54,55] A different strategy may be used to design a related Durham precursor pathway to polyacetylene.[56] The synthesis of crystalline CMPs and expanded adamantane analogues is a challenge. C–C bond forming reactions to make conjugated polymers (*e.g.*, Sonogashira–Hagihara cross-coupling)[29,31,35,57] are irreversible. Hence, strategies that accommodate reversible routes, like molecular preorganization, template pathways or a combination of both, need to be considered.

3.2.1.3 Crystalline Triazine Frameworks

The synthesis of triazine-based networks follows nitrile self-condensation with ZnCl$_2$ to give a series of highly porous networks. The 1,4-dicyano-benzene-based network was among the first triazine networks to show a high

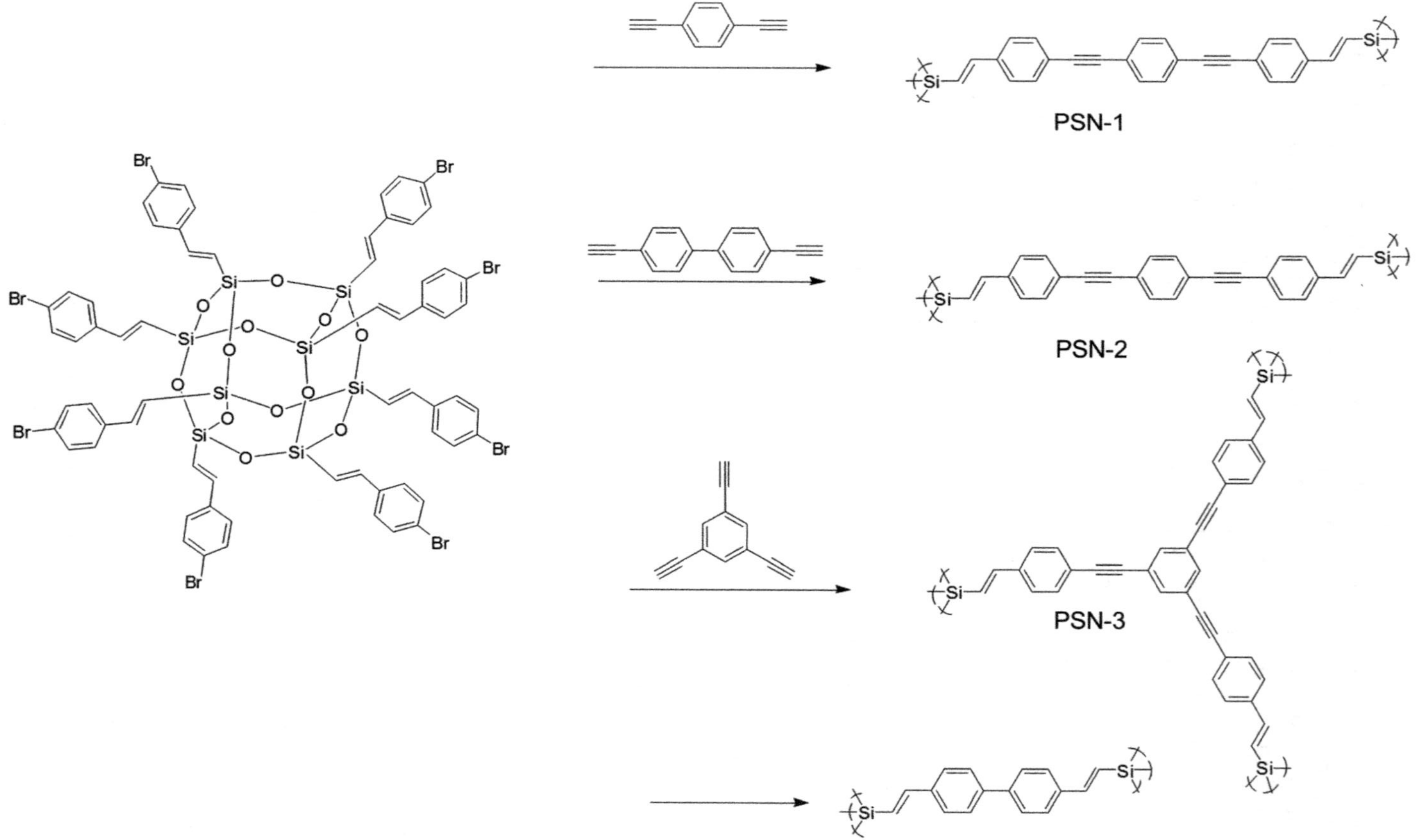

Figure 3.5 Porous networks synthesized from siloxane-centered monomers (PSN, polyorganosiloxane networks).[6] Reprinted with permission from R. Dawson, A. I. Cooper and D. J. Adams, *Prog. Polym. Sci.*, 2012, **37**, 530. Copyright 2011 Elsevier Ltd. All rights reserved.

degree of crystallinity with an apparent BET surface area of 791 $m^2\ g^{-1}$.[58] Too much $ZnCl_2$ led to an increased surface area of 1123 $m^2\ g^{-1}$ but with no crystallinity. Thiophenes, pyridines and related monomers give amorphous products. Desired characteristics of the Type IV isotherm and a maximum BET surface area of 2475 $m^2\ g^{-1}$ in an amorphous network may be achieve through synthesis from a biphenyl analogue (4,4′-dicyanobiphenyl).[59] The porosity and surface area increase from 2.0 nm to 3.6 nm and 920 $m^2\ g^{-1}$ to 2530 $m^2\ g^{-1}$ as the temperature of the reaction is increased from 400 °C to 700 °C, respectively. A maximum surface area of 3270 $m^2\ g^{-1}$ is produced by using a two-step heating method. Here, the reaction first is performed at 400 °C for 20 h and at 600 °C for 4 days. New networks are synthesized by the use of an adamantane-centered tetrahedral network and related cyano monomers.[60] Here also the surface area and mesoporosity increase with an increase in temperature. Coordinating groups are introduced for application in catalysis.[60] Materials with triazine rings have been prepared by condensation of melamine and related aldehydes using Schiff base chemistry.[61] The maximum BET surface area achieved from this route using chip monomers and with a high nitrogen percentage was around 1377 $m^2\ g^{-1}$.[62] An aromatic nitrile-based high-performance network with varying porosity has been reported. Ionothermal[63] synthesis with molten zinc chloride involving dynamic trimerization at high temperature gave triazine-based networks comparable to zeolites, MOFs or boron-oxide-type COFs (Figure 3.6).[58,64,65]

Crystalline polytriazines can be obtained using molten $ZnCl_2$ at 400 °C. The reasons are as follows. (i) In ionic melts, nitriles have good solubility due to strong Lewis acid–base interactions and (ii) for the trimerization reaction, $ZnCl_2$ is a good catalyst. Note that although such nitriles are stable at high temperature, they still start to decompose at temperatures above 400 °C due to C–H bond cleavage with loss of hydrogen.[66] These polymers were made in quartz ampules at 400 °C using a mixture of nitrile and $ZnCl_2$ in almost quantitative yield. Polymerization is incomplete at higher temperature and with a shorter reaction time (*e.g.*, 400 °C for 10 h). New classes of microporous conjugated networks using thiophene linkages are made by oxidative polymerization of designer molecules like 2,2′,7,7′-tetrakis(2-thienyl)-9,9′-spirobifluorene and 1,3,5-tris(2-thienyl)benzene[34] (Figure 3.7). In this synthesis the monomers are used as molecular tectons for creating 3D star-shaped oligothiophene architectures.[31,67,68]

Bithiophene linkers are formed due to C–C coupling of thiophene moieties at the 5-position in the course of oxidative polymerization. Such linkers are stiff but, from a topological viewpoint, it is important to have flexible cross-linking points that can be polymerized. Hence, oxidative polymerization of stiff bithiophene linkers gives a rigid network of molecular motifs that are contorted. Such a reaction is carried out with stoichiometric $FeCl_3$ in acetonitrile, giving a quantitative yield of the polymer, indicating a high degree of polymerization. A covalent triazine framework is prepared from 1,4-dicyanobenzene with $ZnCl_2$ in a molar ratio of 1 : 5 in a sealed quartz ampule heated to 400 and 600 °C. The triazine network is formed in the

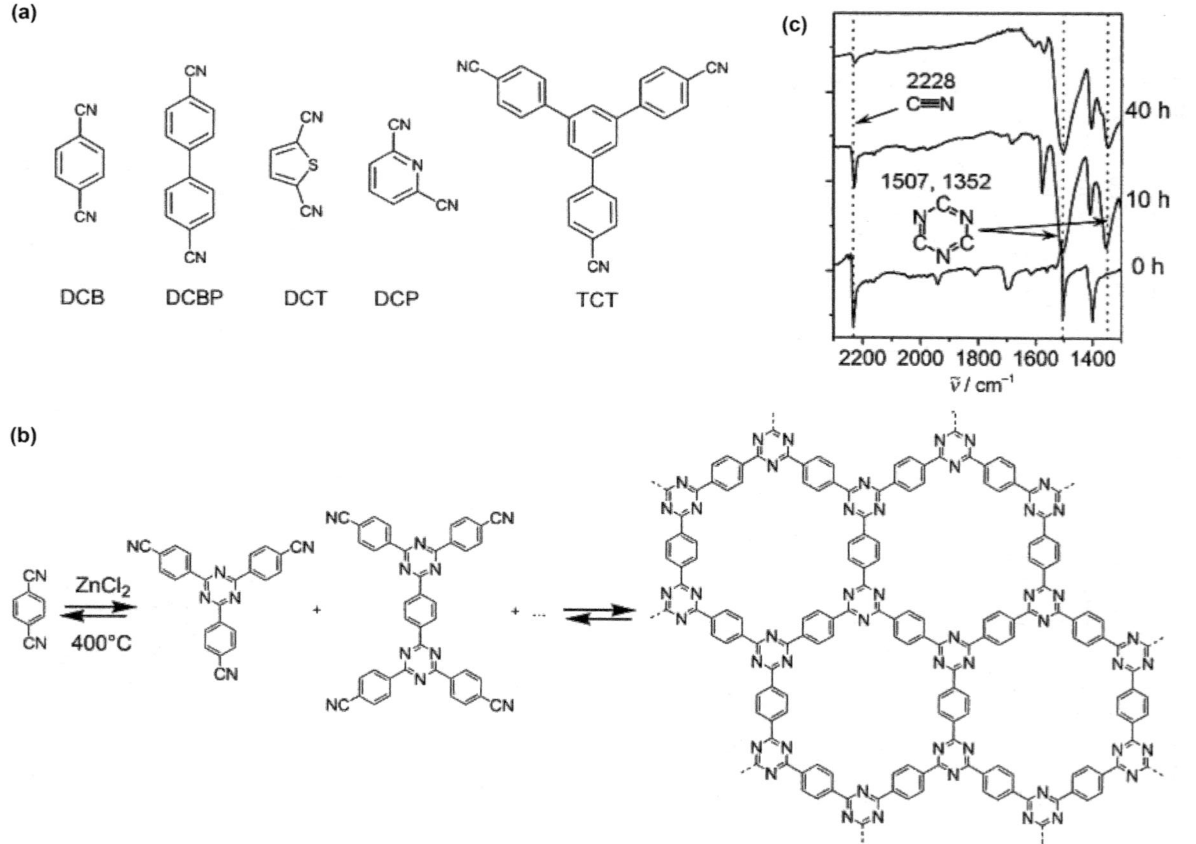

Figure 3.6 (a) Nitrile monomers used for the ionothermal synthesis of polytriazine networks. (b) Trimerization of dicyanobenzene in molten $ZnCl_2$ to trimers and oligomers, and then to a covalent triazine-based framework (CTF-1). (c) Fourier transform infrared spectroscopy (FTIR) spectra following the progress of the condensation.[58]
Reprinted with permission from P. Kuhn, M. Antonietti and A. Thomas, *Angew. Chem., Int. Ed.*, 2008, **47**, 3450. Copyright 2008 WILEY-VCH Verlag GmbH & Co. KGaA, Weinheim.

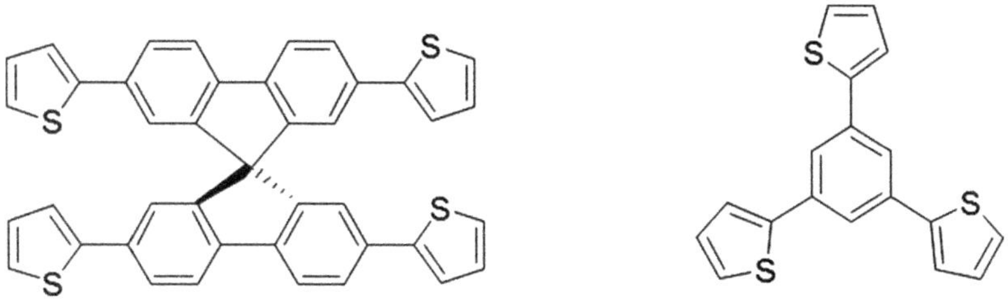

Figure 3.7 Structures of monomers used for the generation of microporous polythiophene networks.[34]
Reprinted with permission from J. Schmidt, J. Weber, J. D. Epping, M. Antonietti and A. Thomas, *Adv. Mater.*, 2009, **21**, 702. Copyright 2009 WILEY-VCH Verlag GmbH & Co. KGaA, Weinheim.

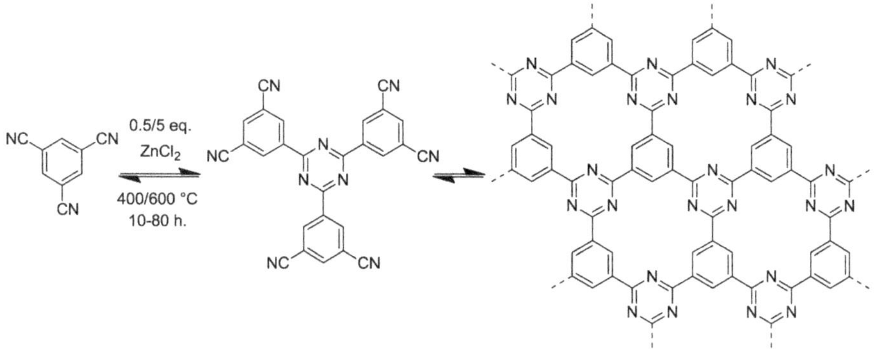

Figure 3.8 Schematic illustration of the trimerization of 1,3,5-tricyanobenzene in molten $ZnCl_2$.[70]
Reprinted with permission from P. Katekomol, J. Roeser, M. Bojdys, J. Weber and A. Thomas, *Chem. Mater.*, 2013, **25**, 1542. Copyright 2013, American Chemical Society.

first step and reorganized at the higher temperature with CN elimination and condensation leading to the formation of an amorphous network.[69] The reorganization in the second step at high temperature is coupled with (i) expansion of the network and (ii) a decrease in nitrogen content from 20 to 10%. In the reorganized structure, a hierarchical porous network with mesopores and micropores is formed, coupled with an increase in surface area and pore volume due to expansion of the network. The CTF is applied with metal nanoparticles and in the carefully ground state after being washed with water to remove $ZnCl_2$. The reaction time, temperature and monomer loading influence the porosity and structure of the network (as shown in Figure 3.8).[70] In such work, preparation of crystalline CTF using trichlorobenzene (TCB) as a monomer under ionothermal conditions can be attempted.

A framework structure with smaller pores can be obtained *via* an ideal reversible polymerization of the monomer. Such a CTF being of smaller pore size and unit cell as compared to CTF1 and CTF2 can be named CTF0. Moreover, frameworks with small pores have a higher nitrogen content as compared to other CTFs, which is useful in catalysis.[54a,71]

3.2.1.4   *Porous Aromatic Frameworks*

Homocoupling of tetrahedral monomers, such as tetrakis(4-bromophe-nyl)methane, by the Yamamoto reaction[67] forms porous aromatic frame-works (PAFs; Figure 3.9). These materials have a Langmuir surface area of 7100 m^2 g^{-1} and are gaining importance. In addition to the high surface area, PAF-1 is very stable under thermal and hydrothermal conditions, and shows high hydrogen (10.7 wt% at 77 K, 48 bar) and carbon dioxide (1300 mg g^{-1} at 298 K, 40 bar) uptakes. Moreover, the high surface area and super hydrophobicity of PAF-1 give rise to high uptake capacities of benzene and toluene vapors at room temperature.[67] PAF-1 synthesis by Yamamoto coupling shows unexpected elimination of the halogen of the end group. Such an option is unique to prepare ultrahigh porous solids with heavy ending halogen[72] atoms that decrease the surface area. This perhaps is the reason why Sonogashira–Hagihara[73–82] and Suzuki[83–92] coupling lead to low surface area when PAF-1 is synthesized following this routes. Si (PAF-3) and Ge (PAF-4) have been prepared following Yamamoto-type Ullmann coupling. Recognition of the gas molecule at 273 K shows that only greenhouse gases, like methane and carbon dioxide, could be adsorbed into these PAFs.[93] The 1,3,5-tris(4-bromophenyl)-benzene monomer has been used to make a new PAF with phenyl rings, giving rise to PAF-5 by a Yamamoto-type Ullmann reaction.[94] Tetrakis(bromophenyl)silane, the silicon-based monomer re-ported by Holst *et al.*,[95] reacts under appropriate conditions to give a PPN-4 network (also called PAF-3) with a surface area of 6461 m^2 g^{-1}.[96] Such ma-terials of high surface area have been prepared by Yamamoto coupling.[95,97] They also include the direct monomer of PAF-1 synthesized from tetrakis(4-bromophenyl)methane.[67] Yamamoto homocoupling under a conventional pathway is effective for the synthesis of PAFs using DMF as a solvent at 80 °C. Applying the same reaction conditions to tetrakis(4-bromophenyl)silane/tetra-kis(4-bromophenyl)-admantane, it was found that materials with lower surface area are formed. Assuming that a reduction of the temperature slows down the reaction to give polymers of higher molecular weight with a view to avoid un-wanted side reactions, Yamamoto homocoupling may be optimized and used. In this process, PPN-3, PPN-4 and PPN-5 with very high surface area are prepared.

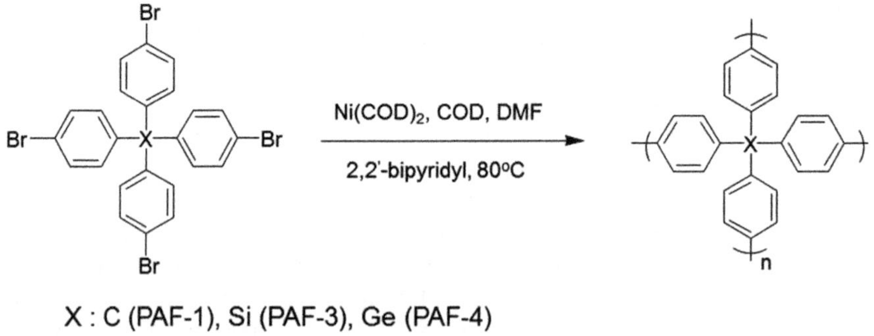

Figure 3.9   Synthesis of PAF-1, PAF-3 and PAF-4.

This procedure is performed at room temperature in a DMF–tetrahydrofuran (THF) solvent mixture. The surface area of PPN-4 approximates the molecular modelling simulation value. This clearly shows the superiority of this process.[98] The 3D monomer coupling with tetrakis(bromophenyl)adamantane by a Yamamoto reaction has been done independently by Holst *et al.*[95] and Lu *et al.* (PPN-3).[97] The surface areas were 3180 and 2840 $m^2 g^{-1}$, respectively. Using the same process to homocouple tetrakis(ethynylphenyl)adamantane (PPN-2) gives a network exhibiting a BET surface area of 1764 $m^2 g^{-1}$.[97]

Using the Suzuki reaction, preparation of 3D networks have also been reported.[99] Tetra-(4-anilyl)methane and tetrakis(4-ethynylphenyl)methane *via* 'click' reactions lead to surface areas of 1440 $m^2 g^{-1}$.[95,100] Tetrakis(4-ethynylbenzene)methane and 1,4-diazidobenzene under similar conditions give rise to a comparable network.[101] In a similar fashion, the adamantane-centered tetrahedral monomer gives rise to a network with a BET surface area around 494 $m^2 g^{-1}$ only.[101] Homocoupling of tetrahedral monomers has led to the synthesis of three porous polymer networks (PPNs). These networks are not conventionally soluble in solvents and show high thermal and chemical stability like other hypercrosslinked networks.[97] Yamamoto coupling (TBPA) and oxidative Eglinton coupling of terminal alkynes (TEPM and TEPA) has also been performed (Figure 3.10).[102] Here, a tetrahedral adamantane core is introduced to spread peripheral phenyl rings further away and to eliminate the inaccessible space.

Tetraphenylmethane block and triangular triazine-ring-based 3D PAFs with Langmuir surface areas of 1109 $m^2 g^{-1}$ selectively adsorb benzene.[103] The synthesis is performed under ionothermal conditions and was elaborated by Kuhn *et al.*[58] In this process, anhydrous $ZnCl_2$ is mixed and heated in a sealed Pyrex ampoule for 48 h at 400 °C. The product obtained from the reaction is stirred for 24 h in 1 M HCl and cleaned with water and THF several times. Three aromatic nitrile groups are trimerized to form a C_3N_3 ring by the catalytic action of $ZnCl_2$. This ring and tetraphenylmethane form a covalently connected open organic framework. This reaction, when followed by FTIR, demonstrates the completion of the trimerization reaction as denoted by the loss of intensity of the C≡N vibration at 2234 cm^{-1} (PAF-2; Figure 3.11).[104]

PAF-16 is assembled by tetrahedral (TCPSi) units and triangular C_3N_3 rings by high temperature trimerization of aromatic nitriles to form the latter catalysed by $ZnCl_2$. The PAF-16 surface area is controlled by (i) the reaction temperature and (ii) monomer loading on the catalyst. The 3D framework of the as synthesized PAF-16 has high capacity for CO_2 adsorption. To form this network, each triangular C_3N_3 ring is polymerized with TCPSi. Here, TCPSi is replaced by TCPM. The reasons are (i) their shapes are almost identical and (ii) TCPSi synthesis occurs under safe conditions.[105] A typical PAF-16 synthesis procedure thus involves heating for 40 h in a sealed Pyrex ampoule with varying mixtures of TCPSi and anhydrous $ZnCl_2$. Varying the temperature from 400 to 600 °C gives different batches of PAF-16[105] (Figure 3.12).

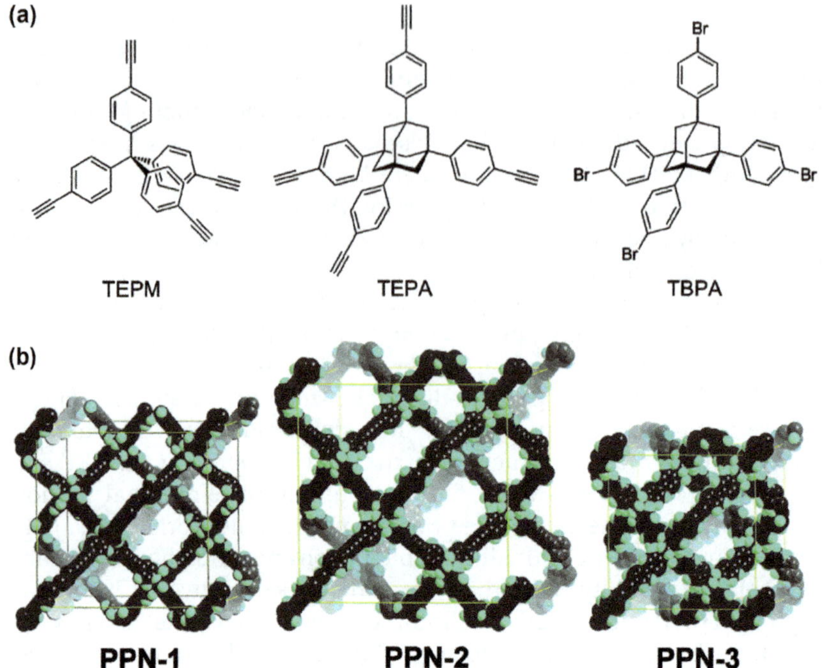

Figure 3.10 (a) Tetrahedral monomers and (b) the default non-interpenetrated diamondoid networks of the PPNs generated by coupling reactions (TEPM, PPN-1; TEPA, PPN-2; TBPA, PPN-3).
Reprinted with permission from W. Lu, D. Yuan, D. Zhao, C. I. Schilling, O. Plietzsch, T. Muller, S. Bräse, J. Guenther, J. Blümel, R. Krishna, Z. Li and H. C. Zhou, *Chem. Mater.*, 2010, **22**, 5964. Copyright 2010, American Chemical Society.

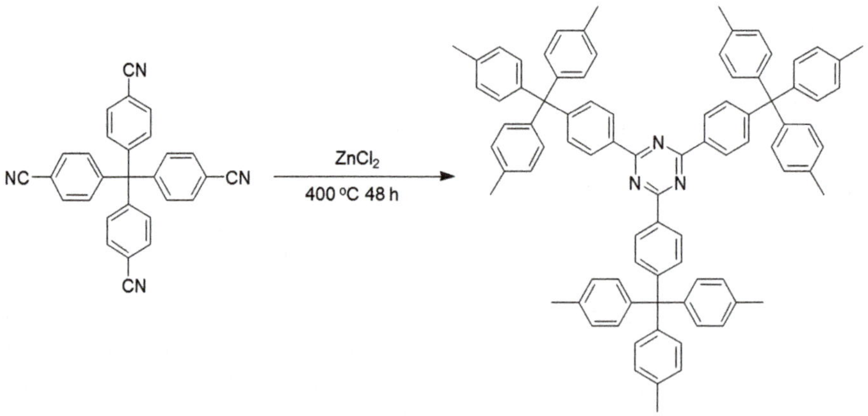

Figure 3.11 The synthesis of PAF-2.

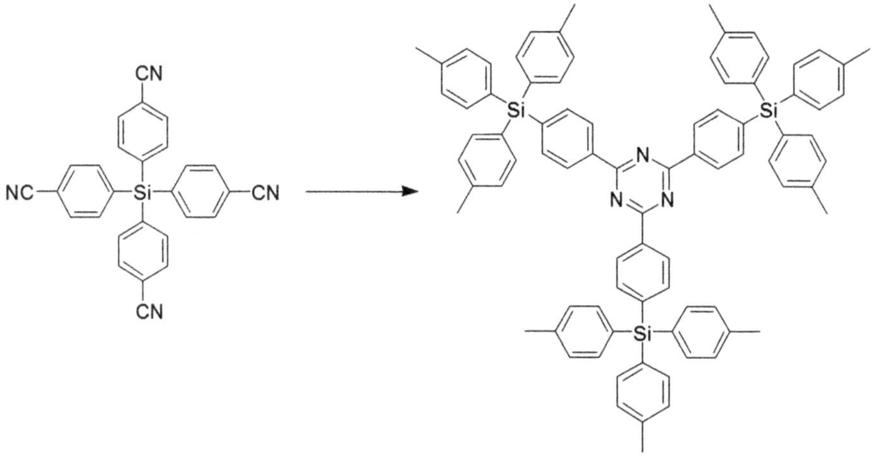

Figure 3.12   Synthesis of PAF-16.

Another novel PAF with tetra-(4-anilyl)-methane and cyanuric chloride, and high selectivity towards CO_2 and CH_4 has been synthesized.[106] Note that cyanuric chloride is highly reactive towards amines and alcohols, and forms triazines at a high nitrogen content, a feature desirable for CO_2 gas adsorption.[107] PAF-30 is such a network formed by the condensation of tetra-(4-anilyl)-methane and cyanuric chloride in dimethylacetyl amide with *N,N'*-diisopropylethylamine. Note here that tetraphenylmethane derivatives, being tetrahedral, serve as an effective building block for making the PAF. PAF-50 with cationic quaternary pyridinium ionic centers with antibacterial properties, although non-toxic to mammalian cells, are synthesized by condensation of cyanuric chloride and 4-pyridinylboronic acid at 120 °C in toluene for 48 h[108] (Figure 3.13).

Well-defined covalently-linked microporous organic–inorganic hybrid frameworks, *e.g.*, polyoctaphenylsilsesquioxane (JUC-Z1), have been effectively synthesized by a Yamamoto-type coupling reaction. JUC-Z1 is prepared using block *para*-iodo-octaphenylsilsesquioxane (I8OPS) with a yield of *ca.* 100%.[109] The synthesis of organic molecular sieves (JUC-Z2) from *para*-tribromotribenzylaniline monomers has also been attempted in a targeted fashion.[110] Yamamoto-type Ullmann chemistry, unlike traditional chemical or electrochemical oxidation to generate hyperbranched nonporous networks, has been develop to prepare 2D polymer sheets controllably. This is achieved *via* the targeted cross-coupling of *para*-tribromotribenzylaniline at the 4-position and the stacked sheets of the lamellar organic framework with an hcb topology *via* successive self-assembly. JUC-Z4-Cl, another novel electroactive PAF, has been prepared by Yamamoto-type Ullmann cross-coupling using tris(4-chlorophenyl)phosphine.

Polytri(*p*-phenyl)phosphine (JUC-Z4) and polytri(*p*-phenyl)phosphine oxide (JUC-Z5), which are stable, reductive PAFs, are obtained as off-white powders[111] following this route (Figure 3.14).

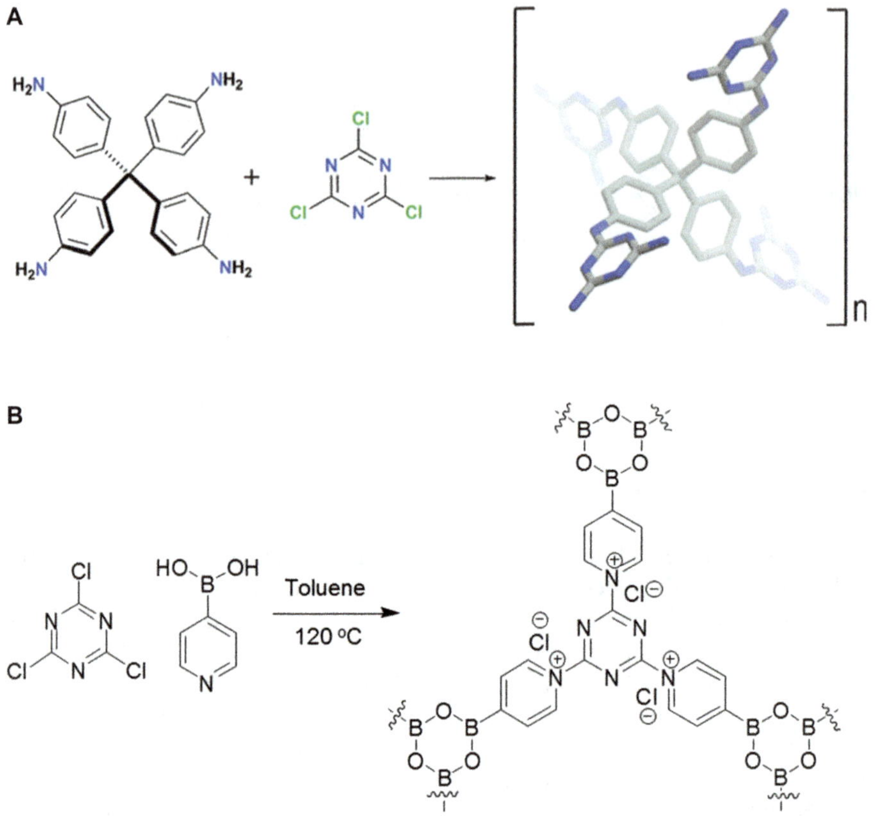

A

B

Figure 3.13 (A) Scheme of PAF-30 synthesis. Reproduced from ref. 106 with permission from The Royal Society of Chemistry. (B) Scheme of PAF-50 synthesis. Condensation reactions are used to produce discrete molecules and extended PAFs. Reprinted with permission from Y. Yuan, F. Sun, F. Zhang, H. Ren, M. Guo, K. Cai, X. Jing, X. Gao and G. Zhu, *Adv. Mater.*, 2013, **25**, 6619. Copyright 2013 WILEY-VCH Verlag GmbH & Co. KGaA, Weinheim.

3.2.2 Soluble Porous Polymers

3.2.2.1 *Polymers with Intrinsic Microporosity*

Polymers with intrinsic microporosity, abbreviated as PIMs, were developed for the first time by McKeown and Budd. The general design strategy involves incorporation of extended aromatic components within a rigid polymer network in order to imitate the structural layout of graphene sheets of activated carbons.[54b,112] The principle behind the synthesis of PIMs employs non-reversible condensation reactions to form polymers. These polymers cannot efficiently fill the space and pack together due to components that are rigid and ladder-like, and thus force the backbone of the polymer to twist or turn.

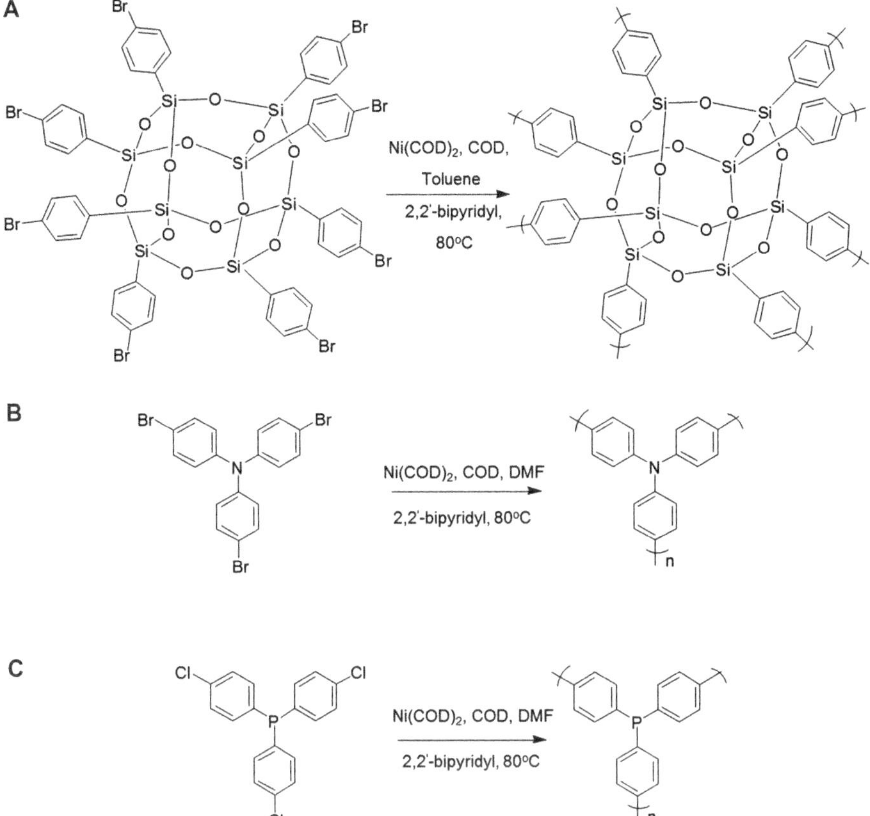

Figure 3.14 Synthesis of JUC-Z1 (A), JUC-Z2 (B) and JUC-Z4-Cl (C) by Yamamoto coupling.

These components in the polymer are the bent monomers containing a tetrahedral carbon, also known as the site of contortion, which results in intrinsic microporosity (Figure 3.15). The spiro center, which is a tetrahedral carbon shared by two rings, can be a site of contortion. The reaction to form dioxane by an *ortho*-dihydroxy monomer and a dihalide, thereby forming a linear network, is a general strategy adopted for the synthesis of such polymers.[112,113] The first PIMs were obtained using phthalocyanines and porphyrins with metal ions/$2H^+$ in the cavities[114,115] (Figure 3.16). Due to the extended planarity of the phthalocyanine macrocycle as the aromatic unit (diameter ~ 1.5 nm) and the oxidative catalytic properties of the transition metal ions in the cavities,[116] several phthalocyanine networks have been designed with different metal centers, such as Zn^{2+}, Cu^{2+} and Co^{2+}. Even $2H^+$ centers have been developed and these materials exhibit porosity and BET surfaces areas of 450–950 $m^2\ g^{-1}$.

In a previously prepared network, the aromatic units tend to form columnar stacks, owing to the strong non-covalent interactions (primarily

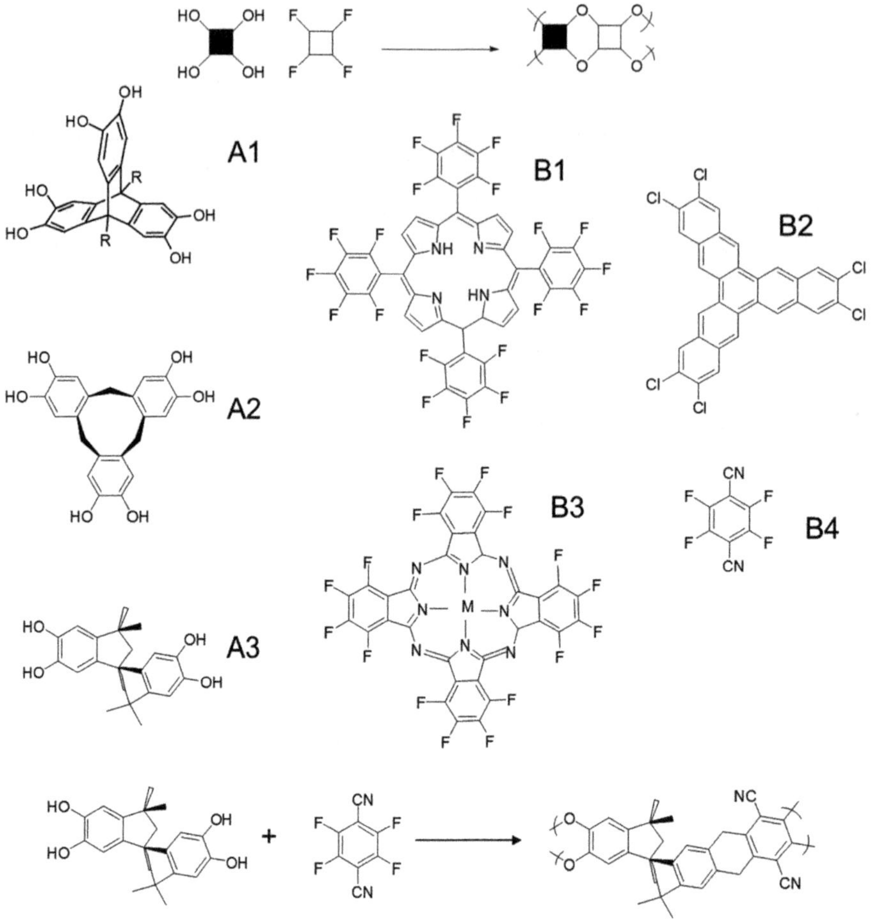

Figure 3.15 Some monomers used in the synthesis of PIMs. A1–3 are hydroxy monomers and B1–4 are halide monomers used in the synthesis of PIMs.

π–π interactions) that form nonporous solids.[117] This has necessitated the use of a highly rigid and nonlinear linking groups, such as the one derived from the 5,5′,6,6′-tetrahydroxy-3,3,3′,3′-tetramethyl-1,1′-spirobisindane monomer (A3), between such subunits, which ensure space-inefficient packing and prevent easing of the structure and loss of porosity (Figure 3.15). The spiro center of this linking group gives the nonlinear shape, whereas the fused ring gives rigidity. Such polymer networks have been prepared from a precursor, bis(phthalonitrile), by dioxane, resulting in cyclotetramerization helped by a metal template. The same process gives highly colored network polymers with A1 and 4,5-dichlorophthalonitrile (an efficient double aromatic nucleophilic substitution reaction—SNAr).[54b] Metal-containing porphyrins are considered as another suitable candidate for PIM synthesis due to their importance in catalysis; for example,

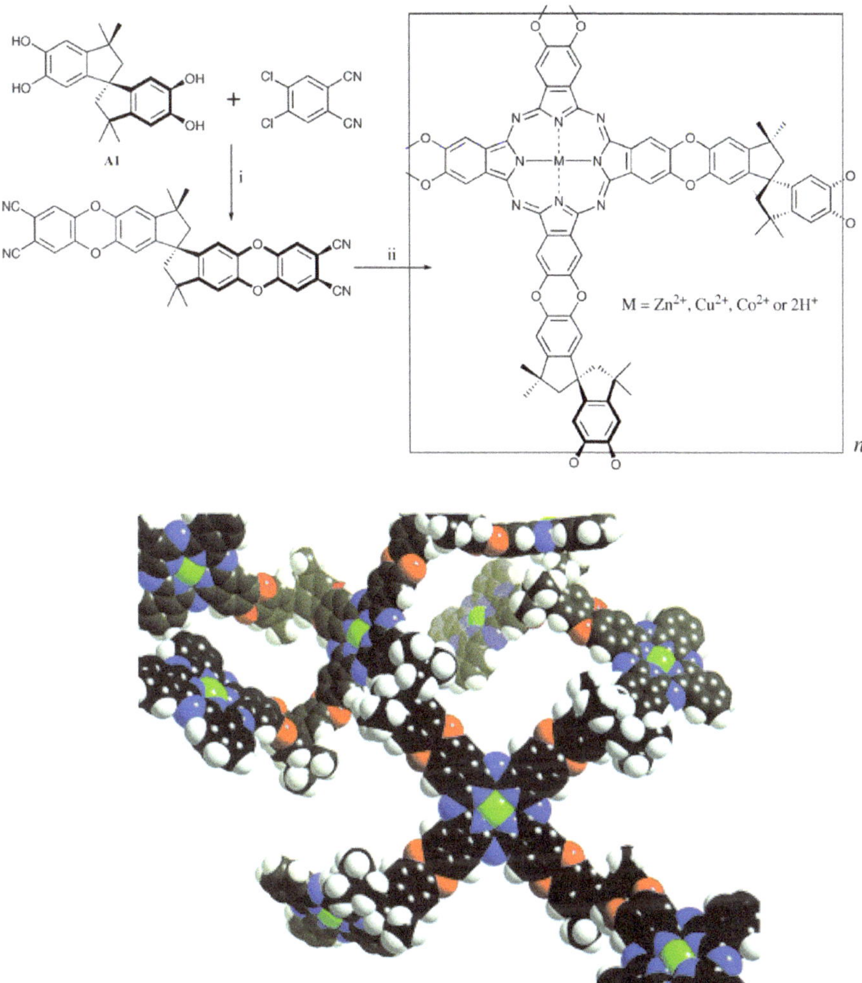

Figure 3.16 The preparation of phthalocyanine-based microporous network polymers (Pc-network-PIM) from the readily available spirocyclic monomer A1. Reproduced from ref. 54 with permission from The Royal Society of Chemistry.

iron–porphyrin derivatives are quite effective in the catalysis of hydrocarbon hydroxylations and alkene epoxidations are almost comparable to the activity of the cytochrome P450 enzymes.[118] These reactions are accomplished in the presence of oxidizing agents, such as O_2 or H_2O_2, and therefore possibly act as useful heterogeneous catalysts. Thus, porphyrins are desirable components of this material. Such a synthesis is a low-yielding reaction, which is very unfavorable for polymer network assembly. In order to resolve this issue, nonflexible groups of spirocyclic are introduced directly between preformed porphyrin units. This is done by a dioxane-forming reaction between the porphyrin and monomer A1. Here, the linking group is not completely made of fused rings. There should be much steric hindrance, which restricts rotation about the single C–C bond at the meso positions of the porphyrins and thus averts structural easing.[112] The synthesis of the porphyrin-network PIM based on a dioxane formation reaction lays out a blueprint for the chemical preparation of PIMs from appropriate rigid monomers and fluorinated (or chlorinated) monomers of aromatic compounds. This general reaction has been used to react monomer A1 with hexa-azatrinaphthylene to form an insoluble HATN-PIM network (HAPN, hexachlorohexaazatrinaphthylene; Figure 3.17), which can bind to Pd metal

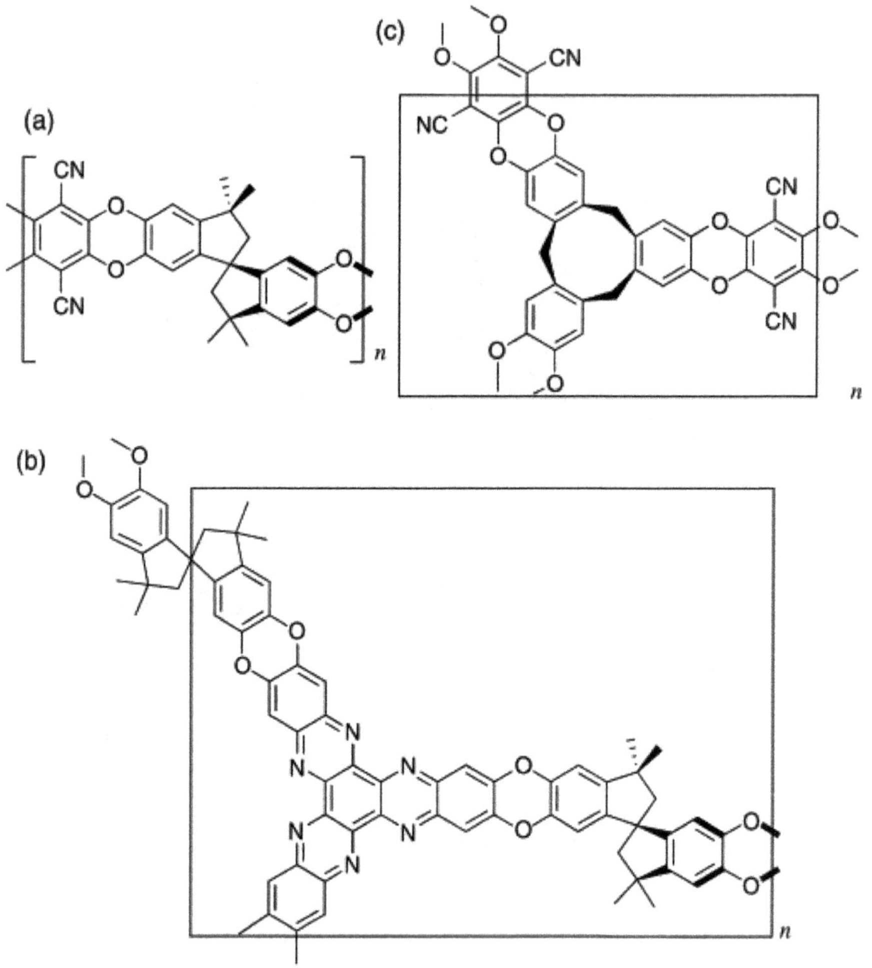

Figure 3.17   Structure of (a) PIM-1, (b) HATN-network PIM and (c) CTC-network PIM.[55]
Reprinted with permission from N. B. McKeown, B. Gahnem, K. J. Msayib, P. M. Budd, C. E. Tattershall, K. Mahmood, S. Tan, D. Book, H. W. Langmi and A. Walton, *Angew. Chem., Int. Ed.*, 2006, **45**, 1804. Copyright 2006 WILEY-VCH Verlag GmbH & Co. KGaA, Weinheim.

owing to the built-in ligands for complexation, forming a metal-coordinating PIM network that acts as an excellent heterogeneous catalyst for Suzuki coupling reactions.[119] The presence of a spiro center provides the required site of contortion essential for porosity and is evident in the synthesis of polyimides, with characteristics like PIMs, through reaction of various aromatic diamines with bis(carboxylic anhydride), which in turn incorporate a spiro center.[120] It has also been found that polymers with greater intrinsic microporosity (IM) can be designed when the rotation about the carbon–nitrogen imide bonds is disturbed in the polyimides by steric congestion accomplished by appropriate substitutions on the diamine co-monomer, which in turn is reacted with the rigid bis(anhydride) spiro(bisindane) monomer.[120–124] In contrast, the imide linking groups and concave monomers with relatively unhindered rotation (*e.g.*, bisnaphthylene)[122] give rise to non-network polymers with lower limits of IM.[120,124] Hence, limited rotational freedom is an essential requirement for IM in non-network polymers.[113]

By manipulating the number of aromatic *ortho*-dihalide groups and catechols in the monomers, a PIM can be prepared either as an insoluble or soluble polymer network. For instance, a soluble PIM (PIM-1) is prepared from the reaction between 5,5′,6,6′-tetrahydroxy-3,3,3′,3′-tetramethyl-1,1′-spirobisindane and tetrafluoroterephthalonitrile. On the other hand, HATN-network PIM is synthesized by the reaction between the readily available spiro-cyclic bis(catechol) monomer and hexachlorohexaazatrinaphthylene.[119]

PIM-1 was one of the first extensively studied materials in the class of soluble polymers. It is synthesized by the tetrahydroxy monomer 5,5′,6,6′-tetrahydroxy-3,3,3′,3′-tetramethyl-1,10-spirobisindane (TTSBI) and the fluoro-monomer 1,4-dicyanotetrafluorobenzene (DTFB) using a step polymerization method, which involves a double aromatic nucleophilic substitution. It is soluble in THF, chloroform, dichloromethane and *o*-dichlorobenzene.[125] PIM-1 can be cast from solution and forms microporous films, which show excellent potential for gas separation[126] and solution-phase membrane separations too.[71] Different monomers have been introduced into soluble PIMs, such as spirobisindanes,[121,127] bis(phenazyl) monomers, 2,2′,3,3′-tetrahydroxy-1,10-dinaphthyl,[128] heptafluoro-*p*-tolylphenylsulfone[129] and disulfone-based monomers.[130] Film-forming copolymers with high free volumes have been synthesized by substituting a fixed ratio of the spiro units of PIM-1 with units derived from 9,10-dimethyl-9,10-dihydro-9,10-ethanoanthracene-2,3,6,7-tetrol (CO1).[131] This co-unit has the shape of a roof due to the ethanol bridge over the middle ring. Thus, CO1 creates a 'bend' in the polymeric structure and induces greater inflexibility than the single tetrahedral carbon spiro center of TTSBI, which in turn modifies the polymer chains packing and increases the free volume.

To synthesize a non-soluble polymer network, the average functionality of the monomer combination must be greater than two. Each pair of adjacent hydroxyl or fluorine substituents may be counted as a single functional group for dioxane formation. The 'pick-and-mix' region of combinable monomers includes pre-formed fluorinated phthalocyanine, the tridentate ligand HATN and rigid hydroxylated monomers, such as cyclotricatechylene (CTC) or calixresorcarene, that possess hosting cavities for molecules.[54b]

Polymers with intrinsic porosity have also been synthesized from triptycene monomers, which contain alkyl groups connected to their bridge-head positions.[143] It has been found that branched (*e.g.*, isopropyl) or shorter (*e.g.*, methyl) alkyl chains form networks of maximum microporosity. On the other hand, longer alkyl chains tend to obstruct the microporosity produced by the rigid organic framework. This is analogous to preferring smaller cross-linkers in HCPs as smaller, rigid monomeric units ensure large surface area. Trip PIM, which has one of the largest BET surface areas among PIMs at 1064 m^2 g^{-1}, is synthesized from hexahydroxytriptycene[132] and cyclotricatechylene[55] by the same dioxane-forming reaction mentioned before.[133] The enhanced microporosity in triptycene-based PIMs, in comparison to other PIMs, can be traced to the polymeric framework, as guided by the triptycene units. These building blocks help to minimize intermolecular connections between the extended planar bridges of the rigid framework and thus give rise to 'loose packing' within the material. Swager and Long devised the concept of internal molecular free volume (IMFV).[134] This concept draws attention in this regard as it proposes that the width and shape of the triptycene unit gives a significant amount of unoccupied space.[135] The choice of triptycene units as monomers can be justified by the enhanced mechanical strength in triptycene-based polymers, which in turn can be attributed to the interlocking of these units leading to greater polymer cohesion.[136–138] The choice of monomers is also influenced by the presence of any pre-formed cavities in the precursors, which can provide sites for hydrogen adsorption and thereby tune pore size distribution.[55] One such example is the incorporation of CTC, where the bowl-shaped monomer acts as a receptor[139] in a network PIM by a reaction forming benzodioxane between CTC and tetrafluoroterephthalonitrile, resulting in an ultramicroporous network[55] (Figure 3.17). In case of non-network soluble PIMs, structural building blocks with well-formed cavities are provided by 1,1′-spirobisindanes,[71,120–122,140,141] 9,9′-spirobisfluorenes,[123] bisnaphthalenes,[124,128,140] 1,1′-spirobis-2,3,4-tetrahydronaphthalenes[127] and 9,10-ethanoanthracene.[141] The soluble polymer that shows the greatest intrinsic microporosity (apparent BET surface area = 850 m^2 g^{-1}) has a 1,1′-spirobisindane framework to which two fluorene units have been fused with spiro centers to give additional concavities.[121] In these PIMs, using an efficient double aromatic nucleophilic substitution reaction, the fused ring linking group is formed between a catechol (1,2-dihydroaryl) and an 1,2-difluoro- (or 1,2-dichloro-) aryl unit.[54a] Extension into building blocks, such as bis(phenazyl) monomers, fused fluorenes and imides, leads to the fabrication of PIM networks with BET surface areas in the range of 470–900 m^2 g^{-1}. This has facilitated further tuning of the separation properties and permeability of PIM networks.[121,141,142] Rigid and twisted spirobifluorenes serve as suitable monomers, providing sites of contortion upon product formation. These form PIM networks *via* imide- and amide-forming reactions.[123] The imide-forming reaction has also been employed to synthesize binaphthalene-based PIMs, where the monomer is used as a site of contortion. In fact it has also been used to design chiral PIMs.[124] Another chemical synthesis strategy devised to synthesize insoluble polymers is based on molecular rearrangement of soluble precursors at high temperatures. For instance, benzoxazole–phenylene- or benzithiazole–phenylene-based polymers are insoluble and infusible in spite of their soluble building units owing to their molecular rearrangement that occurs at about 350–450 °C.[143]

To conclude, a general outline to induce microporosity is as follows: for the lowest microporosity, one of the rigid monomers must have a site of contortion, which can be a spiro center, a single covalent bond around which rotation is blocked or a nonplanar rigid skeleton. When two planar monomers react the result is a nonporous material.[54b]

3.2.2.2 Soluble Conjugated Microporous Polymers

As mentioned in Section 3.2.1.4, the synthesis of these polymers is based on hyperbranching, *e.g.*, to synthesize soluble polyphenylenes that are hyperbranched.[144] In the beginning, 1,3,6,8-tetrabromopyrene was used as an A4 monomer to build insoluble pyrene CMP networks[145] (Figure 3.18). The *t*-butyl-functionalized B2 monomer is incorporated to control the molecular weight of the network and to introduce solubilizing alkyl groups.[146] To synthesize soluble CMPs, a two-step (A4 + B2)-type Suzuki-catalyzed aryl–aryl coupling copolymerization is done. At first, palladium acetate $(Pd(OAc)_2)$[147] catalyzes an aryl halide or diboron coupling to produce arylboronates of both the A4 monomer (1,3,6,8-tetrabromopyrene) and the B2 monomer (1,3-dibromo-7-*tert*-butylpyrene)[148] in a single step 'pre-polymerization'. Statistical copolymerization of the two monomers is then carried out without isolating the arylboronate species and by adding $Pd(PPh_3)_4$ and K_2CO_3 in the second step. The polymer was isolated as a deep yellow film after purification by antisolvent reprecipitation.

3.3 Crystalline Covalent Organic Frameworks

Covalent organic frameworks (COFs) are primarily crystalline in nature, although some amorphous COFs have also been synthesized. The field of COF synthesis has opened up new arenas of the design of tailor-made functionalities imbedded in the building units to facilitate the formation of complex architectural networks with extremely high surface areas. In polymer synthesis, generally, kinetically controlled irreversible reactions are employed. However, it is quite challenging in COF synthesis to crystallize solids through irreversible reactions to link organic polymers. In this regard,

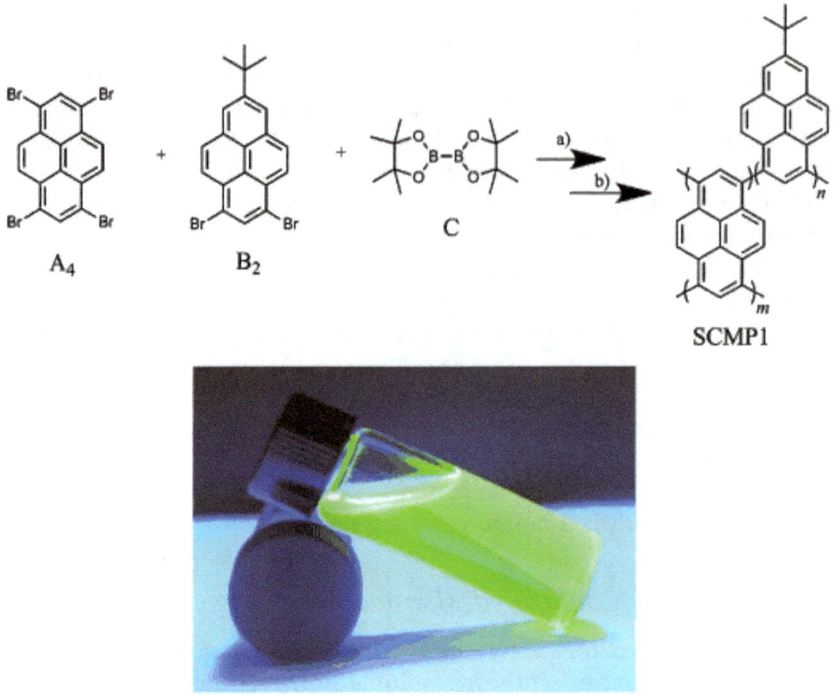

A$_4$ B$_2$ C   a)   b)   SCMP1

Figure 3.18   Top: two-step, one-pot synthesis of a soluble conjugated microporous polymer, SCMP-1 (A4, 1,3,6,8-tetrabromopyrene; B2, 1,3-dibromo-7-*tert*-butylpyrene). The resulting material is a statistical hyperbranched copolymer that is soluble in common organic solvents. Bottom: a solution of SCMP-1 in THF shows green luminescence under ultraviolet (UV) irradiation.[146]
Reprinted with permission from G. Cheng, T. Hasell, A. Trewin, D. J. Adams and A. I. Cooper, *Angew. Chem., Int. Ed.*, 2012, **51**, 12727. Copyright 2012 WILEY-VCH Verlag GmbH & Co. KGaA, Weinheim.

dynamic covalent chemistry (DCC) offers a solution as it deals with the reversible formation, breaking and reformation of covalent bonds.[149] DCC is thermodynamically controlled and, unlike conventional covalent bond formation, it provides reaction systems that are reversible with 'error checking' and 'proof-reading' characteristics. It hence facilitates the formation of the most thermodynamically stable structures. It is this application of the DCC concept that ensures construction of COFs in a manner so that the crystallization process occurs alongside the formation of the polymer skeleton. On the other hand, the self-healing feedback diminishes the structural defects and hence assists in the formation of a structure that is ordered. The final COF product, as a result, has a crystalline structure that has high thermodynamic stability. Two key factors or parameters should be considered in order to establish thermodynamic control in reactions that are reversible: (i) the structure of the monomeric components and (ii) the method of synthesis, the medium of reaction and the reaction conditions.

Different reaction conditions and methods have been tried out for COF synthesis, among which two methods stand out, namely solvothermal and rapid microwave syntheses. A typical solvothermal method for synthesizing COFs involves mixing monomers and a combination of solvents in a Pyrex tube followed by degassing *via* several cycles of freeze–pump–thaw. The sealed tube is then heated for a certain time to a designated temperature. The precipitate is then collected later, washed with appropriate solvents and dried in vacuum to obtain the COF as a solid powder form. While considering the synthesizing media and conditions for solvothermal synthesis it is imperative to take factors like the solubility, crystal nucleation, reaction rate, crystal growth rate and 'self-healing' capacity into account. COFs have been successfully synthesized and designed with the help of phenyl diboronic acid $\{C_6H_4[B(OH)_2]_2\}$ and hexahydroxytriphenylene $[C_{18}H_6(OH)_6]$ condensation reactions.[64] The first COFs, namely COF-1 and COF-5, were produced using a mixture of dioxane/mesitylene under solvothermal conditions.[64] The COF-1 synthesis is based on the molecular dehydration reaction, where a planar six-membered B_3O_3 (boroxine) ring is formed from three boronic acid molecules with the elimination of three water molecules. Such molecular structures of boronic acids, which are cyclotrimerized, are held in planar conformations by $C-H_{(o-C_6H_4)} \cdots O_{(BO)}$ (2.975(2) Å) hydrogen bonds.[150] With the help of expertise, this reaction is applied to 1,4-benzenediboronic acid (BDBA), which upon dehydration is expected to form a layered hexagonal framework. An analogous condensation reaction is used for COF-5. A five-membered BO_2C_2 ring is formed by the dehydration reaction between phenylboronic acid and 2,3,6,7,10,11-hexahydroxytriphenylene (HHTP), which is a trigonal building block. COF-1 is produced in a sealed Pyrex tube by the heating of BDBA at 120 °C for 72 h under a mesitylene-dioxane solution.[64] The dehydration of BDBA is allowed to proceed slowly under these conditions. The nucleation of a crystalline compound is facilitated by the controlled diffusion of the building blocks into solution by the solvent system due to the sparing solubility of BDBA. On the other hand, reversible conditions conducive to crystallite growth are sustained in a closed reaction system that ensures the availability of H_2O. COF-1 is obtained as a white powder in 71% yield based on BDBA after being heated and washed with acetone. By means of similar reaction conditions, COF-5 is produced in 73% yield from a 3:2 stoichiometric ratio of BDBA.[64] It has been shown that such materials can also be synthesized *via* rapid microwave synthesis, without compromising the crystallinity of the resulting networks (Figure 3.19).[151–153]

In fact Cooper and others have created a high-throughput protocol in a microwave reactor for dynamic covalent reactions to rapidly synthesize boronate-ester-linked COFs.[151,154] From the experimental point of view, microwave synthesis is more advantageous than solvothermal methods due, primarily, to three reasons: (i) rapid synthesis of COF favors large scale production, (ii) there is no need for a sealed vessel in microwave synthesis and (iii) the microwave solvent extraction process ensures better porosity by removing impurities and residues in the frameworks more efficiently.

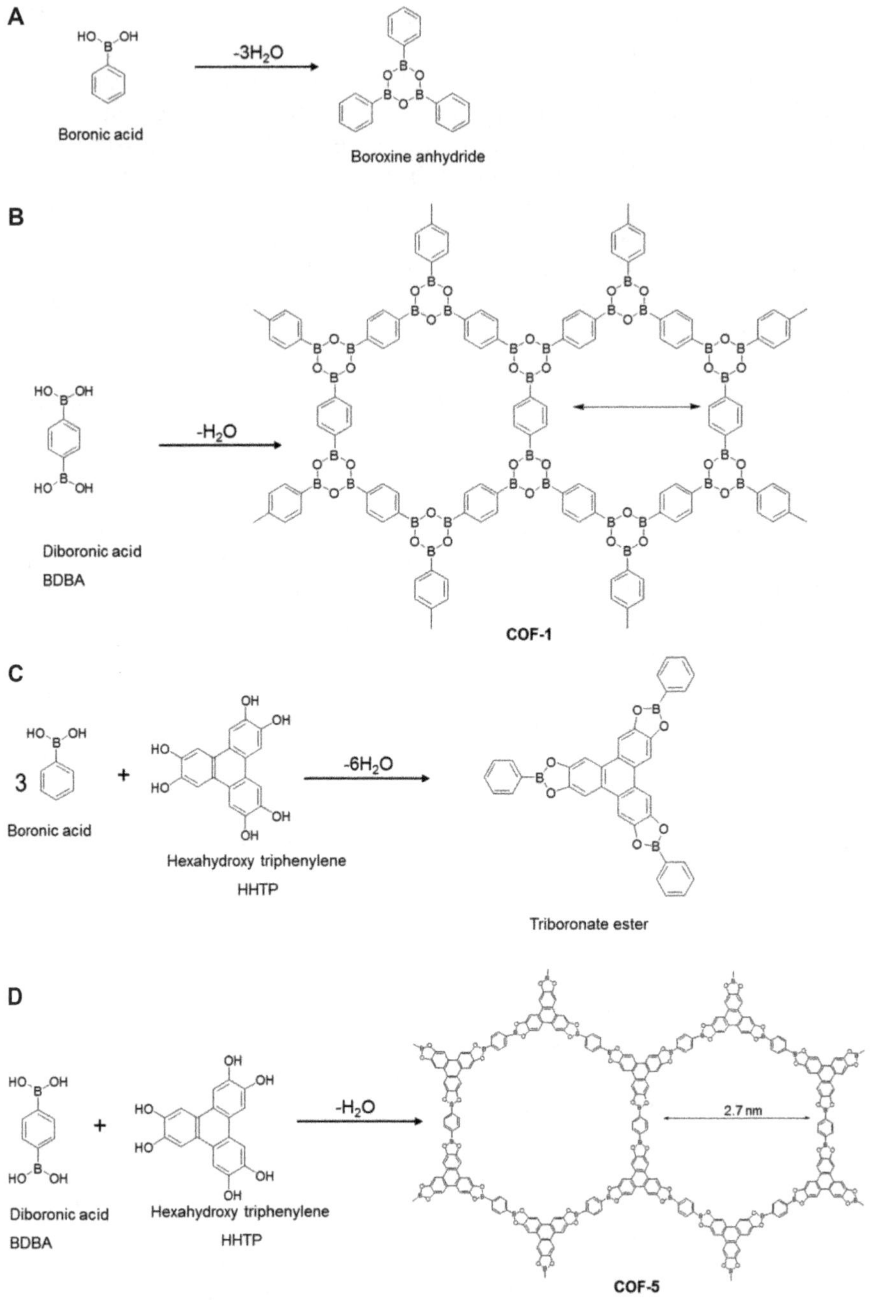

A

B

C

D

Figure 3.19   (A to D) Condensation reactions of boronic acids used to produce discrete molecules and extended COFs.[16]
Reprinted with permission from A. P. Cote, A. I. Benin, N. W. Ockwig, M. O'Keeffe, A. J. Matzger and O. M. Yaghi, *Science*, 2005, **310**, 1166. Copyright 2005, American Association for the Advancement of Science.

COF-1 is the simplest of the compounds among covalent organic frameworks. It is formed by the elimination of water in the self-condensation of benzene-1,4-diboronic acid. On the other hand, COF-5 is synthesized by employing the same type of condensation reaction between benzene-1,4-diboronic acid and HHTP, forming a BC_2O_2-linking ring. Crystalline COFs are formed by condensation of 1,2,4,5-tetrahydroxybenzene and benzene-1,3,5-triboronic acid by simple reflux in a THF and methanol mixture. This forms a five membered ring, as observed in COF-1, COF-102 and COF-103.[153] COFs having larger pores have also been reported. For instance, BTP-COF is prepared using a solvothermal route that employs 1,3,5-benzenetris(4-phenylboronic acid) (BTPA) and 2,3,6,7-tetrahydroxy-9,10-dimethylanthracene to give a pore size of 40 Å. Two-dimensional COFs have also been synthesized using the HHTP linker (COF-6-$C_8H_3BO_2$, COF-8-$C_{14}H_7$–BO_2 and COF-10-C_6H_3BO).[155] Co-condensation reactions between 1,3,5-benzenetriboronic acid (BTBA), HHTP, BTPA and 4,4′-biphenyldiboronic acid (BPDA) leads to formation of COF-6, COF-8 and COF-10, respectively. Upon condensation, C_2O_2B boronate esters are formed. This links HHTP with BTPA (COF-8), BTBA (COF-6) and BPDA (COF-10)[155] (Figure 3.20).

Two-dimensional covalent organic frameworks have also been synthesized *via* condensation of hydrazine with 1,3,6,8-tetrakis(4-formylphenyl)pyrene

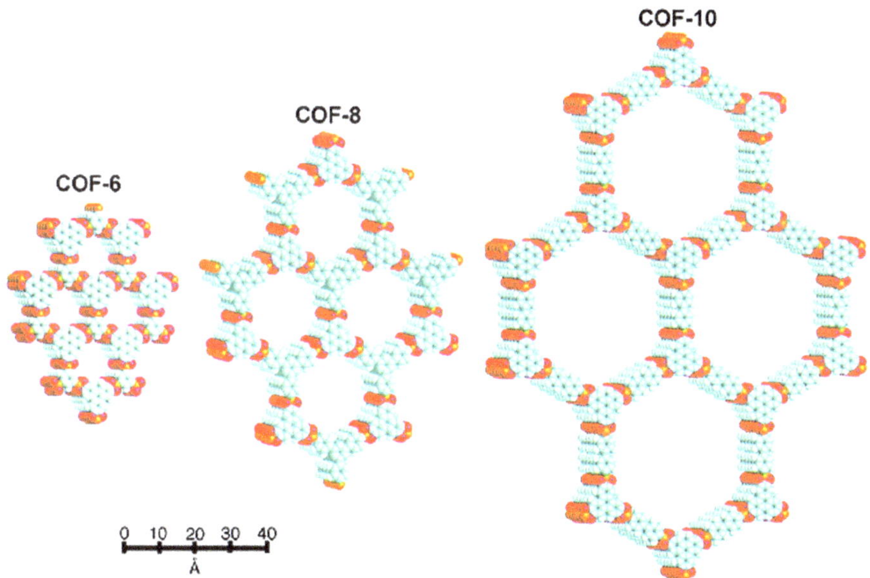

Figure 3.20 Co-condensation of boronic acid building blocks (BTBA, BTPA and BPDA) with HHTP to give 2D COFs (COF-6, COF-8 and COF-10) having systematically designed porous structures. Coloring scheme: C, grey; H, white, B, orange/yellow, O, red/pink.[155]
Reprinted with permission from A. P. Cote, H. M. El-Kaderi, H. Furukawa, J. R. Hunt and O. M. Yaghi, *J. Am. Chem. Soc.*, 2007, **129**, 12914. Copyright 2007, American Chemical Society.

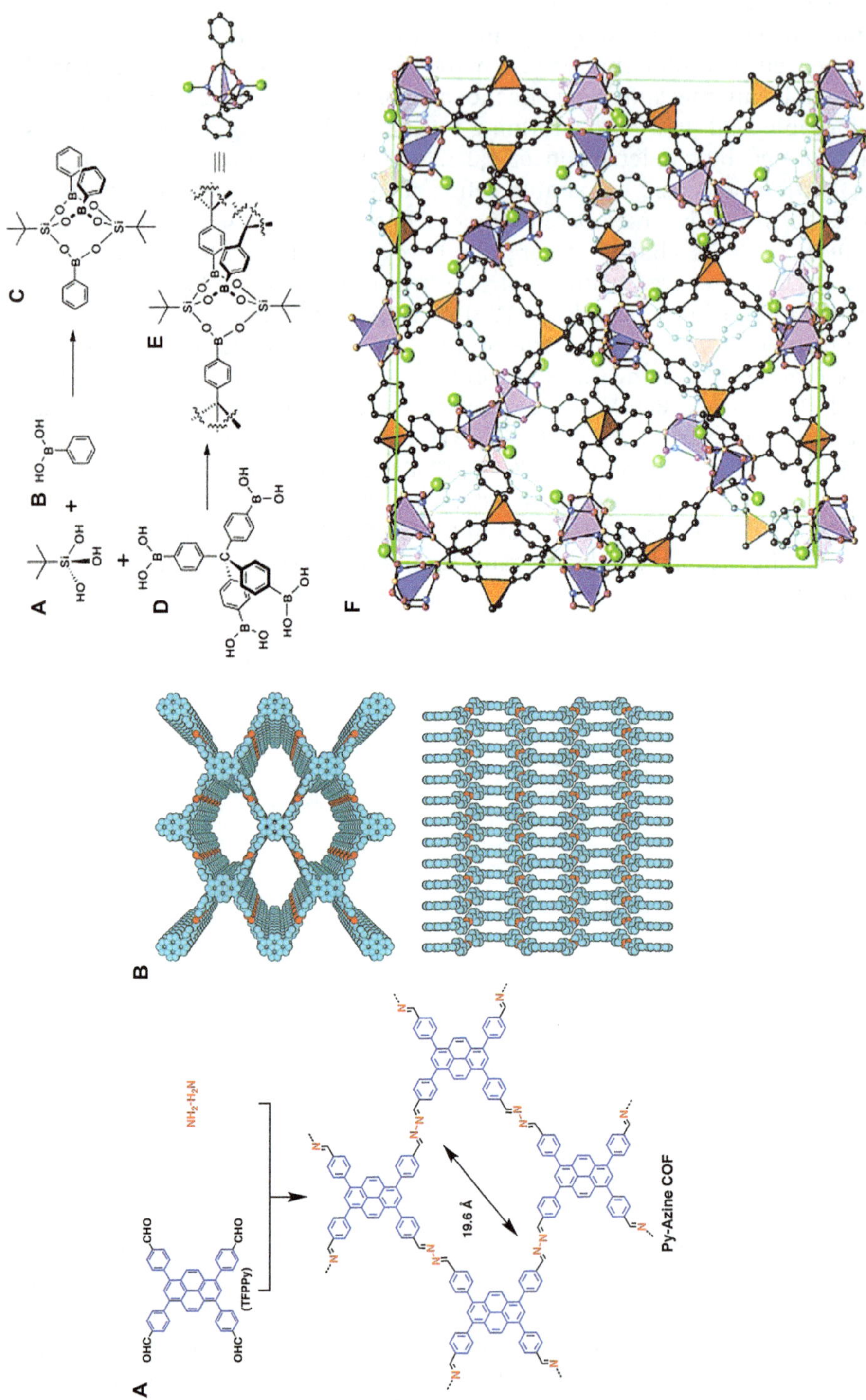
C
B
A
D
E
F
B
A
(TFPPy)
OHC
CHO
CHO
OHC
NH₂·H₂N
N
19.6 Å
Py-Azine COF

under solvothermal conditions yielding highly crystalline networks.[156] Another COF known as COF-202 is designed based on molecular borosilicate cluster chemistry, where the condensation of *tert*-butylsilane triol[157] (A) with monotopic boronic acid (B) forms a high symmetry borosilicate cage (C6). The same condensation reaction with a divergent unit of organic tetraboronic acid is used to also in a covalently linked organic framework, *i.e.*, the borosilicate cage.[158] The building units of this framework are geometrically similar to those used in the design of 3D COFs (*e.g.*, COF-102) that are boroxine based. Here, the tetrahedra and planar triangular rings are connected to give extended structures[158] (Figure 3.21).

Apart from bulk synthesis, COFs can also be designed as monolayers on surfaces. This has been achieved by the condensation of building blocks within a metal surface forming monolayers of COF-5 and COF-1.[159] Sublimation of the building blocks from heated Mo evaporator crucibles onto a clean Ag[160] surface under an ultrahigh vacuum produces the surface covalent organic frameworks (SCOFs) SCOF-1 and SCOF-2, which are covalently bound nano-architectures. However, the synthesis of monolayers on a metal surface that are defect-free demands further refining of the reaction conditions for higher purity of the building blocks and also requires a proper metal substrate interface that is single-crystalline with direct alignment of the building block. The challenge of defect-free monolayer synthesis has been overcome by using a highly ordered pyrolytic graphite (HOPG) interface to synthesize COF monolayers.[160] Biphenyldiboronic acid (BPDA), 1,4-benzene diboronic acid (BDBA) and 9,9′-dihexylfluorene-2,7-diboronic acid have been precipitated onto a HOPG interface from their THF solutions, and HOPG samples so prepared are autoclaved at 150 °C for 1 h to form monolayers of the COF. This method is capable of designing defect-free monolayers due to the presence of $CuSO_4 \cdot 5H_2O$, which serves as a 'reservoir' of water and regulates the equilibrium. This in turn is the key to obtaining a monolayer of high quality. The water molecules released from $CuSO_4 \cdot 5H_2O$ during the process of heating act as an agent to manipulate the equilibrium and the dehydration reaction is pushed backward. It facilitates the process

---

Figure 3.21   (Left) A: schematic representation of the synthesis of the azine-linked COF (py-azine COF); B: top and side views of the AA stacking structure of the py-azine COF. Reprinted with permission from S. Dalapati, S. Jin, J. Gao, Y. Xu, A. Nagai and D. Jiang, *J. Am. Chem. Soc.*, 2013, **135**, 17310. Copyright 2013, American Chemical Society. (Right) Condensation of *tert*-butylsilane triol (A) with monotopic boronic acid (B) forms the molecular borosilicate cage (C). Condensation of A with divergent boronic acid D leads to the building unit E in which the boron atoms occupy the vertices of a triangle and join together the tetrahedral building blocks D to give COF-202 (F). Coloring scheme: C, black; Si, blue; O, red; B, yellow (H atoms have been omitted and *tert*-butyl groups are represented as green spheres for clarity).
Reprinted with permission from J. R. Hunt, C. J. Doonan, J. D. LeVangie, A. P. Cote and O. M. Yaghi, *J. Am. Chem. Soc.*, 2008, **130**, 11872. Copyright 2008, American Chemical Society.

of remedying the defect and leads to the formation of a COF network with a highly ordered structure. During the cooling process, decomposition of the SCOFs is prevented by the reabsorption of water molecules by $CuSO_4$. COFs designed on templates using the above mentioned methods are either monolayers or insoluble powders, which cannot be reliably placed on electrode interfaces or integrated into devices. Hence, there is a need to fabricate COF thin films on a substrate for industrial-scale applications. In this regard, Dichtel and co-workers have prepared and characterized layers of 2D COF films on surfaces of single-layer graphene (SLG) on SiO_2.[160] In solvothermal reaction systems, SLG/SiO_2 COF thin films of high orientation are produced on SLG. Various COF films of different varieties, such as COF-5, TP-COF, NiPc-COF, HHTP-DPB-COF and ZnPc-PPE-COF, have been successfully synthesized on graphene and their thicknesses were controlled by manipulating the time of the reaction.[160]

Metallophthalocyanine co-condensation with an electron-deficient benzothiadiazole (BTDA) block leads to 2D covalent organic framework (2D-NiPc-BTDA COF) formation with excellent electron mobility and panchromatic photoconductivity properties[161a] (Figure 3.22). The 2D NiPc-BTDA COF takes a belt shape and is made from AA stacking of 2D polymer layers. It has been observed that at the edges of tetragonal metallophthalocyanine, COF integration of BTDA blocks results in dramatic changes in the carrier-transport mode. It switches to an electron-transporting framework from a hole-transporting skeleton, giving rise photoconductivity attributes.

A new squaraine-based reaction has been strategized for the preparation of novel types of 2D COF. The condensation of squaric acid (SA) with aromatic, heteroaromatic/olefinic compounds are mixed together to form squaraines.[173] For instance, the SA condensation with *p*-toludine gives a squaraine (SQ) with a planar, zigzagged zwitterionic structure. By using this linkage, a conjugated COF with a zigzagged skeleton has been formed. First of all copper(II) 5,10,15,20-tetrakis(4-aminophenyl)porphyrin (TAP–CuP) is synthesized and used as a building block for the SA condensation in order to construct a 2D conjugated COF (CuP–SQ COF), which is crystalline in nature with a tetragonal mesoporous skeleton (Figure 3.23).[162] The synthesis of 3D COFs is achieved by targeting triangular and tetrahedral node-based nets: bor and ctn. The below mentioned 3D COFs are prepared as crystals by tetrahedral tetra(4-dihydroxyborylphenyl) methane or tetra(4-dihydroxyborylphenyl)silane condensation and triangular 2,3,6,7,10,11-hexahydroxytriphenylene co-condensation.[65] Co-condensation and self-condensation reactions of rigid, molecular-like building blocks, tetrahedral tetra(4-dihydroxyborylphenyl)methane (TBPM) and its silane analogue (TBPS), and triangular HHTP provide 3D COFs (COF-102, COF-103, COF-105 and COF-108), which are crystalline.

In the synthesis plan, building blocks that are tetrahedral and triangular are chosen. This is because they are unlikely to deform as they are rigid. Hence, they do not distort during the reaction for assembly. Dehydration reactions with these units make triangular rings of B_3O_3 and C_2O_2B. Hence,

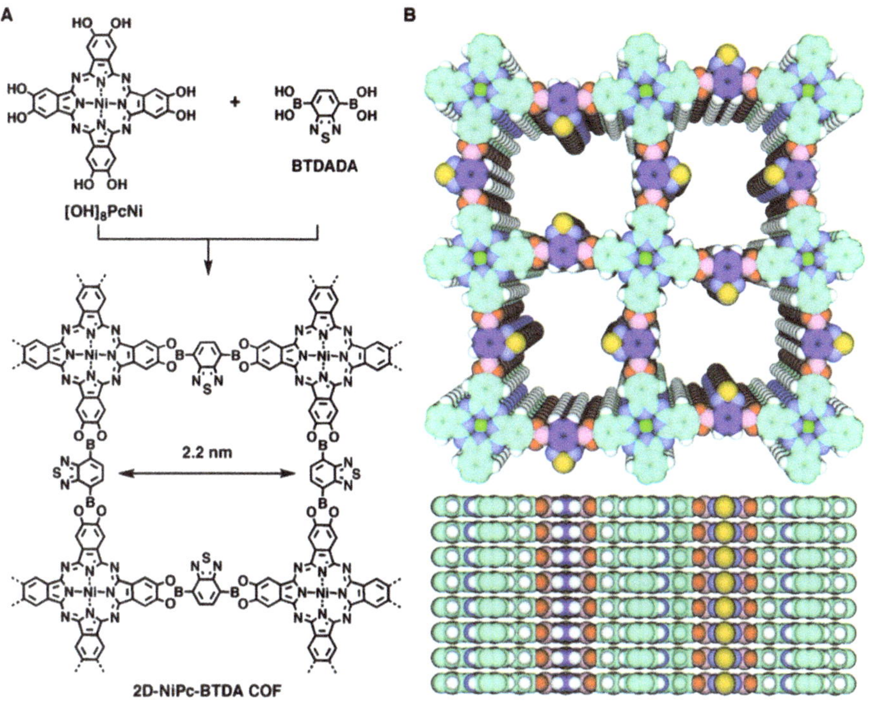

Figure 3.22 (A) Schematic representation of the synthesis of 2D NiPc-BTDA COF with metallophthalocyanine at the vertices and BTDA at the edges of the tetragonal framework and (B) top and side views of a graphical representation of a 2×2 tetragonal grid showing the eclipsed stacking of 2D polymer sheets (Pc, cyan; BTDA, violet; Ni, green; N, blue; S, yellow; O, red; B, pink; and H, white).[161a]
Reprinted with permission from X. Ding, L. Chen, Y. Honsho, X. Feng, O. Saengsawang, J. Guo, A. Saeki, S. Seki, S. Irle, S. Nagase, V. Parasuk and D. Jiang, *J. Am. Chem. Soc.*, 2011, **133**, 14510. Copyright 2011, American Chemical Society.

following the building blocks approach, two kinds of reactions can be envisioned: (i) self-condensation of tetrahedral blocks and (ii) their co-condensation with the triangular unit to form nets of COF structures with tetrahedral and triangular nodes. In principle, there are an infinite number of possibilities of nets resulting from the combination of tetrahedra and triangles. However, analysis of previous assembly reactions shows that the most symmetric nets are the most likely to form in an unbiased system. On the other hand, the nets that have just one kind of link are preferred and are thus easiest to target. The only known nets where the linking of tetrahedral and triangular building blocks meets the above criteria are with ctn and bor. The molecular building units with tetrahedral and triangular shapes can replace the nets. The use of nonflexible, triangular planar units, like B_3O_3 rings, requires rotational freedom at the tetrahedral nodes for the 3D structures of ctn and bor. Cerius2 software has been employed to construct

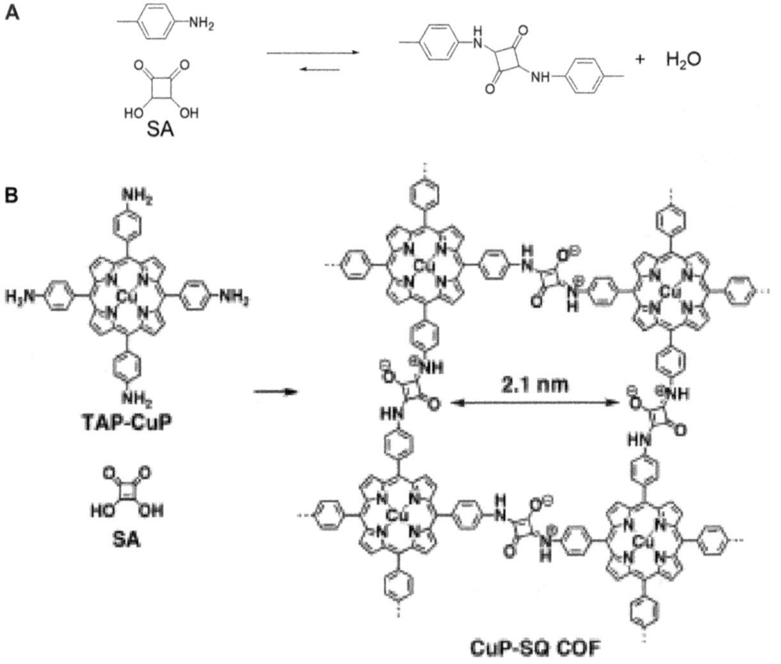

Figure 3.23   (A) Synthesis of SQ through the condensation of SA with *p*-toludine. (B) Synthesis of the CuP-SQ COF through the condensation of SA with TAP–CuP.
Reprinted with permission from A. Nagai, X. Chen, X. Feng, X. Ding, Z. Guo and D. Jiang, *Angew. Chem., Int. Ed.*, 2013, **52**, 3770. Copyright 2013 WILEY-VCH Verlag GmbH & Co. KGaA, Weinheim.

the 'blueprints' for COF synthesis based on ctn and bor nets. It does so by fitting molecular building blocks on the tetrahedral nodes and by the triangular B_3O_3 ring on the triangular nodes keeping the space group symmetries: *I43d* (ctn) and *P43m* (bor). Force-field calculations based on energy minimization are then carried out to get models with chemically reasonable values, bond lengths and angles. Synthesis of the COFs is performed by either suspending TBPM or TBPS in mesitylene/dioxane. These suspensions are put in sealed, evacuated (150 mTorr) Pyrex tubes and heated at 85 °C for 4 days to white crystalline COF-102 and COF-103 in 63 and 73% yields, respectively. Like this, co-condensation of TBPM/TBPS with HHTP (3 : 4 molar ratio) yields green solid crystals of COF-108 (55% yield) and COF-105 (58% yield). The colors of these are due to inclusion of oxidized highly colored HHTP in their pores. The use of the solvent system is to influence the solubility of the starting compounds and to increase product crystallinity. Three-dimensional monomers, like tetra(4-dihydroxyborylphenyl)methane and the Si-centered analogue,[65] are either reacted with HHTP to form COF-105 and COF-108 or self-condensed to produce COF-102 and COF-103.[6] Imine formation has also led to crystalline networks such as COF-300.[6] This compound is held together by imine bonds formed *via* a solvothermal reaction between tetra-(4-anilyl)methane and terephthaldehyde. The resulting 3D COF has a BET surface area of 1360 m^2 g^{-1} with a pore of 7.2 Å. An advantage of this network over COFs that are boron-linked is the increased physicochemical stability.

By self-addition polymerization of suitable monomers one can construct monocrystalline covalent organic networks.[161b] This synthetic strategy creates a new link between supramolecular chemistry and polymer science by demonstrating how ordered covalent or non-covalent structures can be built using a single modular strategy.

3.4 Emerging Porous Organic Materials

3.4.1 Porous Organic Cages

Porous organic cages are a recently developed class of porous materials, which are gaining a great deal of interest among researchers. The main synthetic challenge is to synthesize porous organic crystals gathered by non-covalent forces. These are much weaker than the directional bonding in crystalline networks like COFs.[163] It has been shown that pyramid-like organic cages are prepared and subsequently desolvated to make porous crystals that can uptake gas molecules like nitrogen, hydrogen, methane and carbon dioxide. It has been shown that the pore structure and connectivity is strongly controlled by steric groups linking to the vertices of the cages. Thus, it is now possible to connect/disconnect the molecular empty spaces in the cages. This is a new design strategy for making porous organic compounds from designer molecular pores. For instance three imine-linked[163] pyramidal cages were synthesized by a condensation reaction of 1,3,5-triformylbenzene with 1,2-ethylenediamine (cage 1), 1,2-propylenediamine (cage 2) and (*R,R*)-1,2-diaminocyclohexane (cage 3)[24] in a T4 C 6U cycloimination. The preparations are done at room temperature and need no extra template, only solvent. A preparation principle to obtain shape persistence in amine cages by connecting the cage vertices with formaldehyde-like carbonyls has also been reported. Persistence of shape is predicted by the stability of conformer calculations, providing a basis for design strategy. The linked cages show increased porosity together with unprecedented stability toward acid and base (pH 1.7–12.3). Note, under such conditions many other porous crystalline solids degrade.[175] The parent imine cage, namely RCC3, yields a reduced dodecaamine cage derivative, which is a chiral amine cage (CC3) obtained by treatment of CC3 with $NaBH_4$ that has tetrahedral point group symmetry. This is prepared by cycloimination of 1,3,5-triformylbenzene and (1*R*,2*R*)-1,2-diaminocyclohexane. CC3 is readily prepared on a large scale (>100 g) in a single-step condensation reaction. The shape persistence of this cage is gained back with a suitable molecule that ties on the cage vertices. Such a connected porous crystal is highly stable to acid or base. The material has a BET surface area of $\sim$400 m^2 g^{-1}, even when crystalline[164] (Figure 3.24).

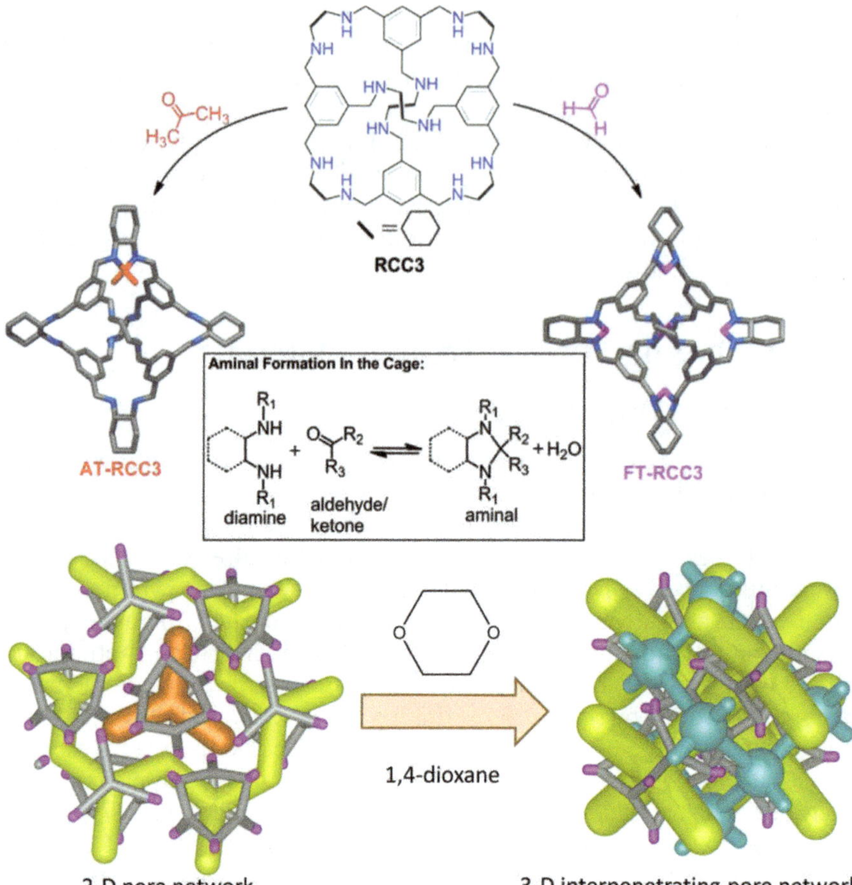

Figure 3.24 (Top) Synthesis of 'tied' porous cages. Reprinted with permission from M. Liu, M. A. Little, K. E. Jelfs, J. T. A. Jones, M. Schmidtmann, S. Y. Chong, T. Hasell and A. I. Cooper, *J. Am. Chem. Soc.*, 2014, **136**, 7583. Copyright 2014, American Chemical Society (Bottom) 1,4-dioxane causes formation of diamondoid pore channels from a 2D pore network. Reprinted with permission from T. Hasell, J. L. Culshaw, S. Y. Chong, M. Schmidtmann, M. A. Little, K. E. Jelfs, E. O. Pyzer-Knapp, H. Shepherd, D. J. Adams, G. M. Day and A. I. Cooper, *J. Am. Chem. Soc.*, 2014, **136**, 1438. Copyright 2014, American Chemical Society.

The solvents also play a crucial role in the porous cages. For instance, an organic solvent, 1,4-dioxane, has a strong effect on the lattice energy for a number of organic cage molecules. Dioxane inclusion directs the packing of the cages to form isostructural, 3D adamantoid pore channels. Such unique function of size, chemistry and geometry of 1,4-dioxane, together with a non-covalent auxiliary interaction results in directional coordination bonding/covalent bonding in this framework. CC13, a new cage with a dual interpenetrating pore structure, has double the uptake of gas and surface area compared to related crystals[165] (Figure 3.24).

A new way to design highly porous organic compounds by targeted assembly of prefabricated molecular pores has been developed. Three imine-linked tetrahedral cages were synthesized by the condensation reaction of 1,3,5-triformylbenzene with 1,2-ethylenediamine (cage 1, 792 g mol^{-1}), 1,2-propylenediamine (cage 2, 876 g mol^{-1}) and (R,R)-1,2-diaminocyclohexane (cage 3, 1118 g mol^{-1}) in a T4 C 6U cycloimination. The preparations were carried out at room temperature and required no extra template.[163] An improved preparation for a [4 + 6] imine-linked cage was shown to exhibit porosity when crystalline. Such a method has been strategized in order to obtain 100% yields and high chemical purity. Moreover, such porous cages are prepared by the [4 + 6] reaction of benzene-1,3,5-tricarbaldehyde with a vicinal 1,2-diamine. Cage 1 is synthesized by reaction of ethylenediamine with benzene-1,3,5-tricarbaldehyde. The starting preparation of the imine-linked cages is enabled by crystallization from the solvent. This reaction takes place by slow diffusion of aldehyde from the solid state into a concentrated liquid state with the amine, thereby effectively diluting the aldehyde. A product is formed in a supersaturated solution that leads to crystallization of cage 1. It is also likely that the poor yields and side reactions come from the need to reach the right conformation for enhanced intramolecular cyclization. However, it can be affected by freezing the conformation of the two amine groups. For instance, a much better yield (>85%) of the cycloimination reaction is reported for [3 + 3] macrocycles prepared from (1R,2R)-cyclohexanediamine. These reactions are carried out at concentrations between 50 and 670 mmol L^{-1}, without any oligomerization. Similarly, the more complex [4 + 6] tetrahedral cage is formed using the same conformationally restricted cyclohexanediamine and benzene-1,3,5-tricarbaldehyde at a concentration of 167 mmol L^{-1}. Thus, a new approach has been developed for the synthesis of cage 1, where the solution of the aldehyde is added slowly to an agitated solution of the amine. The advantage of this process is the possibility to control key parameters like reactant concentration, rate of addition and mixing of the process. Dichloromethane is considered as a suitable solvent for further study and for optimization of up-scaling. This is because its conversion is clean and it has a low boiling point, which helps isolation of the product.[166] A supramolecular approach to assemble porous organic cages has been reported that involves large steric crowding that prevents crystal packing. This creates, in a designed way, additional 'extrinsic' porosity between the porous cages. One such crystal shows a BET surface area of 854 m^2 g^{-1}, which is more than comparable to unfunctionalized cages[167] (Figure 3.25). The design strategy is that these steric groups may disturb packing and create additional porosity, as compared to previous networks. Cages CC9 and CC10 are synthesized by cycloimination of 1,3,5-triformylbenzene with (R,R)-1,2-diphenylethylene-diamine and (R,R)-1,2-bis(4-fluorophenyl)ethane-1,2-diamine, respectively, in dichloromethane (DCM) using trifluoroacetic acid (TFA) as a catalyst. Under such reaction conditions, the formation of the cage competes with a type of Diaza-Cope rearrangement used to prepare vicinal diamines.

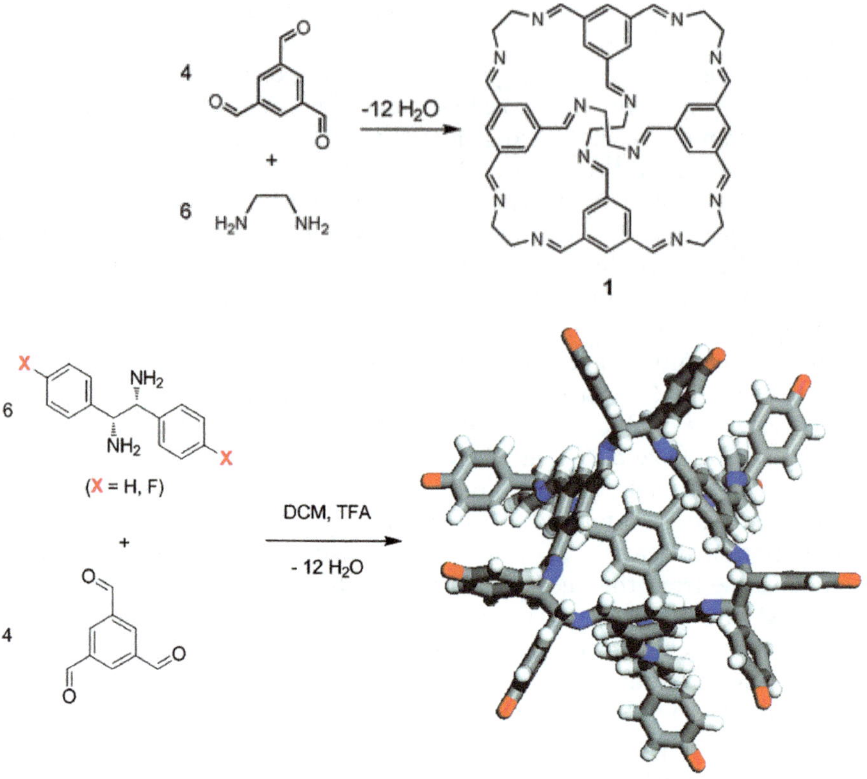

Figure 3.25   (Top) An improved synthesis for a $[4+6]$ imine-linked cage. (Bottom) Reaction scheme for the $[4+6]$ Schiff-base condensation yielding CC9 (X = H) and CC10 (X = F) cage molecules, as shown facing one of the triangular pore windows. C, N and H atoms are colored grey, blue and white, respectively.[167]
Reprinted with permission from M. J. Bojdys, M. E. Briggs, J. T. A. Jones, D. J. Adams, S. Y. Chong, M. Schmidtmann and A. I. Cooper, *J. Am. Chem. Soc.*, 2011, **133**, 16566. Copyright 2011, American Chemical Society.

This competing Diaza-Cope reaction needs a chair-structured geometry to produce the unwanted rearranged by-product. The key to produce these cages in a reasonable amount is to slow down the rate of the Diaza-Cope rearrangement by cooling the reaction and, for CC10, to remove the condensation by product, such as water.[167]

3.4.2   Mastalerz Compounds

Organic cage compounds formed by covalent bonds alone have been synthesized with the help of dynamic covalent bond formation. Cage compounds through formation of imine bonds (hemicarcerands, cavitands, adamantoid nano-cages and chiral nano-cubes based on chiral cyclotriveratrylenes), through formation of boronic esters, through combination of boronic esters and imine bond formation (iminoboronates), though formation of disulfide bridges, through alkene metathesis and through resorcinol or aldehyde condensation have been synthesized and reported.[168]

Pore modification is also a salient feature in these kinds of cage compounds. The Davis group in 2005 demonstrated that steroidal bisphenyl ureas form one-dimensional hydrophilic channels with average diameters between 11.6 and 14.3 Å in the crystalline state. They identified that the R1 and R2 groups are oriented toward the formed channels and thus do not contribute to the self-assembly. Independent of the steric nature of R1, almost the same molecular arrangement is followed by the formation of nano-porous channels. They also showed that the molecular framework can be functionalized to regulate the crystal porosity.[169] Hence, keeping R1 unchanged, R2 (the aromatic group) is changed in size.[170]

Exploiting the formation of dynamic covalent bonds, *e.g.*, aldehydes and amines to give imines, shape-consistent cage compounds can be prepared in high yield. Atwood and co-workers in 2009 first reported a permanent porous but amorphous organic cage compound.[171] Recently, Cooper and co-workers have synthesized an organic cage compound having a high surface area and crystalline form.[172] The key feature of these new types of porous materials is that their smallest units are molecular and, furthermore, intrinsically porous themselves. This is in sharp contrast to most of the other porous materials mentioned before and thus creates new opportunities. The Cooper group have demonstrated that porous organic alloys can be generated by exploiting the varying solubility of different cage compounds.[173] It was possible to precipitate crystalline microporous cage materials inside macro- and meso-porous materials.[174] Their solubility can be used for depositing the cage molecules as thin films on quartz crystal microbalances (QCMs).[175] To date, the cage-decorated QCMs show the best ability for sensing xylenes.[176]

References

1. O. W. Webster, F. P. Gentry, R. D. Farlee and B. E. Smart, *Makromol. Chem., Macromol. Symp.*, 1992, **54**, 477.
2. C. Urban, E. F. McCord, O. W. Webster, L. Abrams, H. W. Long, H. Gaede, P. Tang and A. Pines, *Chem Mater.*, 1995, 7, 1325.
3. V. A. Davankov and M. P. Tsyurupa, *React. Polym.*, 1990, **13**, 27.
4. M. P. Tsyurupa and V. A. Davankov, *React. Funct. Polym.*, 2002, **53**, 193.
5. J. Y. Lee, C. D. Wood, D. Bradshaw, M. J. Rosseinsky and A. I. Cooper, *Chem. Commun.*, 2006, **25**, 2670.
6. R. Dawson, A. I. Cooper and D. J. Adams, *Prog. Polym. Sci.*, 2012, 37, 530.
7. M. P. Tsyurupa and V. A. Davankov, *React. Funct. Polym.*, 2006, **66**, 768.

8. C. D. Wood, B. Tan, A. Trewin, H. J. Niu, D. Bradshaw, M. J. Rosseinsky, Y. Z. Khimyak, N. L. Campbell, R. Kirk, E. Stockel and A. I. Cooper, *Chem. Mater.*, 2007, **19**, 2034.

9. Y. Luo, S. Zhang, Y. Ma, W. Wang and B. Tan, *Polym. Chem.*, 2013, **4**, 1126.

10. J. H. Ahn, J. E. Jang, C. G. Oh, S. K. Ihm, J. Cortez and D. C. Sherrington, *Macromolecules*, 2006, **39**, 627.

11. B. Li, X. Huang, L. Liang and B. Tan, *J. Mater. Chem.*, 2010, **20**, 7444.

12. C. F. Martin, E. Stockel, R. Clowes, D. J. Adams, A. I. Cooper, J. J. Pis, F. Rubiera and C. Pevida, *J. Mater. Chem.*, 2011, **21**, 5475.

13. W. Chaikittisilp, M. Kubo, T. Moteki, A. Sugawara-Narutaki, A. Shimojima and T. Okubo, *J. Am. Chem. Soc.*, 2011, **133**, 13832.

14. N. Fontanals, J. Cortes, M. Galia, R. M. Marce, P. A. G. Cormack, F. Borrull and D. C. Sherrington, *J. Polym. Sci., Part A: Polym. Chem.*, 2005, **43**, 1718.

15. S. Z. Zeng, L. Guo, Q. He, Y. Chen, P. Jiang and J. Shi, *Microporous Mesoporous Mater.*, 2010, **131**, 141.

16. A. P. Cote, A. I. Benin, N. W. Ockwig, M. O'Keeffe, A. J. Matzger and O. M. Yaghi, *Science*, 2005, **310**, 1166.

17. J. Germain, J. M. J. Frechet and F. Svec, *J. Mater. Chem.*, 2007, **17**, 4989.

18. J. Germain, J. M. J. Frechet and F. Svec, *Chem. Commun.*, 2009, **12**, 1526.

19. J. Germain, F. Svec and J. M. J. Fréchet, *Chem. Mater.*, 2008, **20**, 7069.

20. J. Germain, F. Svec and J. M. J. Fréchet, *PMSE Prepr.*, 2007, **97**, 272.

21. R. Gujadhur, D. Venkataraman and J. T. Kintigh, *Tetrahedron Lett.*, 2001, **42**, 4791.

22. A. Ito, H. Ino, K. Tanaka, K. Kanemoto and T. Kato, *J. Org. Chem.*, 2002, **67**, 491.

23. B. H. Yang and S. L. Buchwald, *J. Organomet. Chem.*, 1999, **576**, 125.

24. T. Agou, J. Kobayashi and T. Kawashima, *Org. Lett.*, 2006, **8**, 2241.

25. J. P. Wolfe, H. Tomori, J. P. Sadighi, J. J. Yin and S. L. J. Buchwald, *Org. Chem.*, 2000, **65**, 1158.

26. *Polymer Handbook*, ed. J. Brandrup, E. H. Immergut, E. A. Grulke, A. Abe and D. R. Bloch, Wiley, New York, 4th edn, 2005.

27. T. E. Reich, K. T. Jackson, S. Li, P. Jena and H. M. El-Kaderi, *J. Mater. Chem.*, 2011, **21**, 10629.

28. R. Chinchilla and C. Najera, *Chem. Rev.*, 2007, **107**, 874.

29. J. X. Jiang, F. Su, A. Trewin, C. D. Wood, N. L. Campbell, H. Niu, C. Dickinson, A. Y. Ganin, M. J. Rosseinsky, Y. Z. Khimyak and A. I. Cooper, *Angew. Chem., Int. Ed.*, 2007, **46**, 8574.

30. C. Weder, *Angew. Chem., Int. Ed.*, 2008, **47**, 448.

31. J. X. Jiang, F. Su, A. Trewin, C. D. Wood, H. Niu, J. T. A. Jones, Y. Z. Khimyak and A. I. Cooper, *J. Am. Chem. Soc.*, 2008, **130**, 7710.

32. J. X. Jiang, F. Su, H. Niu, C. D. Wood, N. L. Campbell, Y. Z. Khimyak and A. I. Cooper, *Chem. Commun.*, 2008, **4**, 486.

33. J. X. Jiang, A. Trewin, F. Su, C. D. Wood, H. Niu, J. T. A. Jones, Y. Z. Khimyak and A. I. Cooper, *Macromolecules*, 2009, **42**, 2658.

34. J. Schmidt, J. Weber, J. D. Epping, M. Antonietti and A. Thomas, *Adv. Mater.*, 2009, **21**, 702.
35. J. Weber and A. Thomas, *J. Am. Chem. Soc.*, 2008, **130**, 6334.
36. J. Schmidt, M. Werner and A. Thomas, *Macromolecules*, 2009, **42**, 4426.
37. J. Brandt, J. Schmidt, A. Thomas, J. D. Epping and J. Weber, *Polym. Chem.*, 2011, **2**, 1950.
38. Q. Chen, M. Luo, T. Wang, J. X. Wang, D. Zhou, Y. Han, C. S. Zhang, C. G. Yan and B. H. Han, *Macromolecules*, 2011, **44**, 5573.
39. K. Zhang, B. Tieke, F. Vilela and P. J. Skabara, *Macromol. Rapid Commun.*, 2011, **32**, 825.
40. Y. C. Zhao, D. Zhou, Q. Chen, X. J. Zhang, N. Bian, A. D. Qi and B. H. Han, *Macromolecules*, 2011, **44**, 6382.
41. H. Ren, T. Ben, F. Sun, M. Guo, X. Jing, H. Ma, K. Cai, S. Qui and G. Zhu, *J. Mater. Chem.*, 2011, **21**, 10348.
42. L. Chen, Y. Honsho, S. Seki and D. L. Jiang, *J. Am. Chem. Soc.*, 2010, **132**, 6742.
43. L. Chen, Y. Yang and D. Jiang, *J. Am. Chem. Soc.*, 2010, **132**, 9138.
44. S. W. Yuan, B. Dorney, D. White, S. Kirklin, P. Zapol, L. P. Yu and D. J. Liu, *Chem. Commun.*, 2010, **46**, 4547.
45. W. Chaikittisilp, A. Sugawara, A. Shimojima and T. Okubo, *Chem. Eur. J.*, 2010, **16**, 6006.
46. W. Chaikittisilp, A. Sugawara, A. Shimojima and T. Okubo, *Chem. Mater.*, 2010, **22**, 4841.
47. Y. Kim, K. Koh, M. F. Roll, R. M. Laine and A. J. Matzger, *Macromolecules*, 2010, **43**, 6995.
48. M. F. Roll, J. W. Kampf, Y. Kim, E. Yi and R. M. Laine, *J. Am. Chem. Soc.*, 2010, **132**, 10171.
49. H. Lim and J. Y. Chang, *Macromolecules*, 2010, **43**, 6943.
50. (a) J. Roncali, *Chem. Rev.*, 1992, **92**, 711; (b) H. Sirringhaus, P. J. Brown, R. H. Friend, M. M. Nielsen, K. Bechgaard, B. M. W. Langeveld-Voss, A. J. H. Spiering, R. A. J. Janssen, E. W. Meijer, P. Herwig and D. M. de Leeuw, *Nature*, 1999, **401**, 685.
51. A. G. MacDiarmid, *Synth. Met.*, 1997, **84**, 27.
52. J. F. Morin, M. Leclerc, D. Ades and A. Siove, *Macromol. Rapid Commun.*, 2005, **26**, 761.
53. U. Scherf and E. J. W. List, *Adv. Mater.*, 2002, **14**, 477.
54. (a) P. M. Budd, B. S. Ghanem, S. Makhseed, N. B. McKeown, K. J. Msayib and C. E. Tattershall, *Chem Commun.*, 2004, **2**, 230; (b) N. B. McKeown and P. M. Budd, *Chem. Soc. Rev.*, 2006, **35**, 675.
55. N. B. McKeown, B. Gahnem, K. J. Msayib, P. M. Budd, C. E. Tattershall, K. Mahmood, S. Tan, D. Book, H. W. Langmi and A. Walton, *Angew. Chem., Int. Ed.*, 2006, **45**, 1804.
56. J. H. Edwards and W. J. Feast, *Polymer*, 1980, **21**, 595.
57. U. H. F. Bunz, *Chem. Rev.*, 2000, **100**, 1605.
58. P. Kuhn, M. Antonietti and A. Thomas, *Angew. Chem., Int. Ed.*, 2008, **47**, 3450.

59. P. Kuhn, A. Forget, D. Su, A. Thomas and M. Antonietti, *J. Am. Chem. Soc.*, 2008, **130**, 13333.

60. P. Kuhn, A. Thomas and M. Antonietti, *Macromolecule*, 2009, **42**, 319.

61. M. G. Schwab, B. Fassbender, H. W. Spiess, A. Thomas, X. Feng and K. Mullen, *J. Am. Chem. Soc.*, 2009, **131**, 7216.

62. G. P. Hao, W. C. Li, D. Qian and A. H. Lu, *Adv. Mater.*, 2010, **22**, 853.

63. K. Nishi, R. W. Thompson, Synthesis of classical zeolites, in *Handbook of Porous Solids*, ed. F. Schüth, K. S. W. Sing and F. Weitkamp, Wiley-VCH Verlag GmbH, Weinheim, 2002.

64. A. P. Cote, A. I. Benin, N. W. Ockwig, M. O'Keeffe, A. J. Matzger and O. M. Yaghi, *Science*, 2005, **310**, 1166.

65. H. M. El-Kaderi, J. R. Hunt, J. L. Mendoza-Cortes, A. P. Cote, R. E. Taylor, M. O'Keeffe and O. M. Yaghi, *Science*, 2007, **316**, 268.

66. F. Rodríguez-Reinoso, Production and applications of activated carbons, in *Handbook of Porous Solids*, ed. F. Schüth, K. S. W. Sing and F. Weitkamp, Wiley-VCH Verlag GmbH, Weinheim, 2002.

67. T. Ben, H. Ren, S. Q. Ma, D. P. Cao, J. H. Lan, X. F. Jing, W. C. Wang, J. Xu, F. Deng, J. M. Simmons, S. L. Qiu and G. S. Zhu, *Angew. Chem., Int. Ed.*, 2009, **48**, 9457.

68. A. Trewin and A. Cooper, *Angew. Chem., Int. Ed.*, 2010, **49**, 1533.

69. C. E. Chan-Thaw, A. Villa, P. Katekomol, D. Su, A. Thomas and L. Prati, *Nano Lett.*, 2010, **10**, 537.

70. P. Katekomol, J. Roeser, M. Bojdys, J. Weber and A. Thomas, *Chem. Mater.*, 2013, **25**, 1542.

71. P. M. Budd, E. S. Elabas, B. S. Ghanem, S. Makhseed, N. B. McKeown, K. J. Msayib, C. E. Tattershall and D. Wang, *Adv. Mater.*, 2004, **16**, 456.

72. U. H. F. Bunz, *Chem. Rev.*, 2000, **100**, 1605.

73. K. Sonogashira, Y. Tohda and N. Hagihara, *Tetrahedron Lett.*, 1975, **16**, 4467.

74. D. L. Trumbo and C. S. Marvel, *J. Polym. Sci., Part A: Polym. Chem.*, 1986, **24**, 2311.

75. E. Hittinger, A. Kokil and C. Weder, *Angew. Chem., Int. Ed.*, 2004, **43**, 1808.

76. J. D. Mendez, M. Schroeter and C. Weder, *Macromol. Chem. Phys.*, 2007, **208**, 1625.

77. Z. J. Donhauser, B. A. Mantooth, K. F. Kelly, L. A. Bumm, J. D. Monnell, J. J. Stapleton, D. W. Price, A. M. Rawlett, D. L. Allara, J. M. Tour and P. S. Weiss, *Science*, 2001, **292**, 2303.

78. Q. Zhou and T. M. Swager, *J. Am. Chem. Soc.*, 1995, **117**, 12593.

79. D. Venkataraman, S. Lee, J. S. Zhang and J. S. Moore, *Nature*, 1994, **371**, 591.

80. W. Zhang and J. S. Moore, *Angew. Chem., Int. Ed.*, 2006, **45**, 4416.

81. Y. H. Kiang, G. B. Gardner, S. Lee, Z. T. Xu and E. B. Lobkovsky, *J. Am. Chem. Soc.*, 1999, **121**, 8204.

82. N. Miyaura, K. Yamada and A. Suzuki, *Tetrahedron Lett.*, 1979, **36**, 343.
83. N. Miyaura, T. Yanagi and A. Suzuki, *Synth. Commun.*, 1981, **11**, 513.
84. H. Zhang, F. Xue, T. C. W. Mak and K. S. Chan, *J. Org. Chem.*, 1996, **61**, 8002.
85. H. Zhang and K. S. Chan, *Tetrahedron Lett.*, 1996, **37**, 1043.
86. D. Badone, M. Baroni, R. Cardamone, A. Ielmini and U. Guzzi, *J. Org. Chem.*, 1997, **62**, 7170.
87. A. L. Casalnuovo and J. C. Calabrese, *J. Am. Chem. Soc.*, 1990, **112**, 4324.
88. Y. M. A. Yamada, K. Takeda, H. Takahashi and S. Ikegami, *Org. Lett.*, 2002, **4**, 3371.
89. R. Frenette and R. W. Friesen, *Tetrahedron Lett.*, 1994, **35**, 9177.
90. T. Ishiyama, M. Murata and N. Miyaura, *J. Org. Chem.*, 1995, **60**, 7508.
91. T. Ishiyama, Y. Itoh, T. Kitano and N. Miyaura, *Tetrahedron Lett.*, 1997, **38**, 3447.
92. T. Ben and S. Qiu, *CrystEngComm*, 2013, **15**, 17.
93. T. Ben, C. Pei, D. Zhang, J. Xu, F. Deng, X. Jing and S. Qiu, *Energy Environ. Sci.*, 2011, **4**, 3991.
94. H. Ren, T. Ben, F. Sun, M. Guo, X. Jing, H. Ma, K. Cai, S. Qiu and G. Zhu, *J. Mater. Chem.*, 2011, **21**, 10348.
95. J. R. Holst, E. Stöckel, D. J. Adams and A. I. Cooper, *Macromolecules*, 2010, **43**, 8531.
96. D. Yuan, W. Lu, D. Zhao and H. C. Zhou, *Adv. Mater.*, 2011, **23**, 3723.
97. W. Lu, D. Yuan, D. Zhao, C. I. Schilling, O. Plietzsch, T. Muller, S. Bräse, J. Guenther, J. Blümel, R. Krishna, Z. Li and H. C. Zhou, *Chem. Mater.*, 2010, **22**, 5964.
98. D. Yuan, W. Lu, D. Zhao and H. Zhou, *Adv. Mater.*, 2011, **23**, 3723.
99. M. Rose, N. Klein, W. Bohlmann, B. Bohringer, S. Fichtner and S. Kaskel, *Soft Matter*, 2010, **6**, 3918.
100. P. Pandey, O. K. Farha, A. M. Spokoyny, C. A. Mirkin, M. G. Kanatzidis, J. T. Hupp and S. T. Nguyen, *J. Mater. Chem.*, 2011, **21**, 1700.
101. O. Plietzsch, C. I. Schilling, T. Grab, S. L. Grage, A. S. Ulrich, A. Comotti, P. Sozzani, T. Muller and S. Bräse, *New J. Chem.*, 2011, **35**, 1577.
102. G. Eglinton and A. R. Galbraith, *J. Chem. Soc.*, 1959, 889.
103. H. Ren, T. Ben, E. Wang, X. Jing, M. Xue, B. Liu, Y. Cui, S. Qiu and G. Zhu, *Chem. Commun.*, 2010, **46**, 291.
104. G. Manecke and D. Woehrle, *Makromol. Chem.*, 1968, **120**, 176.
105. W. Wang, H. Ren, F. Sun, K. Cai, H. Ma, J. Du, H. Zhaoc and G. Zhu, *Dalton Trans.*, 2012, **41**, 3933.
106. H. Zhao, Z. Jin, H. Su, J. Zhang, X. Yao, H. Zhao and G. Zhu, *Chem. Commun.*, 2013, **49**, 2780.
107. N. Y. Du, H. B. Park, G. P. Robertson, M. M. Dal-Cin, T. Visser, L. Scoles and M. D. Guiver, *Nat. Mater.*, 2011, **10**, 372.

108. Y. Yuan, F. Sun, F. Zhang, H. Ren, M. Guo, K. Cai, X. Jing, X. Gao and G. Zhu, *Adv. Mater.*, 2013, **25**, 6619.

109. Y. Peng, T. Ben, J. Xu, M. Xue, X. Jing, F. Deng, S. Qiu and G. Zhu, *Dalton Trans.*, 2011, **40**, 2720.

110. T. Ben, K. Shi, Y. Cui, C. Pei, Y. Zuo, H. Guo, D. Zhang, J. Xu, F. Deng, Z. Tian and S. Qiu, *J. Mater. Chem.*, 2011, **21**, 18208.

111. C. Pei, T. Ben, H. Guo, J. Xu, F. Deng, Z. Xiang, D. Cao and S. Qiu, *Philos. Trans. R. Soc., A*, 2013, **371**, 20120312.

112. N. B. McKeown, P. M. Budd, K. J. Msayib, B. S. Ghanem, H. J. Kingston, C. E. Tattershall, S. Makhseed, K. J. Reynolds and D. Fritsch, *Chem. – Eur. J.*, 2005, **11**, 2610.

113. N. B. McKeown and P. M. Budd, *Macromolecules*, 2010, **43**, 5163.

114. N. B. McKeown, S. Makhseed and P. M. Budd, *Chem. Commun.*, 2002, **23**, 2780.

115. N. B. McKeown, S. Hanif, K. Msayib, C. E. Tattershall and P. M. Budd, *Chem. Commun.*, 2002, **23**, 2782.

116. N. B. McKeown, *Phthalocyanine Materials: Synthesis, Structure and Function*, CUP, Cambridge, 1998.

117. N. B. McKeown, *J. Mater. Chem.*, 2000, **10**, 1979.

118. B. Meunier, *Chem. Rev.*, 1992, **92**, 1411.

119. P. M. Budd, B. Ghanem, K. Msayib, N. B. McKeown and C. Tattershall, *J. Mater. Chem.*, 2003, **13**, 2721.

120. B. S. Ghanem, N. B. McKeown, P. M. Budd, N. M. Al-Harbi, D. Fritsch, K. Heinrich, L. Starannikova, A. Tokarev and Y. Yampolskii, *Macromolecules*, 2009, **42**, 7881.

121. M. Carta, K. J. Msayib, P. M. Budd and N. B. McKeown, *Org. Lett.*, 2008, **10**, 2641.

122. B. S. Ghanem, N. B. McKeown, P. M. Budd and D. Fritsch, *Adv. Mater.*, 2008, **20**, 2766.

123. J. Weber, O. Su, M. Antonietti and A. Thomas, *Macromol. Rapid Commun.*, 2007, **28**, 1871.

124. N. Ritter, M. Antonietti, A. Thomas, I. Senkovska, S. Kaskel and J. Weber, *Macromolecules*, 2009, **42**, 8017.

125. P. M. Budd and N. B. McKeown, *Polym. Chem.*, 2010, **1**, 63.

126. P. M. Budd, K. J. Msayib, C. E. Tattershall, B. S. Ghanem, K. J. Reynolds, N. B. McKeown and D. Fritsch, *J. Membr. Sci.*, 2005, **251**, 263.

127. M. Carta, K. J. Msayib and N. B. McKeown, *Tetrahedron Lett.*, 2009, **50**, 5954.

128. N. Du, G. P. Robertson, I. Pinnau, S. Thomas and M. D. Guiver, *Macromol. Rapid Commun.*, 2009, **30**, 584.

129. N. Du, G. P. Robertson, J. Song, I. Pinnau, S. Thomas and M. D. Guiver, *Macromolecules*, 2008, **41**, 9656.

130. N. Du, G. P. Robertson, I. Pinnau and M. D. Guiver, *Macromolecules*, 2009, **42**, 6023.

131. T. Emmler, K. Heinrich, D. Fritsch, P. M. Budd, N. Chaukura, D. Ehlers, K. Ratzke and F. Faupel, *Macromolecules*, 2010, **43**, 6075.

132. B. S. Ghanem, M. Hashem, K. D. M. Harris, K. J. Msayib, M. Xu, P. M. Budd, N. Chaukura, D. Book, S. Tedds, A. Walton and N. B. McKeown, *Macromolecules*, 2010, **43**, 5287.

133. B. S. Ghanem, K. J. Msayib, N. B. McKeown, K. D. M. Harris, Z. Pan, P. M. Budd, A. Butler, J. Selbie, D. Book and A. Walton, *Chem. Commun.*, 2007, **1**, 67.

134. T. M. Long and T. M. Swager, *Adv. Mater.*, 2001, **13**, 601.

135. C. L. Hilton, C. R. Jamison, H. K. Zane and B. T. King, *J. Org. Chem.*, 2009, **74**, 405.

136. N. T. Tsui, L. Torun, B. D. Pate, A. J. Paraskos, T. M. Swager and E. L. Thomas, *Adv. Funct. Mater.*, 2007, **17**, 1595.

137. T. M. Swager, *Acc. Chem. Res.*, 2008, **41**, 1181.

138. N. T. Tsui, Y. Yang, A. D. Mulliken, L. Torun, M. C. Boyce, T. M. Swager and E. L. Thomas, *Polymer*, 2008, **49**, 4703.

139. A. S. Lindsay, *J. Chem. Soc.*, 1965, 1685.

140. P. M. Budd, B. S. Ghanem, S. Makhseed, N. B. McKeown, K. J. Msayib and C. E. Tattershall, *Chem. Commun.*, 2004, **2**, 230.

141. B. S. Ghanem, N. B. McKeown, P. M. Budd and D. Fritsch, *Macromolecules*, 2008, **41**, 1640.

142. B. S. Ghanem, N. B. McKeown, P. M. Budd, N. M. Al-Harbi, D. Fritsch, K. Heinrich, L. Starannikova, A. Tokarev and Y. Yampolskii, *Macromolecules*, 2009, **42**, 7881.

143. H. B. Park, C. H. Jung, Y. M. Lee, A. J. Hill, S. J. Pas, S. T. Mudie, E. Van Wagner, B. D. Freeman and D. J. Cookson, *Science*, 2007, **318**, 254.

144. (a) Y. H. Kim and O. W. Webster, *J. Am. Chem. Soc.*, 1990, **112**, 4592; (b) Y. H. Kim and O. W. Webster, *Macromolecules*, 1992, **25**, 5561.

145. J. X. Jiang, A. Trewin, D. J. Adams and A. I. Cooper, *Chem. Sci.*, 2011, **2**, 1777.

146. G. Cheng, T. Hasell, A. Trewin, D. J. Adams and A. I. Cooper, *Angew. Chem., Int. Ed.*, 2012, **51**, 12727.

147. L. Zhu, J. Duquette and M. B. Zhang, *J. Org. Chem.*, 2003, **68**, 3729.

148. T. M. Figueira-Duarte, S. C. Simon, M. Wagner, S. I. Drtezhinin, K. A. Zachariasse and K. Muellen, *Angew. Chem., Int. Ed.*, 2008, **47**, 10175.

149. S. J. Rowan, S. J. Cantrill, G. R. L. Cousins, J. K. M. Sanders and J. F. Stoddart, *Angew. Chem., Int. Ed.*, 2002, **41**, 898.

150. C. P. Brock, R. P. Minton and K. Niedenzu, *Acta Crystallogr., Sect. C: Cryst. Struct. Commun..*, 1987, **43**, 1775.

151. N. L. Campbell, R. Clowes, L. K. Ritchie and A. I. Cooper, *Chem. Mater.*, 2009, **21**, 204.

152. L. K. Ritchie, A. Trewin, A. Reguera-Galan, T. Hasell and A. I. Cooper, *Microporous Mesoporous Mater.*, 2010, **132**, 132.

153. R. W. Tilford, W. R. Gemmill, H. C. zur Loye and J. J. Lavigne, *Chem. Mater.*, 2006, **18**, 5296.

154. M. Dogru, A. Sonnauer, A. Gavryushin, P. Knochel and T. Bein, *Chem. Commun.*, 2011, **47**, 1707.

155. A. P. Cote, H. M. El-Kaderi, H. Furukawa, J. R. Hunt and O. M. Yaghi, *J. Am. Chem. Soc.*, 2007, **129**, 12914.

156. S. Dalapati, S. Jin, J. Gao, Y. Xu, A. Nagai and D. Jiang, *J. Am. Chem. Soc.*, 2013, **135**, 17310.

157. N. Winkhofer, H. Roesky, M. Noltmeyer and W. Robinson, *Angew. Chem., Int. Ed.*, 1992, **31**, 599.

158. J. R. Hunt, C. J. Doonan, J. D. LeVangie, A. P. Cote and O. M. Yaghi, *J. Am. Chem. Soc.*, 2008, **130**, 11872.

159. N. A. A. Zwaneveld, R. Pawlak, M. Abel, D. Catalin, D. Gigmes, D. Bertin and L. Porte, *J. Am. Chem. Soc.*, 2008, **130**, 6678.

160. J. W. Colson, A. R. Woll, A. Mukherjee, M. P. Levendorf, E. L. Spitler, V. B. Shields, M. G. Spencer, J. Park and W. R. Dichtel, *Science*, 2011, **332**, 228–231.

161. (a) X. Ding, L. Chen, Y. Honsho, X. Feng, O. Saengsawang, J. Guo, A. Saeki, S. Seki, S. Irle, S. Nagase, V. Parasuk and D. Jiang, *J. Am. Chem. Soc.*, 2011, **133**, 14510; (b) D. Beaudoin, T. Maris and J. D. Wuest, *Nat. Chem.*, 2013, **5**, 830.

162. A. Nagai, X. Chen, X. Feng, X. Ding, Z. Guo and D. Jiang, *Angew. Chem., Int. Ed.*, 2013, **52**, 3770.

163. T. Tozawa, J. T. A. Jones, S. I. Swamy, S. Jiang, D. J. Adams, S. Shakespeare, R. Clowes, D. Bradshaw, T. Hasell, S. Y. Chong, C. Tang, S. Thompson, J. Parker, A. Trewin, J. Bacsa, A. M. Z. Slawin, A. Steiner and A. I. Cooper, *Nat. Mater.*, 2009, **8**, 973.

164. M. Liu, M. A. Little, K. E. Jelfs, J. T. A. Jones, M. Schmidtmann, S. Y. Chong, T. Hasell and A. I. Cooper, *J. Am. Chem. Soc.*, 2014, **136**, 7583.

165. T. Hasell, J. L. Culshaw, S. Y. Chong, M. Schmidtmann, M. A. Little, K. E. Jelfs, E. O. Pyzer-Knapp, H. Shepherd, D. J. Adams, G. M. Day and A. I. Cooper, *J. Am. Chem. Soc.*, 2014, **136**, 1438.

166. D. P. Lydon, N. L. Campbell, D. J. Adams and A. I. Cooper, *Synth. Commun.*, 2011, **41**, 2146.

167. M. J. Bojdys, M. E. Briggs, J. T. A. Jones, D. J. Adams, S. Y. Chong, M. Schmidtmann and A. I. Cooper, *J. Am. Chem. Soc.*, 2011, **133**, 16566.

168. M. Mastalerz, *Angew. Chem., Int. Ed.*, 2010, **49**, 5042.

169. A. L. Sisson, V. del Amo Sanchez, G. Magro, A. M. E. Griffin, S. Shan, J. P. H. Charmant and A. P. Davis, *Angew. Chem., Int. Ed.*, 2005, **44**, 6878.

170. M. Mastalerz, *Angew. Chem., Int. Ed.*, 2012, **51**, 584.

171. J. Tian, P. K. Thallapally, S. J. Dalgarno, P. B. Mcgrail and J. L. Atwood, *Angew. Chem., Int. Ed.*, 2009, **48**, 5492.

172. T. Tozawa, J. T. A. Jones, S. I. Swamy, S. Jiang, D. J. Adams, S. Shakespeare, R. Clowes, D. Bradshaw, T. Hasell, S. Y. Chong, C. Tang, S. Thompson, J. Parker, A. Trewin, J. Bacsa, A. M. Z. Slawin, A. Steiner and A. I. Cooper, *Nat. Mater.*, 2009, **8**, 973.

173. (a) T. Hasell, S. Y. Chong, M. Schmidtmann, D. J. Adams and A. I. Cooper, *Angew. Chem., Int. Ed.*, 2012, **51**, 7154; (b) J. T. A. Jones, T. Hasell, X. Wu, J. Basca, K. E. Jelfs, M. Schmidtmann, S. Y. Chong, D. J. Adams, A. Trewin, F. Schiffman, F. Cora, B. Slater, A. Steiner, G. M. Day and A. I. Cooper, *Nature*, 2011, **474**, 367; (c) T. Hasell, S. Y. Chong, K. E. Jelfs, D. J. Adams and A. I. Cooper, *J. Am. Chem. Soc.*, 2012, **134**, 588.
174. T. Hasell, H. Zhang and A. I. Cooper, *Adv. Mater.*, 2012, **24**, 5732.
175. M. Brutschy, M. W. Schneider, M. Mastalerz and S. R. Waldvogel, *Adv. Mater.*, 2012, **24**, 6049.
176. M. Mastalerz, *Synlett.*, 2013, **24**, 781.

Recent Developments of Hypercrosslinked Microporous Organic Polymers

LIANGXIAO TAN AND BIEN TAN*

School of Chemistry and Chemical Engineering, Huazhong University
of Science and Technology, Wuhan 430074, China
*Email: bien.tan@mail.hust.edu.cn

4.1 Short Overview of Microporous Materials

According to the IUPAC classification, a micropore refers to a pore size less than 2.0 nm. Nowadays, microporous materials have received an increasing level of research interest because of their potential applications, especially in gas storage, separation, heterogeneous catalysis, sensors and energy storage.[1] During the past several decades, scientists and researchers have extensively developed and produced a series of microporous solids, including zeolites,[2] silica,[3] active carbon,[4] metal–organic frameworks (MOFs),[5,6] porous organic cages[7,8] and microporous organic polymers (MOPs).[1,9] Compared with other materials, MOPs, which are comprised of light, nonmetallic elements (C, H, N, O and B),[10] have high specific surface areas and have become one of the most attractive species with special advantages. For instance, the enormous varieties of monomer make it possible to incorporate desired functionalities into the framework and pore surface by introducing various functional groups or exerting control over the characteristics of the pore to tune the pore structure or size.[11,12]

Monographs in Supramolecular Chemistry No. 17
Porous Polymers: Design, Synthesis and Applications
By Shilun Qiu and Teng Ben
© The Royal Society of Chemistry 2016
Published by the Royal Society of Chemistry, www.rsc.org

Microporous organic polymers are usually obtained by the formation of C–C bonds and many types of chemical reactions have been investigated to develop various kinds of microporous organic polymers, such as Sonogashira, Suzuki or Yamamoto coupling, resulting in conjugated microporous polymers (CMPs),[13–15] dioxane-forming polymerization for soluble and crosslinked polymers of intrinsic microporosity (PIMs),[16–18] covalent organic frameworks (COFs)[19–21] through reversible borate chemistry, hypercrosslinked polymers (HCPs)[9,22–24] *via* the Friedel–Crafts reaction, porous aromatic frameworks (PAFs)[25,26] by various cross-coupling reactions of aromatic compounds and covalent triazine-based frameworks (CTFs)[27,28] obtained by the trimerization of aromatic nitrile monomers. In addition to these synthetic methods, MOPs can also be formed by trimerization of ethynyl or nitrile groups,[29,30] by amide, imide or imine formation,[31–33] *via* "click" chemistry[34,35] and oxidative polymerization.[36–38]

All of these synthetic methodologies aimed at developing novel microporous organic polymers with larger specific surface areas, precisely controlled pore sizes and multitudinous functions. However, the transition or noble metal catalysts used in the synthesis of CMPs, PAFs and some other MOPs are expensive and rare, which limit the practical applications.[1] Therefore, the sustainable industrial production of microporous organic polymers is still an unsolved challenge.

HCPs represent a novel, broad class of porous materials; moreover, they are one of the first kinds of pure microporous organic materials and are mainly prepared by Friedel–Crafts alkylation chemistry. This was developed by taking the concepts of "crosslinking", which has been used for the preparation of some other crosslinked materials, while the crosslinking level was further extended, thus making the networks highly rigid and preventing the polymer chains from collapsing into a dense form.[9] Therefore, the resulting materials contain permanent small pores, high surface areas and large micropore volumes. The diversity of the monomer used for synthesis has given the materials a wide range of functionalities and potential applications.[39,40] So, the materials have shown remarkable property enhancement in recent studies compared with research from the past ten years. According to the development of the synthetic methods, there are mainly three approaches to produce HCPs: (1) post-crosslinking functionalized precursors by Friedel–Crafts alkylation, (2) direct one-step polycondensation of a monomer containing multifunctional groups and (3) knitting rigid aromatic building blocks using an external crosslinker. In this chapter, we will describe the recent development of hypercrosslinked polymers, as well as their practical and potential applications.

4.2   Synthetic Strategy to Hypercrosslinked Polymers

4.2.1   Post-crosslinking Procedure

"Davankov-type" resins[41] can be defined as the first kind of hypercrosslinked polymers. The single-phase hypercrosslinked polystyrene

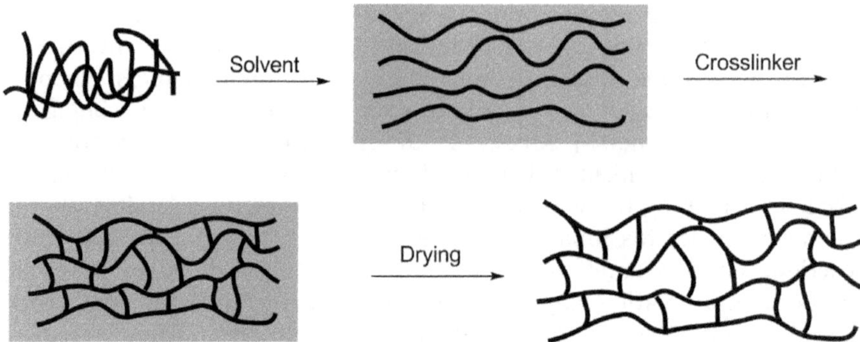

Figure 4.1   Schematic representation of the hypercrosslinking process. Reproduced with permission.[42] Copyright 2007, Royal Society of Chemistry.

networks were prepared by the intensive post-crosslinking of either dissolved linear polystyrene or swollen gel-type styrene–divinylbenzene copolymers. As shown in Figure 4.1, the swelling process separates polymer chains first to create space that is filled with solvent. Then, the pivotal crosslinking step was adopted immediately to lock the polymer chains in an expanded form around the solvent. When the solvent was removed, while rigid crosslinks kept the polymer chains separated, the space that was previously occupied became the pore volume and the interconnected pores were left in the polymer network, resulting in a material with permanent porosity.[42]

Hypercrosslinked polystyrene networks and sorbents were described in the scientific literature for the first time in the early 1970s. From then on, many efforts have been made within the synthesis field. Davankov with his colleagues[43] have prepared an intermolecular hypercrosslinked soluble material from linear polystyrene using monochlorodimethyl ether (MCDE) as the crosslinking agent (which is also called the crosslinker) in the presence of an $SnCl_4$ catalyst. The nanosponges display a high inner surface area of 680–1000 m^2 g^{-1} and show potential applications in chromatography, separation of contaminants from liquid solutions and adsorption of organic vapors.

Compared with MCDE, tetrachloromethane is also a good crosslinker with less toxicity. Hradil and Králová[44] reported a series of crosslinked resins with surface areas up to 1000 m^2 g^{-1}, which were obtained by post-crosslinking styrene–divinylbenzene copolymer precursors with tetrachloromethane *via* a Friedel–Crafts reaction in the presence of aluminium or ferric chloride.

As the polystyrene analogue, poly(divinylbenzene-co-vinylbenzyl chloride) (DVB-VBC) was also proved to be an excellent precursor to get very high surface area HCPs *via* the post-crosslinking process.[45] The synthetic strategy is quite simple and based on the crosslinking of a swollen chloromethylated polymer in the presence of a Friedel–Craft catalyst. This converts the chloromethyl groups into methylene bridges, thus creating a stronger crosslinking in addition to the divinylbenzene part of the original polymer.

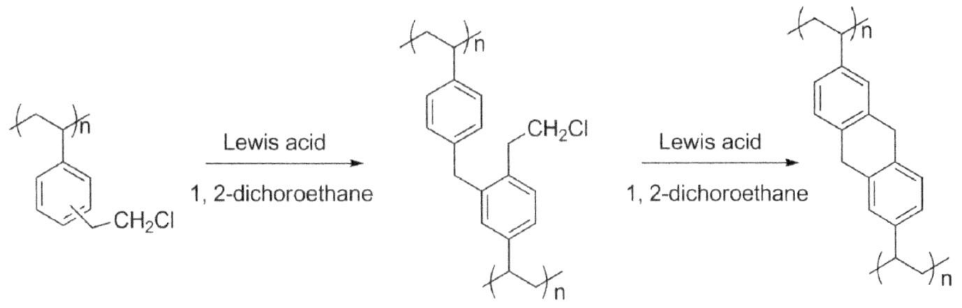

Figure 4.2 Hypercrosslinking of DVB-VBC Precursor resins *via* generation of an "internal" electrophile.
Reproduced with permission.[46] Copyright 2006, American Chemical Society.

As a result, the hypercrosslinked polymers prepared from this gel-type precursor only contain micropores in the networks.

Since then, for a long period these precursors were widely used in the synthesis of HCPs. Sherrington and co-workers[46] investigated the effect of different synthetic conditions (such as the monomer ratios of different vinylbenzyl chlorides, the reaction solvent and the catalyst) with regards to the pore structure and the properties of the resulting materials. It was confirmed that a different polymer structure of the starting precursor and various reaction conditions can lead to huge difference in surface areas (in the range of 300–2000 m^2 g^{-1}) of the resulting materials. Among all the examined Lewis acids, such as $FeCl_3$, $AlCl_3$ and $SnCl_4$, $FeCl_3$ was the most effective catalyst for the Friedel–Crafts reaction. In the case of a gel-type 2 mol% DVB-VBC precursor (Figure 4.2), an extensive permanent microporous structure was generated within only 15 min of initiating the crosslinking reaction, which yielded a surface area of $\sim$1200 m^2 g^{-1}, and the surface area rose steadily to a maximum approaching 2090 m^2 g^{-1} after 18 h.

Hydrogen is considered to be an ideal green renewable energy resource. The advantages, such as environmentally friendliness, renewability and high calorific value, make it one of the most promising new energies in the future. However, even assuming that producing hydrogen can be efficient and scalable, the safe and vast storage of hydrogen is still a big problem for usage in transport. Hydrogen stored as the metal hydride has been widely investigated in the past, but there are problems such as the irreversible process and slow rate. Meanwhile, physisorption using highly porous materials has become an attracting method in recent years because of the large capacity for storage and reversible absorption/desorption. In porous materials, the presence of small micropores is crucial for enhanced storage under practicable conditions, because the multiple sorbent–sorbate interactions would increase the heat of adsorption.[47] Recently, some studies have suggested a "hydrogen spillover" technique to further enhance the hydrogen storage capacity of porous materials.[48] To achieve this goal, the Tan group[49] developed an *"in situ"* method for the synthesis of microporous

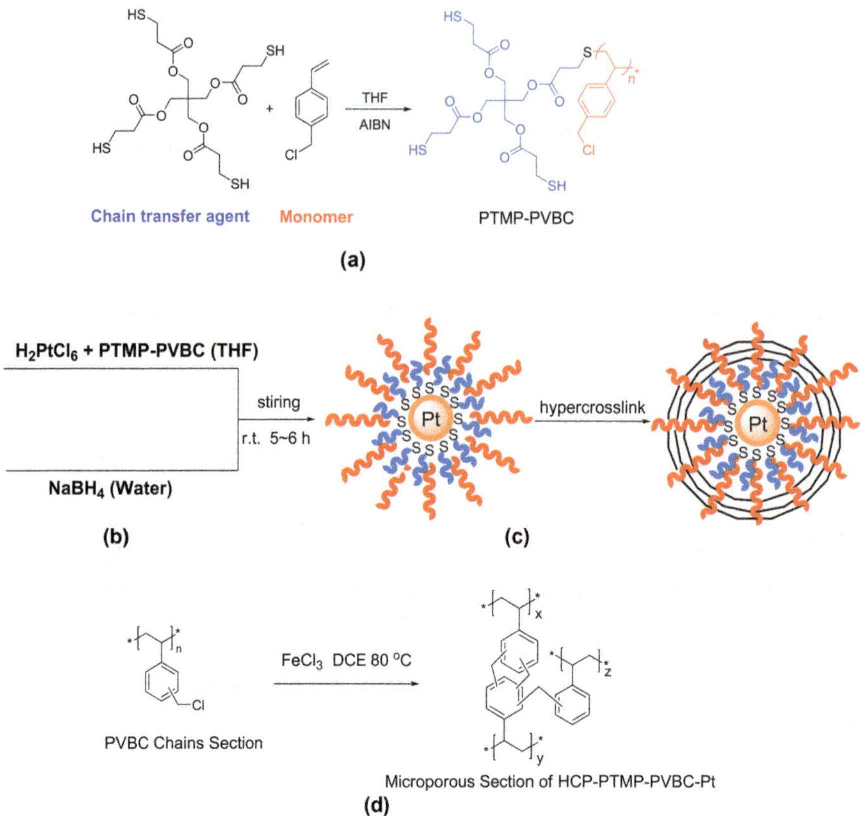

Figure 4.3 (a) Synthesis of functional tridentate thiol polymer ligands; (b) preparation of PTMP-PVBC-Pt; (c) formation of microporous HCP-PTMP-PVBC-Pt; (d) hypercrosslinking reaction details. Note: red chains represent PVBC, blue chains represent functional tridentate thiol polymer ligands. Reproduced with permission.[49] Copyright 2012, International Assoc. of Hydrogen Energy.

hypercrosslinked polymers with highly dispersed Pt nanoparticles (Figure 4.3). By the dissociation of hydrogen molecules on the Pt surface and the subsequent surface diffusion and adsorption of atomic hydrogen on the hypercrosslinked polymer surface, the hydrogen storage capacity of HCPs containing 2 wt% Pt nanoparticles increased to 0.21 wt% with an enhancement factor of 1.75 (from 0.12 wt%) compared to the similar material without Pt nanoparticles (at 298 K/19 bar).

Despite a great deal of work on the synthetic methods of MOPs, the controllable pore size and size distribution has rarely been reported. The Tan group[50] investigated the influence of the precursor structure on the porosity in detail. By adjusting the DVB content in DVB-VBC, the pore size of the hypercrosslinked DVB-VBC (HCP-DVB-VBC) can be precisely tailored from the macropore to micropore scale.

According to the nitrogen sorption analysis, the Brunauer–Emmett–Teller (BET) surface area of HCP-DVB-VBC initially increases and then decreases along with increasing DVB concentration. The highest BET surface area of up to 2064 m^2 g^{-1} was obtained for HCP-DVB-VBC with 2% DVB. With the DVB content increased from 0 to 10%, the pore size decreased and a more uniform and narrower microporous structure was produced effectively. When the DVB concentration was higher than 7%, a typical Type I nitrogen adsorption/desorption isotherm was observed, indicating a pure microporous polymer network. The mechanism was also proposed to explain the variation in the surface area and pore size distribution. In general, a phenyl ring would be more likely to interact with phenyls of distant segments. For the DVB-VBC precursor without DVB, the polymer chains dissolve well in organic solvent, becoming loose and disordered; the distance between two neighboring chains is random. This implied that not every $-CH_2Cl$ group possesses a neighboring benzene ring, which results in random hypercrosslinking density and the appearance of macropores in the polymer networks. However, for the DVB-VBC precursor containing DVB, every chain was slightly crosslinked by DVB and was not so free to disperse in the solvent but was almost fully swollen. For the low DVB concentration, the crosslinking level was also very low and the chains were loose, which means the distance between two polymer chains was not uniform. This is quite similar to the DVB-VBC precursor without DVB in terms of the formation of pore structure. In terms of DVB-VBC polymer precursors with higher DVB content, a more uniform pore structure was obtained because of the non-linear crosslinked texture after the hypercrosslinking reaction. Meanwhile, the precursor with a higher crosslinking level possessed a more rigid network and the distance between each chain tended to be uniform at every location, resulting in a more uniform hypercrosslinking degree in the polymer networks. The fracture section of HCP-DVB-VBC from 2% to 10% DVB thus represents a smoother, narrower and more uniform pore size distribution (Figure 4.4). The gas adsorption properties of these materials are totally different for different samples. For example, HCP-DVB-VBC with 2% DVB possesses the highest nitrogen adsorption, while HCP-DVB-VBC with 5% DVB shows the greatest hydrogen uptake and HCP-DVB-VBC with 10% DVB adsorbs carbon dioxide the most. These various gas adsorption properties indicate that the smaller micropore size and higher microporous volume is beneficial for H_2 and CO_2 uptakes.

To have a better understanding of pore formation in hypercrosslinked polymers, research[51] has been done on a series of poly(styrene-co-vinylbenzyl chloride) hypercrosslinked polymers utilizing the *Polymatic* simulated polymerization algorithm recently. During the virtual synthesis, including the simulated crosslinking process in the swollen state and on moving from the swollen to dried states, the evolution of the porosity was examined (Figure 4.5). These details from the molecular simulations provide necessary insight for elucidating the progression of pore formation in hypercrosslinked polymers, which is currently unobtainable experimentally.

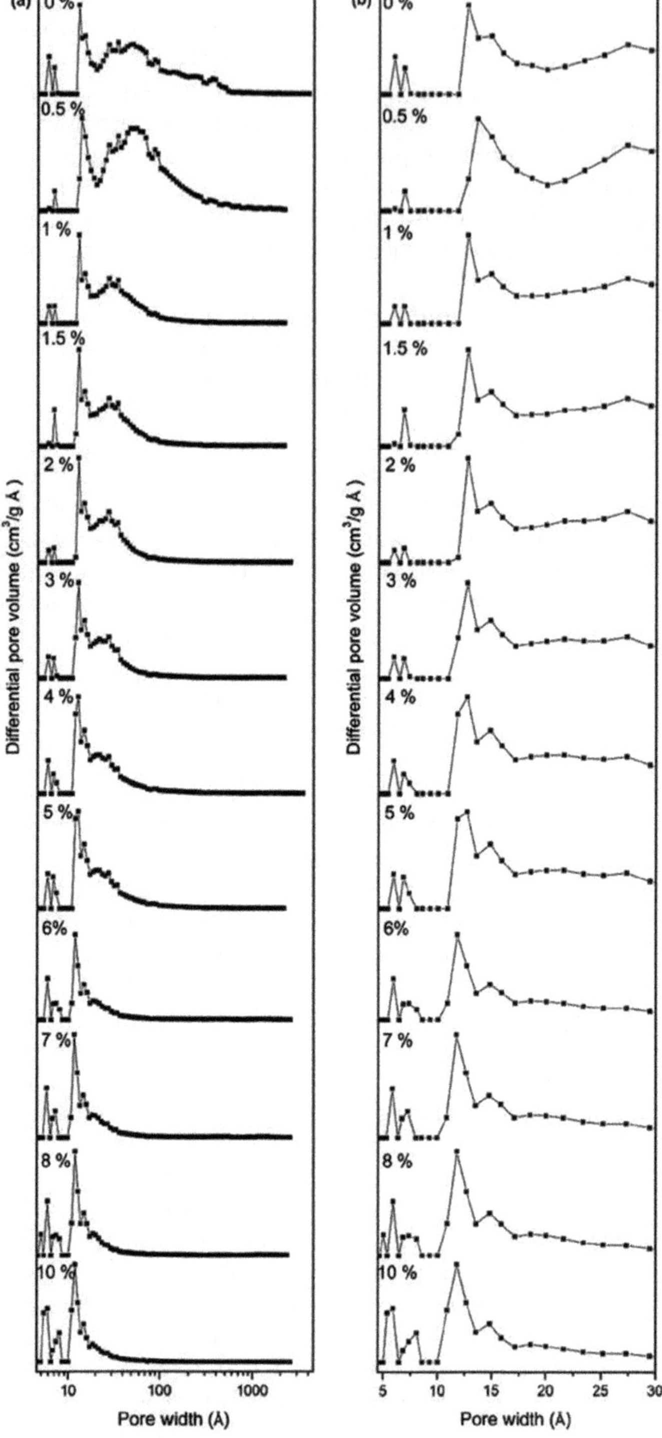

(a)
0 %
0.5 %
1 %
1.5 %
2 %
3 %
4 %
5 %
6%
7 %
8 %
10 %
Differential pore volume (cm³/g Å)
10
100
1000
Pore width (Å)
(b)
0 %
0.5 %
1 %
1.5 %
2 %
3 %
4 %
5 %
6%
7 %
8 %
10 %
Differential pore volume (cm³/g Å)
5
10
15
20
25
30
Pore width (Å)

In particular, the results showed that greater porosity was obtained when the polymers were crosslinked at lower densities and with higher degrees of crosslinking. The trends in the surface areas as a function of the VBC content were consistent with available experimental data. Moreover, measures of the surface areas and pore size distributions throughout the virtual synthesis indicated a gradual formation of pores in the swollen state, as the crosslinks redistributed the dispersion of the chains. Larger degrees of crosslinking were required to prevent collapse of the pores when moving from the swollen to the dried states. Complementary to experimental studies, these simulation results have provided better understanding of the structure and porosity of hypercrosslinked polymers, which will allow for a more purposeful approach to tuning the properties in these types of materials for given applications from the swollen to dried states.

4.2.2 Direct One-step Self-polycondensation

In spite of the post-crosslinking procedure, HCPs can also be synthesized from small molecule monomers that contain functional groups such as the chloromethyl group. By a simple one-step direct condensation polymerization process, small molecules linked together with the elimination of functional groups and formed a highly crosslinked network. This work was first studied by Cooper and his co-workers[52] using three kind of bis(chloromethyl) aromatic monomers, including dichloroxylene (DCX), bis(chloromethyl)biphenyl (BCMBP) and bis(chloromethyl) anthracene (BCMA; Figure 4.6). A series of polymer networks were obtained by changing the monomer ratio between the DCX isomers, BCMBP and BCMA, and the different amount of Lewis acid catalyst could also be an important factor that affects the copolymerization. The resulting networks, to some extent, are similar to those of the Friedel–Crafts-linked polystyrene materials. According to the N_2 adsorption/desorption at 77.3 K, these materials are predominantly microporous and exhibit apparent high BET surface areas of up to 1904 m^2 g^{-1} (Langmuir surface area $= 2992$ m^2 g^{-1}). Among all of the materials, the polymer network based on BCMBP and *p*-DCX exhibits a gravimetric H_2 storage capacity of 3.68 wt% at 77.3 K/15 bar. The high H_2 uptake on these materials is attributed to the low isosteric heat of adsorption (in the range of 6–7.5 kJ mol^{-1}). A molecular model was selected for the *p*-DCX network with the best overall agreement. The particular model was constructed from a combination of poly(*p*-DCX) clusters containing 500 carbon atoms and 60 carbon atoms that simulated certain key physical

---

Figure 4.4   Pore size distributions calculated using density functional theory (DFT) methods (slit pore models, differential pore volumes, pore width) of samples with 0% to 10% DVB. (a) Pore size distribution in the range between 4 and 3000 Å; (b) pore size distribution in the range between 4 and 30 Å.
Reproduced with permission.[50] Copyright 2011, Royal Society of Chemistry.

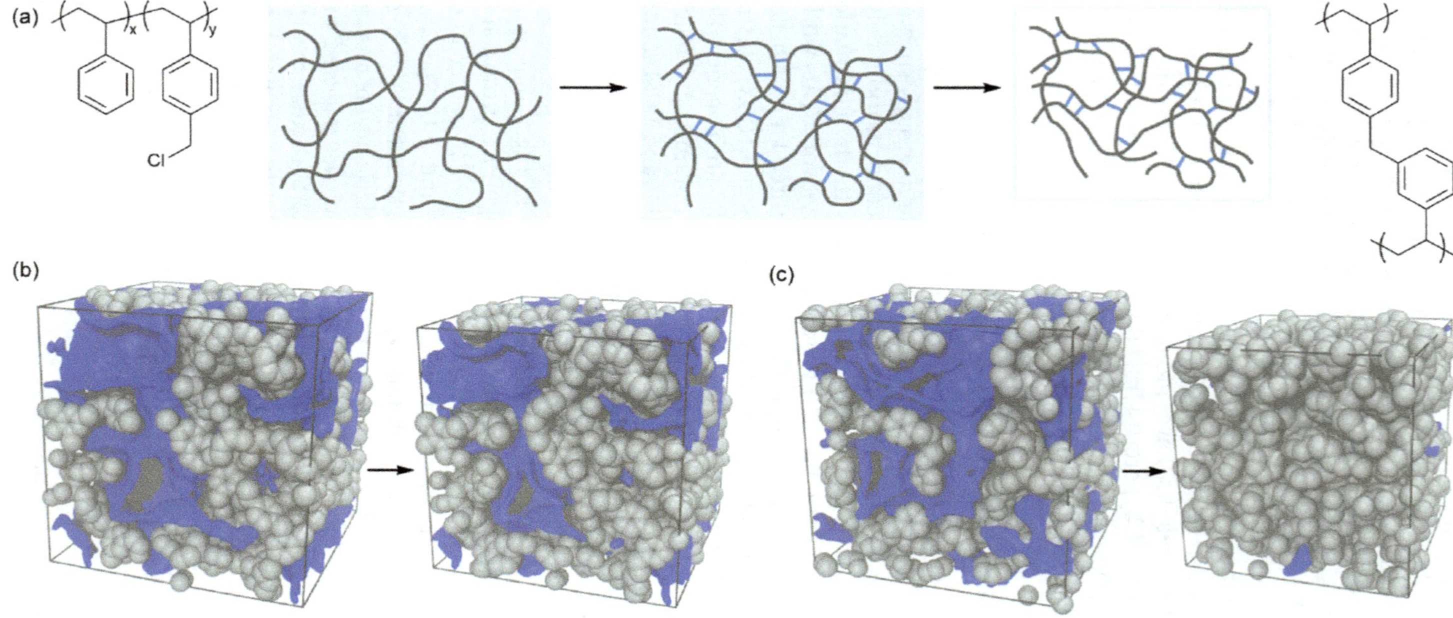

Figure 4.5 (a) Schematic showing the stages of synthesis: uncrosslinked linear chains swollen in solvent (left), crosslinks formed in the swollen state (middle) and the crosslinked structure in the dried state (right). Chemical structures of the poly(styrene-co-vinylbenzyl chloride) (St-VBC) precursor polymer and the hypercrosslinked polymer formed by Friedel–Crafts alkylation are also given. Example snapshots from the simulations are illustrated for (b) VBC and (c) St-VBC with 25% VBC in the swollen (left) and dried (right) states, with pore surfaces mapped out by the center of a nitrogen-sized probe in blue. Reproduced with permission.[51] Copyright 2014, American Chemical Society.

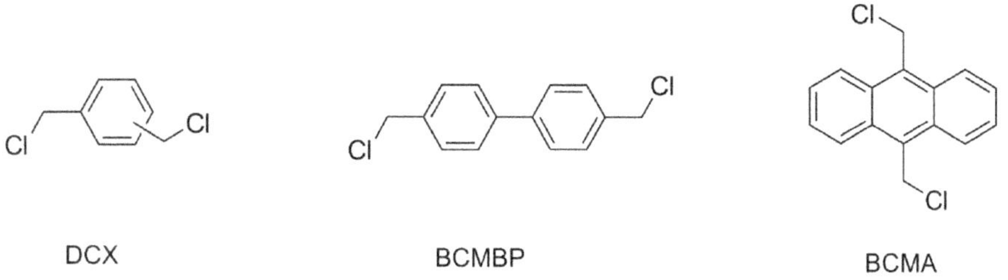

DCX   BCMBP   BCMA

Figure 4.6   Monomers used for the synthesis of the hypercrosslinked polymer networks. DCX: dichloroxylene (*ortho-*, *meta-* and *para-*isomers). BCMBP: 4,4′-bis(chloromethyl)-1,1′-biphenyl. BCMA: bis(chloromethyl)anthracene. Reproduced with permission.[52] Copyright 2007, American Chemical Society.

properties such as pore volume, pore width, absolute density and bulk density. This model also predicts the isotherm shape and isosteric heat for H_2 adsorption at 77.3 K and 87.2 K but overestimates the absolute H_2 uptake capacity. This is most likely because of the overstated surface area and the simulated sorption of H_2 in areas of occluded volume. The methane uptake abilities of these hypercrosslinked organic polymers were also investigated.[53] With a monomer ratio of 1:3, the resulting hypercrosslinked copolymers of DCX and BCMBP showed adsorption capacities up to 5.2 mmol g^{-1} (116 cm^3 g^{-1}) for methane at 298 K/20 bar. The adsorption capacity is similar to many other microporous materials but falls short of materials with higher micropore volumes.[54–56] Moreover, the polymers can be synthesized as continuous monolithic blocks, suggesting a potential route to overcome the volumetric issues associated with the packing of powders. Environmental issues are always a concern. The change in the global climate caused by excessive carbon dioxide emissions has attracted widespread public attention in recent years. CO_2 capture and storage (CCS) technology is a promising method using MOPs as the physical adsorption material. Martin *et al.*[57] investigated the CO_2 adsorption capacity of these hypercrosslinked polymers using a thermogravimetric analyzer (atmospheric pressure tests) and a high-pressure magnetic suspension balance (high pressure tests). The CO_2 capture capacities were related to the porous structural properties of the HCPs and a maximum CO_2 uptake of 1.7 mmol g^{-1} (7.4 wt% at 298 K) was obtained at atmospheric pressure. At higher pressures (30 bar), the polymers show superior CO_2 uptake up to 13.4 mmol g^{-1} (59 wt%) compared to zeolite-based materials and commercial activated carbons. Moreover, the low isosteric heats of adsorption and good selectivity make these materials promising for potential applications as CO_2 adsorbents in pre-combustion capture processes. As a promising solution to the global greenhouse effect, the CCS technique has attracted a large amount attention. And high adsorption capacity materials should be a key factor. Enormous research has shown that the introduction of functional groups, such as carboxyl and amine, may enhance the CO_2 adsorption and selectivity for CO_2/N_2 by increasing the interactions between adsorbent and adsorbate.[58,59]

Yang *et al.*[60] described the synthesis of microporous organic copolymers prepared from triphenylamine and dichloro-*p*-xylene using a combination of oxidative polymerization and Friedel–Crafts alkylation promoted by an anhydrous $FeCl_3$ catalyst. The BET specific surface areas for the obtained samples ranged from 318 to 1530 m^2 g^{-1} with the increasing content of dichloro-*p*-xylene. A high CO_2 uptake of 4.60 mmol g^{-1} was observed, which is comparable to the best reported results for MOPs, activated carbon and MOFs under the same conditions.

Further study about hydrocarbons/water separation was made by Yang *et al.*[61] while using 4,4-bis(chloromethyl)biphenyl incorporated with triphenylamine. By changing the monomer ratio, a series of hypercrosslinked polymers were obtained with high surface areas and a predominantly microporous structure. With apparent BET surface areas of 1362 m^2 g^{-1} for PBP-N-25 and 1338 m^2 g^{-1} for PBP-N-50, the benzene/water vapor selectivity was as high as 53.5 and 63.6, respectively. Moreover, a monolithic polymer (M-PBP-N-25) was prepared with an apparent BET surface area of 551 m^2 g^{-1}. Owing to its hydrophobic nature and low density, the monolith showed the potential for applications in oil spill cleanup operations.

By the self-condensation of benzyl-chloride-terminated cubic siloxane cages *via* Friedel–Crafts alkylation, Chaikittisilp *et al.*[62] described a novel hierarchically micro- and meso-porous hypercrosslinked siloxane-organic hybrid material. During the synthesis, this benzyl-chloride-terminated unit can serve as a good and single precursor for the construction of hypercrosslinked polymers, although the siloxane cages collapse, which eventually resulted in the porous siloxane-organic hybrid with an extremely high BET surface area of ~ 2500 m^2 g^{-1} and large pore volume of ~ 3.3 cm^3 g^{-1}. This finding suggests that simultaneous polymerization of organic functional groups and destruction of the siloxane cages would be a strategic concept in the construction of highly porous siloxane-organic hybrids.

Without chloride-terminated functional groups, Wu *et al.*[63] described the synthesis of siloxane-organic hybrids using octavinylsilsesquioxane (OVS) and benzene as the building blocks a *via* Friedel–Crafts reaction catalyzed by anhydrous aluminum chloride. The resulting polymers were highly porous with both micro- and meso-porous structures, showing relatively high BET surface areas ranging from 400 to 904 m^2 g^{-1} and pore volumes ranging from 0.24 to 0.99 cm^3 g^{-1}. The gas adsorption applications revealed that the porous hybrid possessed comparable H_2 uptake of 3.47 mmol g^{-1} (0.70 wt%) at 77.3 K/1 bar and a CO_2 uptake of 0.62 mmol g^{-1} (2.73 wt%) at 298 K/1 bar.

Despite the traditional monomers, which contain a chloromethyl group, the Tan group[64] has demonstrated that the hydroxymethyl group can also serve as a functional group. Two kinds of aromatic hydroxymethyl monomer, including bishydroxymethyl monomer (1,4-benzenedimethanol, BDM) and monohydroxymethyl monomer (benzyl alcohol, BA), were used for the synthesis process by a Friedel–Crafts reaction catalyzed under self-condensation (Figure 4.7). N_2 adsorption/desorption isotherms for the polymers showed that both polymer networks possessed a predominantly microporous

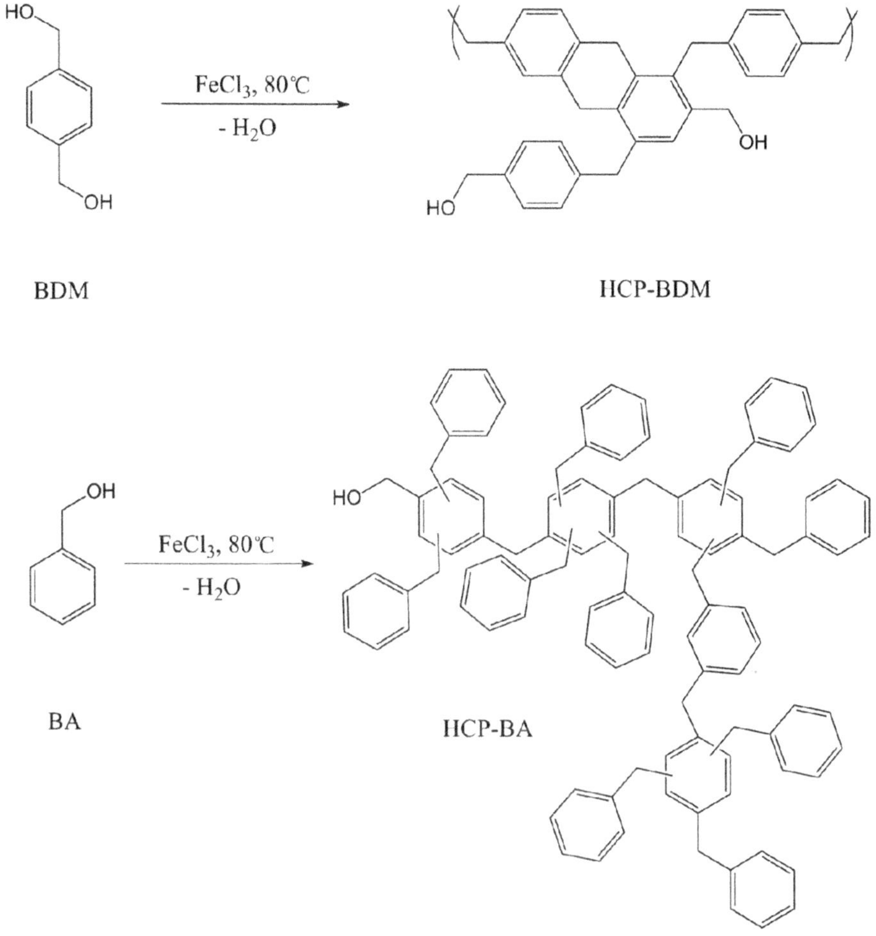

Figure 4.7 Synthesis of HCP-BDM and HCP-BA by a Friedel–Crafts reaction catalyzed under self-condensation.
Reproduced with permission.[64] Copyright 2013, Royal Society of Chemistry.

structure with high surface areas of up to 847 $m^2\ g^{-1}$ for HCP-BDM and 742 $m^2\ g^{-1}$ for HCP-BA. It is worth noting that the networks based on BA can also store a significant amount of CO_2, up to 8.46 wt% at 273 K/1 bar, and H_2, up to 0.97 wt% at 77.3 K/1 bar. This is different from the previous research, where multifunctional monomers are crucial for the construction of porous polymer networks and this study opens up the possibility of synthesizing porous materials using monofunctional monomers.

Following this study, the Tan group developed a novel strategy in which nonfunctional monomers were used for the one-step self-polycondensation reaction.[37] A series of microporous organic polymers (Scholl-coupling microporous polymer, SMP) with high surface areas and highly microporous structures were obtained by directly linking rigid aromatic building blocks (Figure 4.8). This strategy is based on the Scholl reaction, which involves the

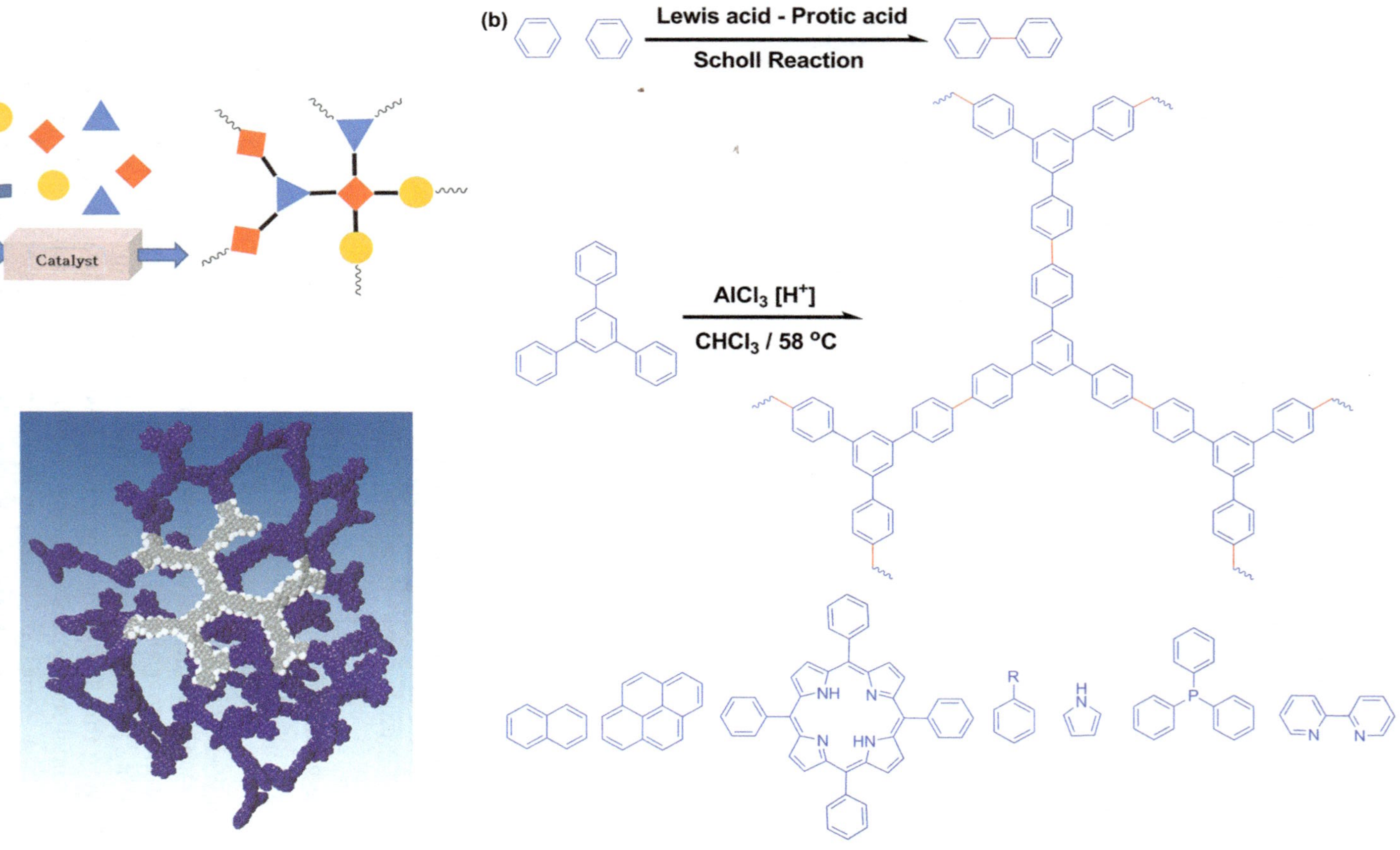

Figure 4.8   A general strategy for coupling various building blocks to form networks. (a) The typical Scholl reaction. (b) A model structure of the SMP-1 network. (c) Other monomers or co-monomers. Reproduced with permission.[37] Copyright 2014, Royal Society of Chemistry.

elimination of two aryl-bound hydrogen atoms accompanied by the formation of a new aryl–aryl bond in the presence of a Friedel–Crafts catalyst. Moreover, this method has many other advantages; for example, monomers with different functional groups [$-Br$, $-\equiv$, $-B(OH)_2$] are not necessary for the construction of networks and the catalysts used are not limited to transition or noble metals, which are expensive and rare. Because of the high activity of the reaction, a variety of monomers are suitable for the synthesis, including monomers with both high and low electron density, acidic or alkaline functional-group-containing monomers and aryl- or fused-ring-based monomers, as well as heterocyclic-ring-containing monomers. Thus, the BET surface areas of the resulting materials, determined by N_2 sorption analysis, range from 636 to 1421 $m^2\,g^{-1}$ according to the different activities of the monomer in the reaction. Moreover, the porous structure and properties of the resulting polymer networks may change a lot. For example, SMP-1 displays the highest H_2 adsorption capacity of 1.74 wt% at 77.3 K/1.13 bar, although it does not possess the highest surface area. The most probable reason should be the abundant ultra-microporous structure in the polymer networks, which is of great importance in gas storage and separation applications. In terms of CO_2 adsorption, the functional groups play the most important role. Compared to SMPs containing acidic groups or no functional groups, the SMPs containing alkaline functional groups possess higher CO_2 heat of adsorption and thus have higher CO_2 adsorption capacities. For instance, SMP-7 consists of a pyrrole monomer and exhibits the best CO_2 adsorption properties of about 20.4 wt% at 273 K/1 bar and 11.5 wt% at 298 K/1 bar. These results are comparable to that of the best reported MOPs. Small molecular ligands can also be incorporated into the polymer matrix as a part of the polymer scaffold providing an interesting bottom-up strategy to design materials for heterogeneous catalysts. SMP-8b containing phosphorus was used to fix Pd nanoparticles, which showed relatively good catalytic activities for the Suzuki–Miyaura cross-coupling reaction. The ultimate π stacking of some components was widely investigated in photo-electronic materials. Copolymerized with pyrrole, SMP-7 polymer networks show excellent conductivity of about 8.12×10^{-6} $S\,m^{-1}$, which is the highest conductivity reported for the MOPs to date. Moreover, this strategy provides a facile way to synthesize MOPs with luminescent properties. The luminescence properties of these SMPs may be due to the extended π-conjugated polymer systems and the three-dimensional network of π–π stacked aryl rings. As a result, the solid state ultraviolet (UV)–vis spectra of SMPs show relatively broad absorptions and SMPs with varied functional groups can exhibit colorful luminescence (Figure 4.9).

By introducing three-dimensional monomers, such as triphenylamine, tetraphenylmethane, tetraphenylsilane and tetraphenylgermane, the Zhu group[65] successfully synthesized a series of PAFs with moderate surface areas ranging from 515 $m^2\,g^{-1}$ to 1119 $m^2\,g^{-1}$ *via* the AlCl$_3$-catalyzed Scholl coupling reaction. The resulting polymer networks exhibited relatively high CH_4 and CO_2 sorption capacities of 1.04 mmol g^{-1} and 3.52 mmol g^{-1} at

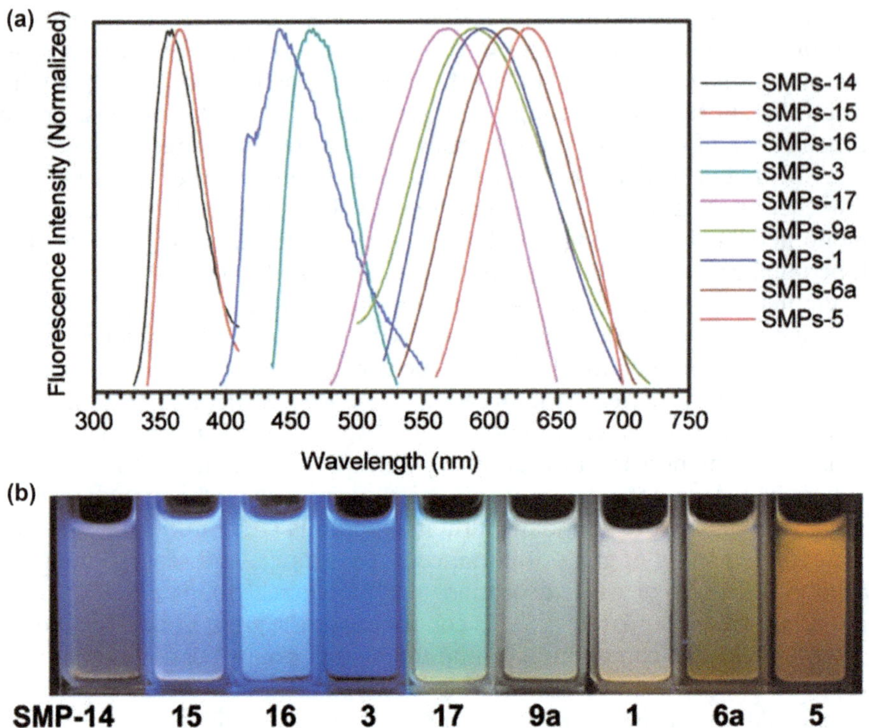

Figure 4.9   (a) Fluorescence emission spectra of SMPs; (b) optical images of SMPs in tetrahydrofuran (THF) under a UV light at 365 nm.
Reproduced with permission.[37] Copyright 2014, Royal Society of Chemistry.

273 K, respectively, as well as comparably high isosteric heat of adsorptions at 34.8 kJ mol^{-1} for CO_2 and 29.7 kJ mol^{-1} for CH_4.

4.2.3   External Crosslinking Strategy

Although numerous kinds of HCPs have been developed during the last few years, the synthetic routes remain very limited. Based on the need for functional groups like the chloromethyl group on the polymeric precursors and monomers, eventually there is a decrease in the diversification of the resulting materials. Moreover, the hydrogen chloride generated from the chloromethyl group during the Friedel–Crafts alkylation also needs to be considered for large scale industrial production and environmental issues. Considering these problems, the Tan group came up with a new strategy to obtain microporous organic polymers by knitting rigid aromatic building blocks using an external crosslinker *via* a simple one-step Friedel–Crafts reaction.[24] In this one-step crosslinking approach, formaldehyde dimethyl acetal (FDA) was used as an external crosslinker to react with various aromatic monomers, resulting in materials with a predominantly microporous

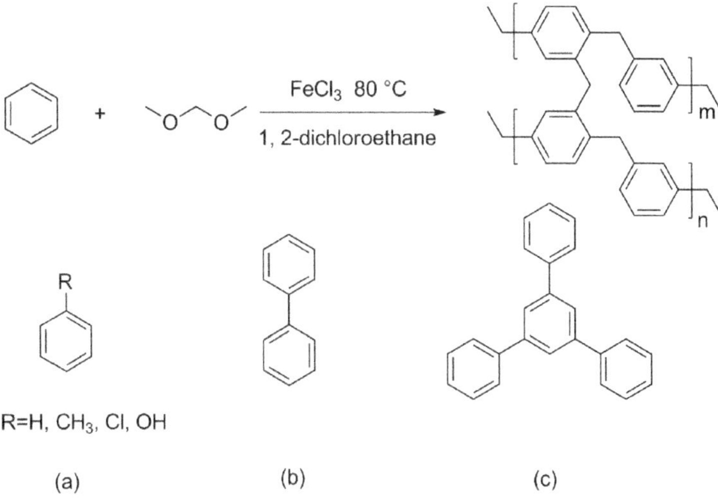

Figure 4.10 The synthetic pathway to the network structure. Structures (a)–(c) are molecular structures of building blocks for the network.
Reproduced with permission.[24] Copyright 2011, American Chemical Society.

structure and high surface area (Figure 4.10). This method needs a very simple monomer and cheap metal catalyst, thus making it economical and efficient. The highest BET surface area of 1391 m^2 g^{-1} was obtained for the benzene-based network. Meanwhile, the surface area, pore volume and porous structure can be roughly controlled by changing the ratio of different monomers to the external crosslinker. The functionality on the polymer networks can easily be introduced, as well as attained by choosing proper monomers containing different functional groups. For instance, when chlorobenzene and phenol were chosen as the starting monomers, polymer networks with –Cl and –OH groups can be easily obtained.

This strategy provides a facile way for the construction of hyper-crosslinked polymers with unique micro-morphology structures or special performances. Dawson *et al.*[66] successfully synthesized a series of MOPs containing –NH_2 groups by the copolymerization of a simple amine-functionalized aromatic monomer (aniline) with benzene (Figure 4.11). Despite the non-porous network synthesized solely by aniline, the strategy of copolymerization with benzene was used to fine tune the porous properties and finally obtain the networks with apparent BET surface areas of up to 1100 m^2 g^{-1}. Moreover, the increasing aniline content in the hypercrosslinked networks led to an improved CO_2/N_2 selectivity of up to 49.2 : 1 but, as a contrast, the 100% benzene network only showed a CO_2/N_2 selectivity of 15.9 : 1.

The Zhu group[67] successfully incorporated two functional groups (amino and hydroxyl) into tetrahedral building blocks, resulting in highly porous materials that possessed surface areas up to 1230 m^2 g^{-1} and 1608 m^2 g^{-1},

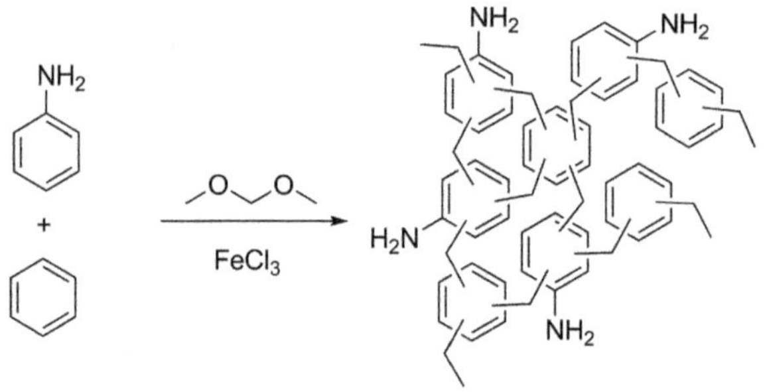

Figure 4.11   Synthesis of aniline/benzene copolymer networks.
Reproduced with permission.[66] Copyright 2012, Royal Society of Chemistry.

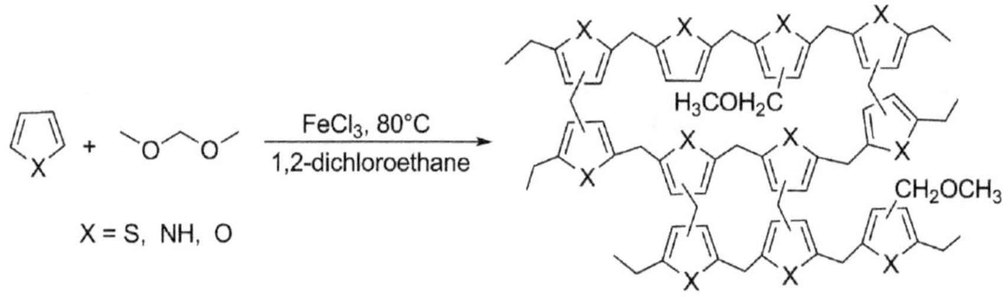

Figure 4.12   Schematic representation of the synthesis of aromatic heterocyclic microporous polymers.
Reproduced with permission.[40] Copyright 2012, John Wiley & Sons, Inc.

respectively. As expected, the corresponding functionalized PAF materials display enhanced CO_2 adsorption capacities and higher heats of adsorption than those of the non-functionalized materials.

Based on the outstanding outcomes of this knitting strategy, the Tan group also undertook further research on the gas (H_2, CO_2) adsorption process by introducing heterocyclic molecules into the microporous networks (Figure 4.12).[40] Three typical heterocyclic molecules (pyrrole, thiophene and furan) were used as monomers and the highest BET surface area was about 726 $m^2\ g^{-1}$ for the thiophene-1 (Th-1) polymer network. As expected, the microporous heterocyclic polymer networks obtained high adsorption capacities for H_2 (1.11 wt%, 77.3 K/1.13 bar) and CO_2 (12.7 wt%, 273 K/1 bar) because of their confined pore structures and heteroatom-rich pore surfaces. The CO_2 uptake capacities are much higher than many of the amine- and carboxylic-acid-functionalized materials. Moreover, pyrrole-1 (Py-1) showed an extraordinarily high selectivity for CO_2/N_2 of about 117 : 1 at 273 K, which was the highest among all the microporous materials reported at that time. These results indicated the significant impact of heterocyclic molecules for CO_2 capture applications and also highlighted the direction for carbon capture material development.

The Jiang group[68] reported a series of hypercrosslinked microporous organic polymer and copolymer networks synthesized *via* the Friedel–Crafts alkylation reaction using tetraphenylethylene (TPE) and/or 1,1,2,2-tetraphenylethane-1,2-diol (TPD) as the monomers. The polymer networks of 100% TPE showed a high BET surface area of up to 1980 $m^2 g^{-1}$ with a CO_2 uptake ability of 3.63 mmol g^{-1} (at 273 K/1 bar), while increased TPD content led to improved CO_2/N_2 selectivity. When increasing the TPD content to 100%, the resulting polymer network exhibited higher CO_2/N_2 selectivity of 119 : 1, while giving a lower CO_2 uptake ability of 1.92 mmol g^{-1} (at 273 K/1 bar) compared to other resulting polymer networks.

Zhu *et al.*[69] described the synthesis of two kinds of porous triazine and carbazole bifunctionalized task-specific polymers (TSPs) (Figure 4.13) with high levels of porosity (BET surface areas up to 913 $m^2 g^{-1}$ for the TSP-2 network). The resultant porous TSP-2 framework also exhibits good CO_2 uptake (18.0 wt% at 273 K/1 bar) and good adsorption selectivity (about 38 : 1 at 273 K), which is competitive with many other CO_2 adsorbents.

In summary, the Friedel–Crafts alkylation reaction by the FDA knitting method provides a simple and efficient synthetic method to produce

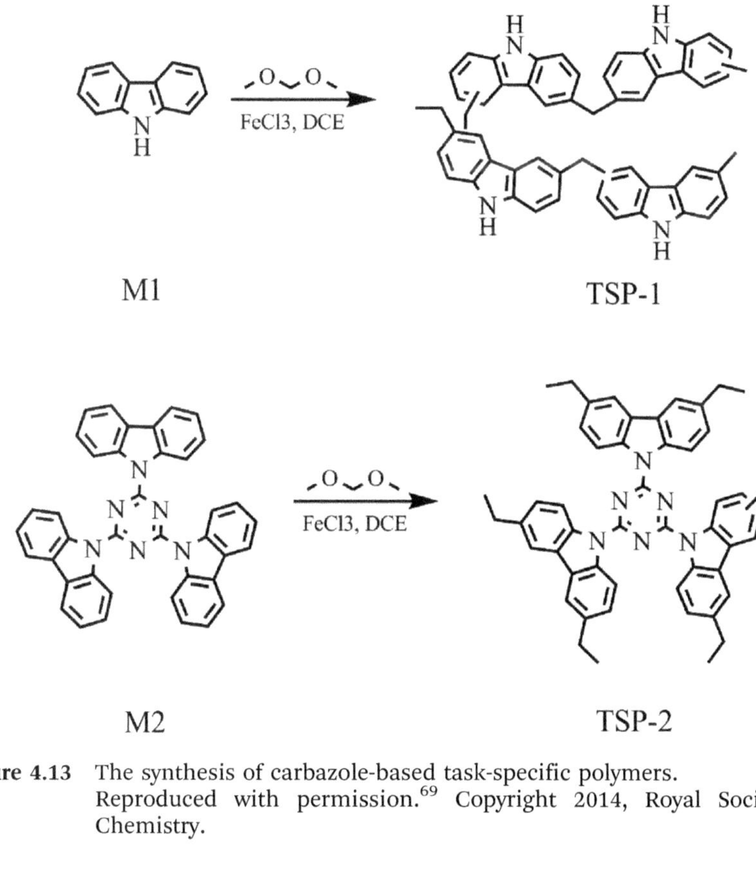

Figure 4.13 The synthesis of carbazole-based task-specific polymers. Reproduced with permission.[69] Copyright 2014, Royal Society of Chemistry.

cost-effective functionalized porous materials. These functionalized materials can be promising solid sorbents for practical application in CCS technologies.

Although lots of methodologies for MOP synthesis have been explored, only little attention has been paid to MOPs with controllable micromorphology. There are only a few reports of MOPs with tunable micromorphology such as nanoparticles,[70] hollow capsules,[71] 2D films[72,73] and monoliths.[61,74]

By using the simple "knitting" method, a kind of novel magnetic microporous polymer nanoparticle (MMPNs) with high surface area and superparamagnetic properties has been designed and synthesized successfully (Figure 4.14).[70] The BET surface area reached 560 m^2 g^{-1} after hypercrosslinking. After coating with polystyrene (PS), the Fe$_3$O$_4$@PS nanoparticles still maintained one third saturation magnetization (11.9 emu g^{-1}), making potential applications in industrial magnetic separation possible. For example, the MMPNs can be used as adsorbents for organic pollutants and are potential candidates for magnetic drug delivery systems.

In recent times, the Tan group[71] also made some progress in fabricating hollow microporous organic capsules (HMOCs; Figure 4.15), which combine the advantages of both microporous organic polymers and non-porous nanocapsules. In this work, a facile traditional emulsion polymerization process was used to prepare SiO$_2$@PS-DVB precursors, which contained an SiO$_2$ core and a PS-DVB copolymer shell. After hypercrosslinking *via* the knitting approach, the HMOCs with specific morphology were obtained by chemical etching of the sacrificial SiO$_2$ cores. The highest BET surface area achieved was 1129 m^2 g^{-1} for 0.5% DVB HMOCs and it displayed a decreasing trend with increasing DVB content, which means the porous structure can be controlled precisely. As expected, the cavity inside the HOMCs makes drug (*e.g.*, ibuprofen) uptake much higher than the solid HCP nanoparticles. The controlled microporous structure made HOMCs possess zero order drug release kinetics, which indicated their attractive applications in the medical field. Moreover, the multifunctional HMOCs also showed almost no cytotoxicity and fluorescence, which suggested promising prospects for biomedical applications.

In addition to the biomedical applications of HMOCs, they were also demonstrated as nanoscale reactors for the preparation of nanoparticles.[75] Pt nanoparticles supported by amine-functionalized condensation polymers were highly dispersed in the micropores of HMOCs *via* a simple chemical reduction reaction of chloroplatinic acid (Figure 4.16). The microporous structure ensured the high dispersion of Pt active sites. As a result, Pt-modified HMOCs (Pt/m-HMOCs) exhibited good activity (97%) and selectivity (99%) even after ten runs for the reduction of nitrobenzene to aniline under mild conditions.

Shi *et al.*[76] also designed a yolk–shell-type porous organic network with gold nanoparticles inside the cavities, which can be used as an efficient nanoreactor for the catalytic decomposition of cyclohexyl hydroperoxide.

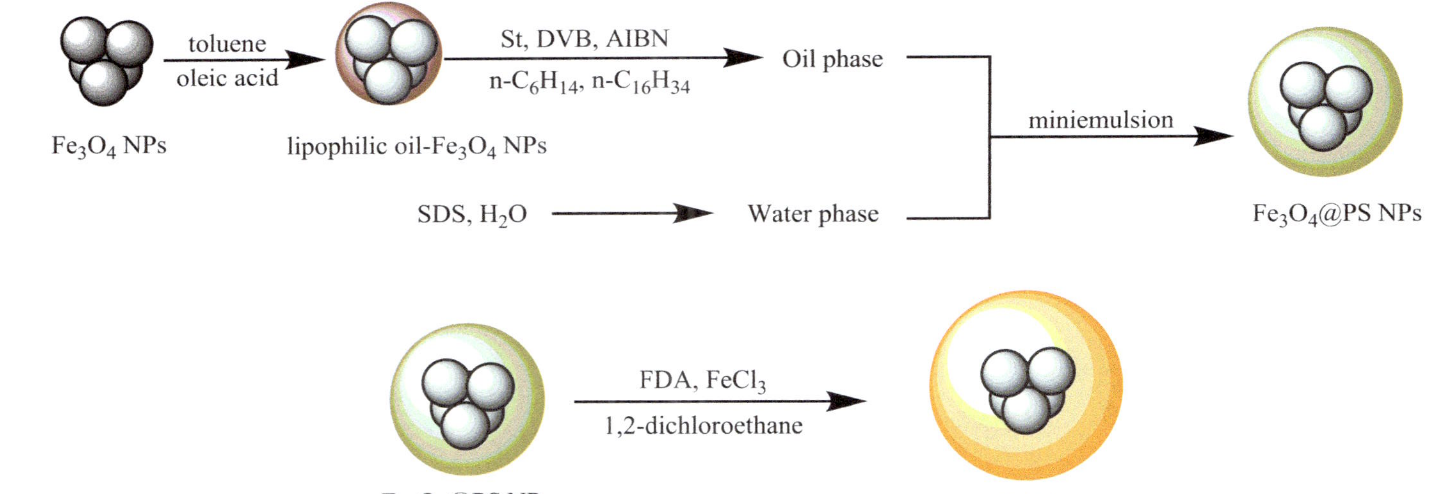

Figure 4.14   The synthesis pathway of magnetic microporous polymer nanoparticles (AIBN, azobisisobutyronitrile; SDS, sodium dodecyl sulfate).
Reproduced with permission.[70] Copyright 2013, Royal Society of Chemistry.

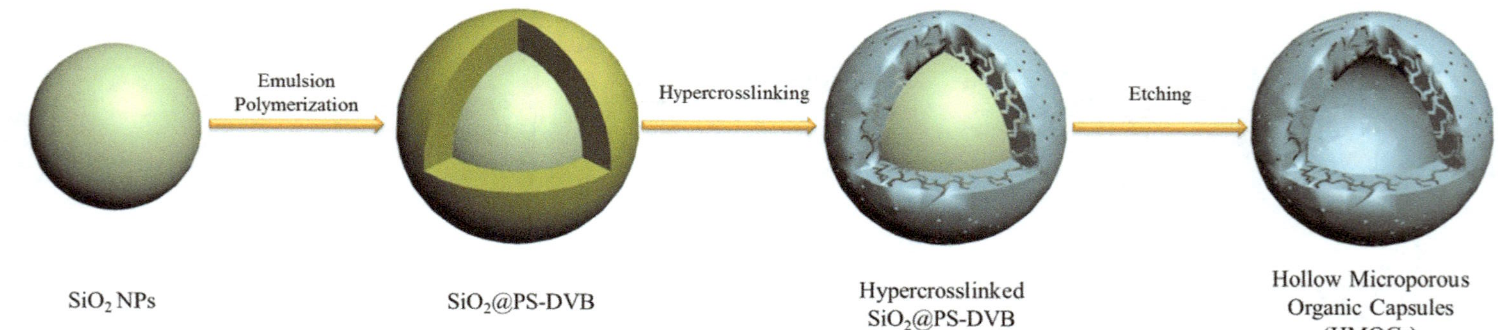

Figure 4.15   Schematic synthetic route to hollow microporous organic capsules. Reproduced with permission.[71] Copyright 2013, Macmillan Publishers Limited.

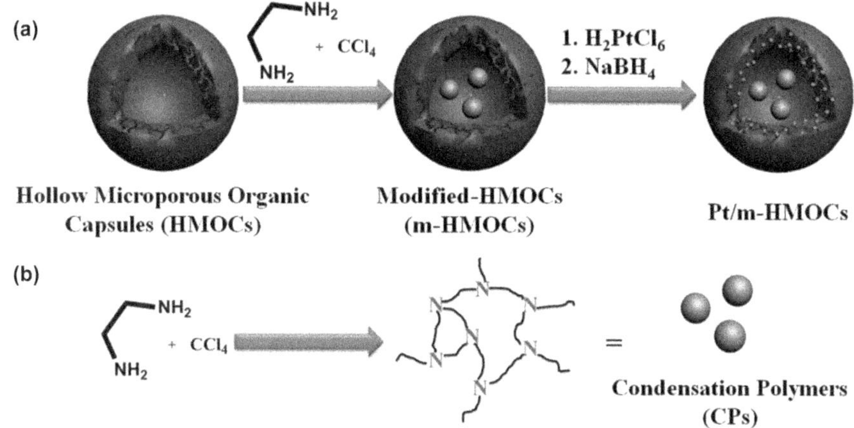

Figure 4.16　The synthesis pathways of (a) a catalyst and (b) condensation polymers. Reproduced with permission.[75] Copyright 2014, John Wiley & Sons, Inc.

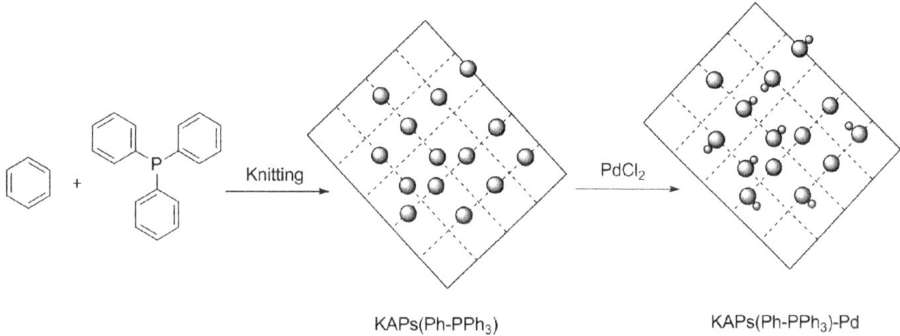

Figure 4.17　Synthetic route to KAPs(Ph-PPh₃)-Pd. Reproduced with permission.[39] Copyright 2012, John Wiley & Sons, Inc.

The porous properties of the samples were analyzed by N_2 adsorption/desorption analysis at 77.3 K and a BET surface area of 1240 $m^2\ g^{-1}$ was reached. The gold nanoparticles were trapped separately in the cavities, thus prevented from aggregation *via* the protection of the polymer shells. As a result, 90% conversion of cyclohexyl hydroperoxide and 86% selectivity for K-A oil (a mixture of cyclohexanone and cyclohexanol) were obtained.

The "knitting" strategy can be directly used to prepare a palladium–phosphine heterogeneous catalyst by the following strategy: knitting PPh_3 with benzene to produce functionalized knitting aryl polymers [KAPs(Ph-PPh₃)] and then binding Pd with PPh_3 groups to form KAPs(Ph-PPh₃)-Pd (Figure 4.17).[39] The BET surface area of KAPs(Ph-PPh₃) was found to be 1036 $m^2\ g^{-1}$ according to the nitrogen sorption analysis and showed no difference after binding with Pd. The microporous structure in the polymer networks ensured the high dispersion of Pd nanoparticles and the heterogeneous system improved the diffusion of small organic reactant molecules. Owing to these significant advantages, KAPs(Ph-PPh$_3$)-Pd exhibited excellent activity and selectivity for the Suzuki–Miyaura cross-coupling reaction of aryl chlorides in an aqueous ethanol solution under mild conditions (80 °C). Moreover, the non-toxic aqueous media, mild reaction conditions (80 °C), stability of the catalyst and the facile synthesis approach may eventually be helpful to use KAPs(Ph-PPh$_3$)-Pd on an industrial scale. This work also highlights that the microporous polymers can not only play a role as support materials, but they also protect the catalyst and positively affect the catalytic activity.

Compared with the 3D HMOC model, the 2D micro-morphology possesses great superiority in the synthesis of high-performance gas separation membranes, which are attractive for molecular-level separations in industrial scale chemical, energy and environmental processes. Wang *et al.*[72] described a novel, positively charged thin film composite nanofiltration (NF) membrane synthesized by interfacial polymerization with a polyacrylonitrile (PAN) support membrane by utilizing highly functional 2,2′,4,4′,6,6′-biphenyl hexaacyl chloride (BHAC) and piperazine (PIP). This series of membranes comprising hypercrosslinked polyamide barrier layers have a mean effective pore radius of 0.44 nm and exhibit a high efficiency for Na$^+$/Mg^{2+} separation with considerable permeate flux.

Most commercial polymer membranes for gas separation have been limited to a small number of polymers, which have low permeability and high selectivity that create the need for large surface areas to compensate for the lack of performance. This increases both cost and space requirements for large scale applications. From this view, Qiao *et al.*[73] developed a facile one-pot approach for the preparation of polymeric molecular sieve membranes with high permeability and selectivity (Figure 4.18). This is a simple method using a non-porous polystyrene (Ps) membrane precursor as the template, which was followed by the *in situ* crosslinking process. The resulting hypercrosslinked polymer membrane showed a sandwich hierarchical porous structure comprising a mesoporous surface composed of small polymer nanoparticles and a macroporous core with a layer of dense micropores, as confirmed by atomic force microscopy (AFM), scanning electron microscopy (SEM) and high-resolution transmission electron microscopy (TEM) images. The procedure can be explained as follows: the PS membrane surface was first reacted with FDA to produce a microporous crosslinked polymer shell outer layer; then FDA and FeCl$_3$ diffused through the microporous layer to react with the inner PS molecules. Meanwhile, the preferred outward PS molecules diffused from the core to the shell, leading to a net material flux across the membrane interface and simultaneously resulting in a flow of fast-moving vacancies to the vicinity of the solid–liquid interface. The macroporous core was formed through coalescence of the vacancies based on a nanoscale Kirkendall effect. Nitrogen sorption analyses were applied to investigate the porous properties of the membranes at 77.3 K. The microporous polymeric membranes exhibit Type I reversible sorption isotherms

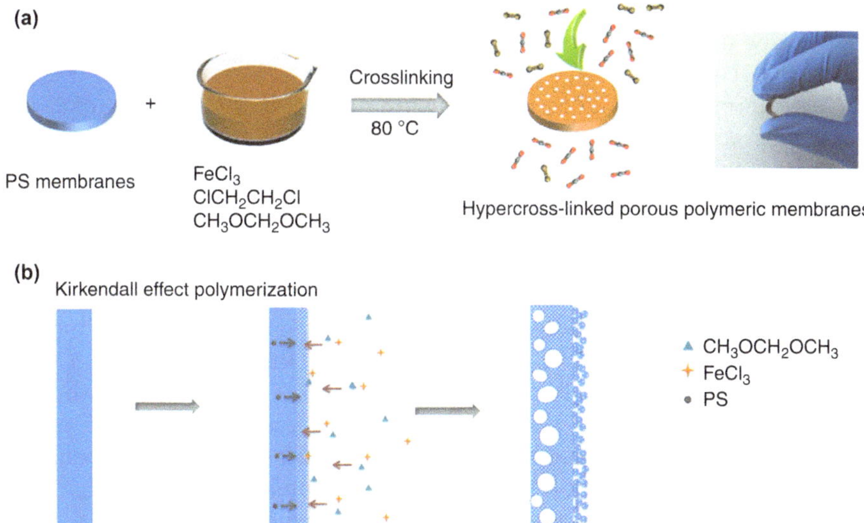

Figure 4.18   (a) Schematic illustration of the preparation procedure for the hypercrosslinked porous polymeric membranes (left) and an optical image of the hypercrosslinked polymeric membrane (right); (b) the nonequilibrium diffusion at the interface in the membrane by Kirkendall effect polymerization.
Reproduced with permission.[73] Copyright 2014, Macmillan Publishers Limited.

with a slight hysteresis loop at higher relative pressures, indicating abundant microporous and mesoporous structures in the membrane surface. The BET surface areas ranged from 218 to 618 m^2 g^{-1} with increasing crosslinking time, while the CO_2 permeability increased from 222.2 to 5261 barrer, but the CO_2/N_2 selectivity decreased from 30 to 18.5. The best O_2/N_2 selectivity and O_2 permeability were 4.2 and 222.2 barrer, respectively, which suggested promising gas separation applications.

Hierarchical porous monoliths possessing well-defined macropores and interconnected mesopores and micropores have attracted significant attention due to the rapid mass transport driven by convection through the internal connecting pores.

Maya and Svec[74] reported the preparation of porous polymer monoliths by hypercrosslinking styrene-type polymers and discussed the factors affecting the properties of the resulting materials by changing the reaction conditions and using three external crosslinkers, including 4,4′-bis(chloromethyl)-1,1′-biphenyl, a,a'-dichloro-p-xylene and formaldehyde dimethyl acetal. Polymer monoliths with extremely large surface areas reaching up to 900 m^2 g^{-1} were obtained using 4,4′-bis(chloromethyl)-1,1′-biphenyl as a crosslinker for only 2.5 h hypercrosslinking. Owing to the high surface area and interconnecting porous structure, the capillary columns containing hypercrosslinked monoliths as the stationary phase for liquid chromatography delivered a significant improvement in efficiency in the reversed phase separation of small molecules.

The external crosslinking strategy using the low-cost FDA crosslinker has become a very practical method for the preparation of hypercrosslinked polymers, while the surge to expand the synthesis strategy has never stopped.

Vinodh *et al.*[77] discovered the formation of hypercrosslinked conjugated quinonoid chromophores during crosslinking of aromatic monomers with 1,3,5-trioxane in the presence of an anhydrous $FeCl_3$ catalyst. As all the polymers bear low band gaps, they strongly absorb light covering the wavelength range of 1000 to 200 nm, which will inspire research on semiconductors, solar cell devices, UV stabilizers/absorbers, infrared absorbers/detectors and molecular sensors. For normal gas storage, most of the hypercrosslinked polymer matrices carry microporous spheres of about 1 to 5 mm in diameter and showed 6–7% CO_2 sorption.

Sun *et al.*[78] came up with a new synthetic methodology for a series of luminescent microporous organic polymers through the palladium-catalyzed tandem Suzuki–Heck C–C coupling reactions of several aromatic halides with potassium vinyltrifluoroborate. The BET surface areas of these polymers ranged from 318 to 693 $m^2\ g^{-1}$. The formation of conjugated polymers with the incorporation of vinyl groups leads to fluorescent properties and the selection of aromatic halides and alternating the ratio of monomers could adjust the emissions from blue to green.

4.3   Conclusion and Outlook

Since they were discovered by Davankov *et al.*, hypercrosslinked polymers have experienced a rapid development in the area of monomer design and synthetic strategy. As compared with other microporous organic polymers, such as PIMs, CMPs and COFs, the HCPs exhibit obvious advantages, including robustness, good thermal and chemical stability, moderate synthetic conditions and low-cost monomers or precursors. In addition, HCP materials can be produced in a monolithic form, which has practical significance with regards to the industrial application. According to the development of the synthetic methods, there are mainly three approaches to produce HCPs:

1) post-crosslinking procedure;
2) direct one-step self-polycondensation;
3) external crosslinking strategy.

Along with the development of these approaches, swollen polystyrene- and poly(chloromethylstyrene)-type polymeric precursors were first investigated for the post-crosslinking procedure. Followed by the discovery of functionalized aromatic benzyl chlorides, groundbreaking progress in this field has been made ever since the use of FDA as a crosslinker, knitting multitudinous low-cost aromatic monomers *via* a simple one-step Friedel–Crafts alkylation. This strategy showed outstanding characteristics compared with other methods: (1) the extensive range of building blocks with possible structures and functionality (there are still countless monomers that have not been used); (2) the mild synthesis conditions, which can be realized for larger scale and economical production in industry; (3) different structural building blocks or copolymerization with multi-functionalized monomers will result in diverse materials for various potential applications. Based on these advantages, we consider the external crosslinking strategy to be a very practical method for the preparation of hypercrosslinked polymers and the knitting of aromatic polymer materials could undoubtedly become the bridge between laboratory methods and industrial production.

Even though there are so many significant advantages for HCPs, they also present some challenges. For example, the structure of HCP materials is highly irregular and the generation of huge amounts of heat during the hypercrosslinking is also an intractable problem when large scale preparation of HCPs is carried out. Looking for new synthetic approaches to obtain novel HCP networks with higher specific surface areas is always highly desired. Further development of HCPs with controllable structures and excellent new functions will be a prolonged research hotspot and it can broaden their application domains such as in photo-electricity, sensors and other semi-conducting devices.

Acknowledgements

The authors would like to thank financial support from the Program for Changjiang Scholars and Innovative Research Teams in University (No. IRT1014) and the National Science Foundation of China (Grant No. 50973037, 51173058).

References

1. R. Dawson, A. I. Cooper and D. J. Adams, *Prog. Polym. Sci.*, 2012, **37**, 530.
2. Y. Li and J. Yu, *Chem. Rev.*, 2014, **114**, 7268.
3. C. Knöfel, J. Descarpentries, A. Benzaouia, V. Zeleňák, S. Mornet, P. L. Llewellyn and V. Hornebecq, *Microporous Mesoporous Mater.*, 2007, **99**, 79.
4. E. L. K. Mui, D. C. K. Ko and G. McKay, *Carbon*, 2004, **42**, 2789.
5. J. Li, J. Sculley and H. Zhou, *Chem. Rev.*, 2011, **112**, 869.
6. A. M. Fracaroli, H. Furukawa, M. Suzuki, M. Dodd, S. Okajima, F. Gándara, J. A. Reimer and O. M. Yaghi, *J. Am. Chem. Soc.*, 2014, **136**, 8863.
7. T. Hasell, J. L. Culshaw, S. Y. Chong, M. Schmidtmann, M. A. Little, K. E. Jelfs, E. O. Pyzer-Knapp, H. Shepherd, D. J. Adams, G. M. Day and A. I. Cooper, *J. Am. Chem. Soc.*, 2014, **136**, 1438.

8. M. Liu, M. A. Little, K. E. Jelfs, J. T. A. Jones, M. Schmidtmann, S. Y. Chong, T. Hasell and A. I. Cooper, *J. Am. Chem. Soc.*, 2014, **136**, 7583.

9. S. Xu, Y. Luo and B. Tan, *Macromol. Rapid Commun.*, 2013, **34**, 471.

10. J. Germain, J. M. J. Fréchet and F. Svec, *Small*, 2009, **5**, 1098.

11. J. Weber, M. Antonietti and A. Thomas, *Macromolecules*, 2008, **41**, 2880.

12. P. Pandey, A. P. Katsoulidis, I. Eryazici, Y. Wu, M. G. Kanatzidis and S. T. Nguyen, *Chem. Mater.*, 2010, **22**, 4974.

13. A. I. Cooper, *Adv. Mater.*, 2009, **21**, 1291.

14. G. Cheng, T. Hasell, A. Trewin, D. J. Adams and A. I. Cooper, *Angew. Chem., Int. Ed.*, 2012, **51**, 12727.

15. J. Jiang, Y. Li, X. Wu, J. Xiao, D. J. Adams and A. I. Cooper, *Macromolecules*, 2013, **46**, 8779.

16. N. B. McKeown and P. M. Budd, *Chem. Soc. Rev.*, 2006, **35**, 675.

17. N. B. McKeown, P. M. Budd, K. J. Msayib, B. S. Ghanem, H. J. Kingston, C. E. Tattershall, S. Makhseed, K. J. Reynolds and D. Fritsch, *Chem. – Eur. J.*, 2005, **11**, 2610.

18. J. Vile, M. Carta, C. G. Bezzu, B. M. Kariuki and N. B. McKeown, *Polymer*, 2014, **55**, 326.

19. H. Ma, H. Ren, S. Meng, Z. Yan, H. Zhao, F. Sun and G. Zhu, *Chem. Commun.*, 2013, **49**, 9773.

20. A. P. Cote, A. I. Benin, N. W. Ockwig, M. O'Keeffe, A. J. Matzger and O. M. Yaghi, *Science*, 2005, **310**, 1166.

21. H. M. El-Kaderi, J. R. Hunt, J. L. Mendoza-Cortes, A. P. Cote, R. E. Taylor, M. O'Keeffe and O. M. Yaghi, *Science*, 2007, **316**, 268.

22. V. A. Davankov and M. P. Tsyurupa, *React. Polym.*, 1990, **13**, 27.

23. C. D. Wood, B. Tan, A. Trewin, H. J. Niu, D. Bradshaw, M. J. Rosseinsky, Y. Z. Khimyak, N. L. Campbell, R. Kirk, E. Stockel and A. I. Cooper, *Chem. Mater.*, 2007, **19**, 2034.

24. B. Li, R. Gong, W. Wang, X. Huang, W. Zhang, H. Li, C. Hu and B. Tan, *Macromolecules*, 2011, **44**, 2410.

25. Y. Li, S. Roy, T. Ben, S. Xu and S. Qiu, *Phys. Chem. Chem. Phys.*, 2014, **16**, 12909.

26. T. Ben and S. Qiu, *CrystEngComm*, 2013, **15**, 17.

27. S. Ren, M. J. Bojdys, R. Dawson, A. Laybourn, Y. Z. Khimyak, D. J. Adams and A. I. Cooper, *Adv. Mater.*, 2012, **24**, 2357.

28. M. J. Bojdys, J. Jeromenok, A. Thomas and M. Antonietti, *Adv. Mater.*, 2010, **22**, 2202.

29. S. Yuan, S. Kirklin, B. Dorney, D.-J. Liu and L. Yu, *Macromolecules*, 2009, **42**, 1554.

30. S. Yuan, B. Dorney, D. White, S. Kirklin, P. Zapol, L. Yu and D.-J. Liu, *Chem. Commun.*, 2010, **46**, 4547.

31. O. K. Farha, A. M. Spokoyny, B. G. Hauser, Y.-S. Bae, S. E. Brown, R. Q. Snurr, C. A. Mirkin and J. T. Hupp, *Chem. Mater.*, 2009, **21**, 3033.

32. Z. Wang, B. Zhang, H. Yu, L. Sun, C. Jiao and W. Liu, *Chem. Commun.*, 2010, **46**, 7730.

33. Y. Luo, B. Li, L. Liang and B. Tan, *Chem. Commun.*, 2011, **47**, 7704.
34. J. R. Holst, E. Stöckel, D. J. Adams and A. I. Cooper, *Macromolecules*, 2010, **43**, 8531.
35. P. Pandey, O. K. Farha, A. M. Spokoyny, C. A. Mirkin, M. G. Kanatzidis, J. T. Hupp and S. T. Nguyen, *J. Mater. Chem.*, 2011, **21**, 1700.
36. J. Schmidt, J. Weber, J. D. Epping, M. Antonietti and A. Thomas, *Adv. Mater.*, 2009, **21**, 702.
37. B. Li, Z. Guan, X. Yang, W. D. Wang, W. Wang, I. Hussain, K. Song, B. Tan and T. Li, *J. Mater. Chem. A*, 2014, **2**, 11930.
38. Q. Chen, M. Luo, P. Hammershøj, D. Zhou, Y. Han, B. W. Laursen, C. Yan and B. Han, *J. Am. Chem. Soc.*, 2012, **134**, 6084.
39. B. Li, Z. Guan, W. Wang, X. Yang, J. Hu, B. Tan and T. Li, *Adv. Mater.*, 2012, **24**, 3390.
40. Y. Luo, B. Li, W. Wang, K. Wu and B. Tan, *Adv. Mater.*, 2012, **24**, 5703.
41. M. P. Tsyurupa and V. A. Davankov, *React. Funct. Polym.*, 2006, **66**, 768.
42. J. Germain, J. M. J. Fréchet and F. Svec, *J. Mater. Chem.*, 2007, **17**, 4989.
43. V. A. Davankov, M. M. Ilyin, M. P. Tsyurupa, G. I. Timofeeva and L. V. Dubrovina, *Macromolecules*, 1996, **29**, 8398.
44. J. Hradil and E. Králová, *Polymer*, 1998, **39**, 6041.
45. P. Veverka and K. Jeřábek, *React. Funct. Polym.*, 1999, **41**, 21.
46. J.-H. Ahn, J.-E. Jang, C.-G. Oh, S.-K. Ihm, J. Cortez and D. C. Sherrington, *Macromolecules*, 2005, **39**, 627.
47. Q. Wang and J. K. Johnson, *Int. J. Hydrogen Energy*, 1999, **110**, 577.
48. D. J. Collins and H.-C. Zhou, *J. Mater. Chem.*, 2007, **17**, 3154.
49. B. Li, X. Huang, R. Gong, M. Ma, X. Yang, L. Liang and B. Tan, *Int. J. Hydrogen Energy*, 2012, **37**, 12813.
50. B. Li, R. Gong, Y. Luo and B. Tan, *Soft Matter*, 2011, 7, 10910.
51. L. J. Abbott and C. M. Colina, *Macromolecules*, 2014, **47**, 5409.
52. C. D. Wood, B. Tan, A. Trewin, H. Niu, D. Bradshaw, M. J. Rosseinsky, Y. Z. Khimyak, N. L. Campbell, R. Kirk, E. Stöckel and A. I. Cooper, *Chem. Mater.*, 2007, **19**, 2034.
53. C. D. Wood, B. Tan, A. Trewin, F. Su, M. J. Rosseinsky, D. Bradshaw, Y. Sun, L. Zhou and A. I. Cooper, *Adv. Mater.*, 2008, **20**, 1916.
54. K. Seki, *Chem. Commun.*, 2001, 1496.
55. M. Eddaoudi, J. Kim, N. Rosi, D. Vodak, J. Wachter, M. O'Keeffe and O. M. Yaghi, *Science*, 2002, **295**, 469.
56. D. Lozano-Castelló, D. Cazorla-Amorós and A. Linares-Solano, *Energy Fuels*, 2002, **16**, 1321.
57. C. F. Martín, E. Stöckel, R. Clowes, D. J. Adams, A. I. Cooper, J. J. Pis, F. Rubiera and C. Pevida, *J. Mater. Chem.*, 2011, **21**, 5475.
58. S. Chen, J. Zhang, T. Wu, P. Feng and X. Bu, *J. Am. Chem. Soc.*, 2009, **131**, 16027.
59. S. Couck, J. F. M. Denayer, G. V. Baron, T. Rémy, J. Gascon and F. Kapteijn, *J. Am. Chem. Soc.*, 2009, **131**, 6326.
60. Y. Yang, Q. Zhang, S. Zhang and S. Li, *Polymer*, 2013, **54**, 5698.
61. Y. Yang, Q. Zhang, S. Zhang and S. Li, *RSC Adv.*, 2014, **4**, 5568.

62. W. Chaikittisilp, M. Kubo, T. Moteki, A. Sugawara-Narutaki, A. Shimojima and T. Okubo, *J. Am. Chem. Soc.*, 2011, **133**, 13832.
63. Y. Wu, D. Wang, L. Li, W. Yang, S. Feng and H. Liu, *J. Mater. Chem. A*, 2014, **2**, 2160.
64. Y. Luo, S. Zhang, Y. Ma, W. Wang and B. Tan, *Polym. Chem.*, 2013, **4**, 1126.
65. L. Li, H. Ren, Y. Yuan, G. Yu and G. Zhu, *J. Mater. Chem. A*, 2014, **2**, 11091.
66. R. Dawson, T. Ratvijitvech, M. Corker, A. Laybourn, Y. Z. Khimyak, A. I. Cooper and D. J. Adams, *Polym. Chem.*, 2012, **3**, 2034.
67. X. Jing, D. Zou, P. Cui, H. Ren and G. Zhu, *J. Mater. Chem. A*, 2013, **1**, 13926.
68. S. Yao, X. Yang, M. Yu, Y. Zhang and J.-X. Jiang, *J. Mater. Chem. A*, 2014, **2**, 8054.
69. X. Zhu, S. M. Mahurin, S.-H. An, C.-L. Do-Thanh, C. Tian, Y. Li, L. W. Gill, E. W. Hagaman, Z. Bian, J.-H. Zhou, J. Hu, H. Liu and S. Dai, *Chem. Commun.*, 2014, **50**, 7933.
70. X. Yang, B. Li, I. Majeed, L. Liang, X. Long and B. Tan, *Polym. Chem.*, 2013, **4**, 1425.
71. B. Li, X. Yang, L. Xia, M. I. Majeed and B. Tan, *Sci. Rep.*, 2013, **3**, 2128.
72. T. Wang, Y. Yang, J. Zheng, Q. Zhang and S. Zhang, *J. Membr. Sci.*, 2013, **448**, 180.
73. Z. Qiao, S. Chai, K. Nelson, Z. Bi, J. Chen, S. M. Mahurin, X. Zhu and S. Dai, *Nat. Commun.*, 2014, **5**, 3705.
74. F. Maya and F. Svec, *Polymer*, 2014, **55**, 340.
75. X. Yang, K. Song, L. Tan, I. Hussain, T. Li and B. Tan, *Macromol. Chem. Phys.*, 2014, **215**, 1257.
76. S. Shi, C. Chen, M. Wang, J. Ma, H. Ma and J. Xu, *Chem. Commun.*, 2014, **50**, 9079.
77. R. Vinodh, P. Hemalatha, M. Ganesh, M. M. Peng, A. Abidov, M. Palanichamy, W. S. Chab and H.-T. Jang, *RSC Adv.*, 2014, **4**, 3668.
78. L. Sun, Y. Zou, Z. Liang, J. Yu and R. Xu, *Polym. Chem.*, 2014, **5**, 471.

CHAPTER 5

Polymers of Instrinsic Microporosity

5.1 Introduction

Polymers with intrinsic microporosity, also abbreviated as PIMs, are soluble porous polymers formed from non-reversible condensation methods. These polymers cannot pack efficiently in the solid state due to rigid ladder-like components. These polymers have a site of contortion in their structure and are known for their properties, including gas permeation and selective separation.

5.2 Design and Synthesis of PIMs

McKeown and Budd have developed PIMs with a design strategy imitating the structural architecture of graphene, *i.e.*, to have an extended network of aromatic groups that form a network-like structure.[1,2] A dioxane-forming reaction between an *ortho*-dihydroxy monomer and an *ortho*-dihalide monomer (difluoride or dichloride) is employed for their synthesis.[3] The very first PIMs were built from phthalocyanines and porphyrins units, and had metal ions incorporated, making them suitable for catalysis.[4,5] In phthalocyanine-based networks, rigid and nonlinear linking groups, such as one derived from the 5,5′,6,6′-tetrahydroxy-3,3,3′,3′-tetramethyl-1,1′-spirobisindane monomer (A1), have been incorporated in between the sub-units in order to prevent aggregation of aromatic units into columnar stacks.[6] Phthalocyanine network polymers were synthesised from bis(phthalonitrile) as a precursor. During this reaction, dioxane is formed by a metal ion template-governed cyclotetramerisation reaction. The similar dioxane reaction gives highly coloured network polymers with monomer A1 and 4,5-dichlorophthalonitrile (a high yielding, double aromatic nucleophilic substitution reaction—SNAr).[1] The dioxane-forming reaction

Monographs in Supramolecular Chemistry No. 17
Porous Polymers: Design, Synthesis and Applications
By Shilun Qiu and Teng Ben
© The Royal Society of Chemistry 2016
Published by the Royal Society of Chemistry, www.rsc.org

has been used to react monomer A1 with the hexaazatrinaphthylene (HATN) monomer to form an insoluble HATN-PIM, which can bind to a Pd metal ion and act as a heterogeneous catalyst.[7]

It is important to note here that for the non-network polymers to have quite significant intrinsic microporosity (IM),[8,9] limited rotational freedom is an essential requirement.[3] It is also interesting that one can manipulate the solubility of PIMs in solvents by varying the ratio of the aromatic *ortho*-dihalide and catechol units in the network.[7,10]

Fluoro-monomer 1,4-dicyanotetrafluorobenzene (DTFB) and a tetrahydroxy monomer 5,5′,6,6′-tetrahydroxy-3,3,3′,3′-tetramethyl-1,10-spirobisindane (TTSBI) participate and polymerise in one step to form soluble PIM-1 and PIM1-CO1-40[7,10] (Figure 5.1).

PIMs have also been synthesised from triptycene monomers with alkyl groups on bridgehead positions.[11] The alkyl groups can be short or branched, which yield high microporosity, or they can be longer chains, in which case the microporosity is reduced. It is also noteworthy that rigid, twisted spirobifluorenes serve as suitable monomers, providing sites of contortion upon product formation. These form PIM networks *via* imide- and amide-forming reactions.[12] The imide forming reaction has also been employed to synthesise binaphthalene-based PIMs, where the monomer is used as a site of contortion. In fact it has also been used to design chiral PIMs.[9]

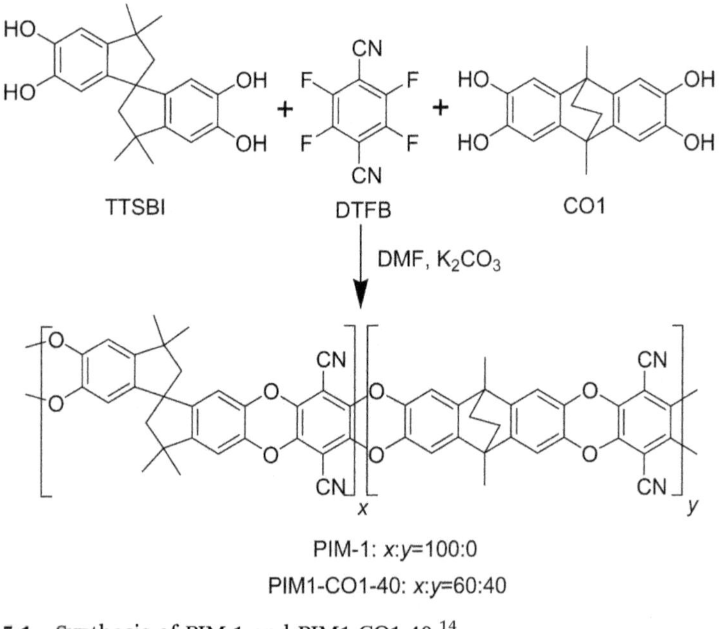

Figure 5.1 Synthesis of PIM-1 and PIM1-CO1-40.[14]
Reprinted with permission from T. Emmler, K. Heinrich, D. Fritsch, P. M. Budd, N. Chaukura, D. Ehlers, K. Ratzke and F. Faupel, *Macromolecules*, 2010, **43**, 6075. Copyright 2010, American Chemical Society.

From the above facts we can generalise the basic design strategy for the synthesis of PIMs as inducing microporosity is the primary target and in order to achieve this at least one of the rigid monomers (*e.g.*, monomers listed in Figure 5.2) should contain a site of contortion, which could be a spiro-centre, a single covalent bond around which rotation is hindered or a nonplanar, rigid backbone. If there is a reaction of two planar monomers it normally gives rise to a non-porous compound. On the contrary, by using the non-planar monomers listed in Figure 5.2, the resulting polymers (herein PIMs) always show high surface area and porosity (Table 5.1).

Very recently, Cooper *et al.* developed a new generation of PIMs called conjugated polymers of intrinsic microporosity (C-PIMs).[43] Generally speaking, PIMs are prepared by nonplanar monomers to increase the intrinsic microporosity. However, such a structure leads to no extended π-conjugation in the PIM backbones. By cross coupling the pyrene or biphenyl moieties, a series of C-PIMs are synthesised with high molecular weights and high apparent BET surface area (Figure 5.3). The porosity may come from the inefficient packing of aromatic main chains and sterically hindered of alkyl side chains. Importantly, these C-PIMs exhibit solubility and microporosity accompanied by a conjugated nature along the main chain, which may be used in electrochemical sensing or printable organic supercapacitors.

5.3   Structure of PIMs

PIMs are twisted polymers with rigidity, which could be due to a nonplanar aromatic ring framework that cannot densely pack in space to give microporosity. PIMs exhibit comparably high specific surface areas (in term of BET 400–1050 m^2 g^{-1}) with linear, synthetic organic macromolecules. Simple and efficient dioxane-like diether-forming reactions are used to prepare PIMs either as insoluble network polymers, like those containing hexaaza-trinaphthylene (HATN-PIM), porphyrin (Porph-PIM), cyclotricatechylene (CTC-PIM) and triptycene (Trip-PIM) moieties, or as soluble non-network polymers, *e.g.*, PIM-1 and PIM-7[13] (Figure 5.4). Later, they were found to give fantastic free-standing films upon casting from normal organic solution, which can be applied in gas permeation membranes[13] (Figure 5.5).

As mentioned before, the components in the PIMs are bent monomers containing a tetrahedral carbon, also known as a site of contortion, which results in intrinsic microporosity. The contortion site can be a spiro-centre with a single sp^3 carbon atom shared by two aromatic rings, a single covalent bond around which rotation is hindered or a rigid, nonplanar skeleton. To understand the PIM structure, one needs to understand the structures of the individual polymer components. For example, introduction of monomer CO1 (9,10-dimethyl-9,10-dihydro-9,10-ethanoanthracene-2,3,6,7-tetrol), which is basically a relatively rigid tetrahydroxy moiety, in increasing amounts introduces a "bend" into the structure of the polymer, but is less flexible than a spiro-centre, so it modifies the total polymer chain packing and thus affects the free volume[14] (Figure 5.1).

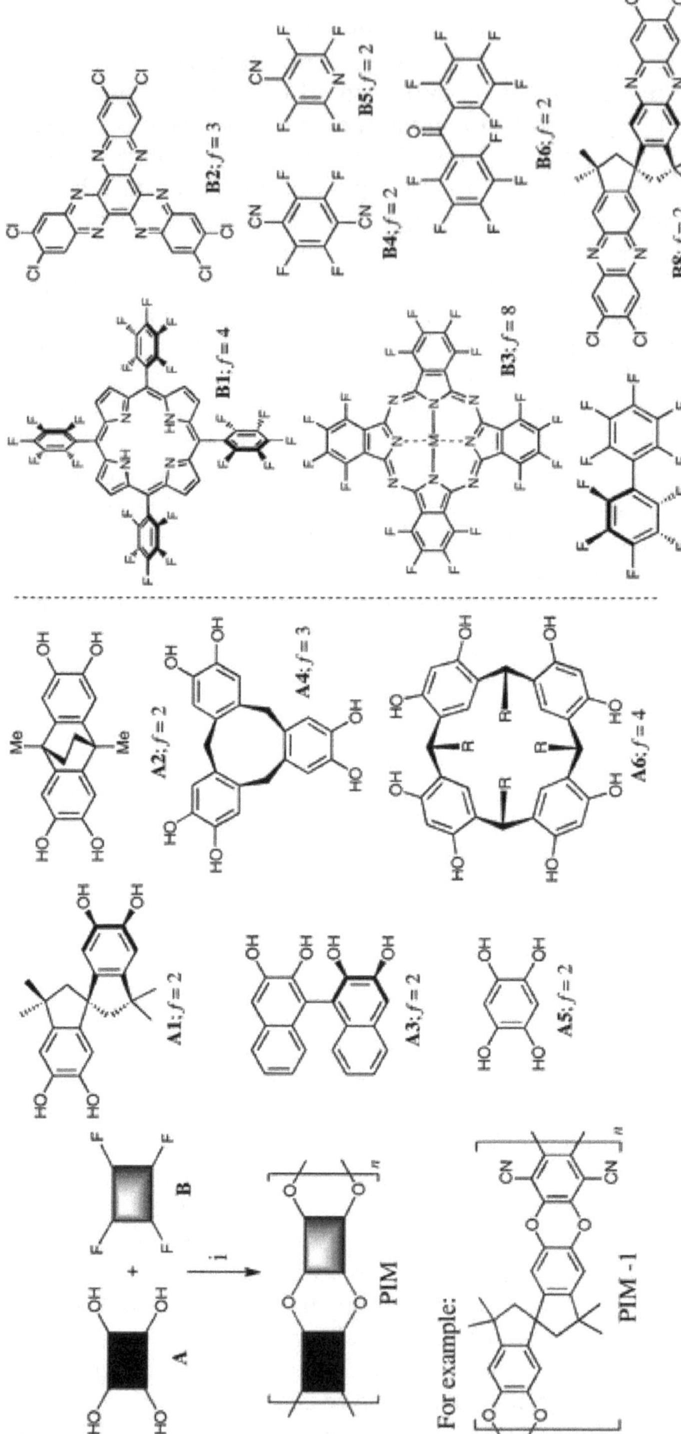

Figure 5.2 Different monomers (A and B) used for PIM synthesis. Reproduced from Ref. 1 with permission from The Royal Society of Chemistry.

Table 5.1 Synthesis and properties of PIMs.[1]

Monomers A B	f_{av}	Insoluble network/ soluble polymer	Molecular mass[a] M_w (g mol^{-1})	Surface area[c] (BET; m^2 g^{-1})	Abbreviation[d]
A1 B1	3	Network	–	1000	Pore-network-PIM
A1 B2	2.5	Network	–	820	HATN-network-PIM
A1 B3	3	Network	–	500	Mpc-network-PIM
A1 B4	2	Soluble	$>200\times10^3$	760	PIM-1
A1 B5	2	Soluble	$>100\times10^3$	750	–
A1 B6	2	Soluble	170×10^3	560	PIM-3
A1 B7	2	Soluble	36×10^3	600	PIM-2
A1 B8	2	Soluble	$>100\times10^3$	750	PIM-7
A2 B2	2.5	Network	–	750	HATN-network-PIM-2
A3 B4	2	Soluble	5×10^3	450	PIM-4
A4 B4	3	Network	–	830	CTC-network-PIM
A5 B7	2	Soluble	[b]	550	PIM-6
A6 B4	3	Network	–	800	Calix-network-PIM
A6 B4	2	Insoluble	–	2	–

[a]As measured by gel permeation chromatography (GPC) *versus* polystyrene standards.
[b]Insoluble in GPC solvents.
[c]BET, Brunauer–Emmett–Teller.
[d]Mpc, metallophthalocyanine; CTC, cyclotricatechylene.

The microporosity of PIMs, like that of other microporous polymers, is composed by a covalent bonding framework.[1] Actually, normal non-network polymers can pack the space densely because the polymer skeleton can twist and bend to maximise inter-chain interactions. However, it should be noted that some polymers can possess large amounts of voids, usually referred to as free volume. It can be anticipated that above a certain amount of free volume, the voids would be interpenetrated, and therefore the polymer will behave as a conventional microporous compound despite the lack of a network structure. Thus, the polymer with interpenetrated voids can be soluble like normal polymers, which could trigger solution processing, providing an advantage over other types of microporous compounds.[1]

It is well known that the introduction of triptycene moieties into a certain polymer backbones can increase mechanical strength. Thus, the polymer shows higher cohesion due to the interlocking of the triptycene units along the chain.[11] Inspired by this, more free volume or voids can be achieved if "expanded" triptycene monomers or "iptycenes" are introduced into the polymer. So-called "iptycenes" are oligomeric systems that contain only bicyclo[2,2,2]octane rings and fused arenes. As shown in Figure 5.6, Mckeown *et al.* synthesised a highly soluble, easily aligned conjugated Trip(Et)-PIM by introducing H-shaped pentiptycene with a huge amount internal molecular free volume.[11] Trip(Et)-PIM is a porous polymer with high free volumes, which exhibit high surface areas and show good properties in hydrogen storage. It has a loose layered structure, which provides space, as indicated by the arrows in Figure 5.6, for nitrogen adsorption and swelling in

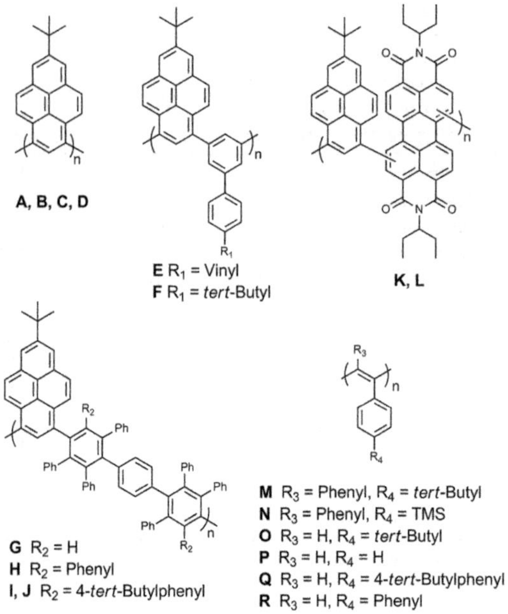

Polymer	Mg [g mol^{-1}]	Mn [g mol^{-1}]	PDI	SA$_{BET}$ [m^2 g^{-1}]
A	2360	1850	1.28	400
B	4670	3910	1.19	573
C	4910	3970	1.24	628
D	8140	6350	1.28	728
E	26300	9900	2.66	646
F	4450	3480	1.28	492
G	31300	14200	2.21	492
H	104000	11200	9.29	724
I	4860	3950	1.23	460
J	14300	9070	1.57	585
K	3940	3170	1.24	322
L	8150	5010	1.63	436
M	1740000	1040000	1.67	436
N	1910000	1410000	1.35	470
O	92700	50900	1.82	438
P	190000	63900	2.97	0
Q	22900	16200	1.41	293
R	23000	17800	1.29	0

Figure 5.3 Structures of the linear, conjugated polymers of intrinsic microporosity that have been prepared. The molecular weight and apparent BET surface area (SA$_{BET}$) are listed in the table (PDI, polydispersity index; TMS, trimethylsilyl).
Reprinted with permission from G. Cheng, B. Bonillo, R. S. Sprick, D. J. Adams, T. Hasell and A. I. Cooper, *Adv. Mater.*, 2014, **24**, 5219. Copyright 2014 WILEY-VCH Verlag GmbH & Co. KGaA, Weinheim.

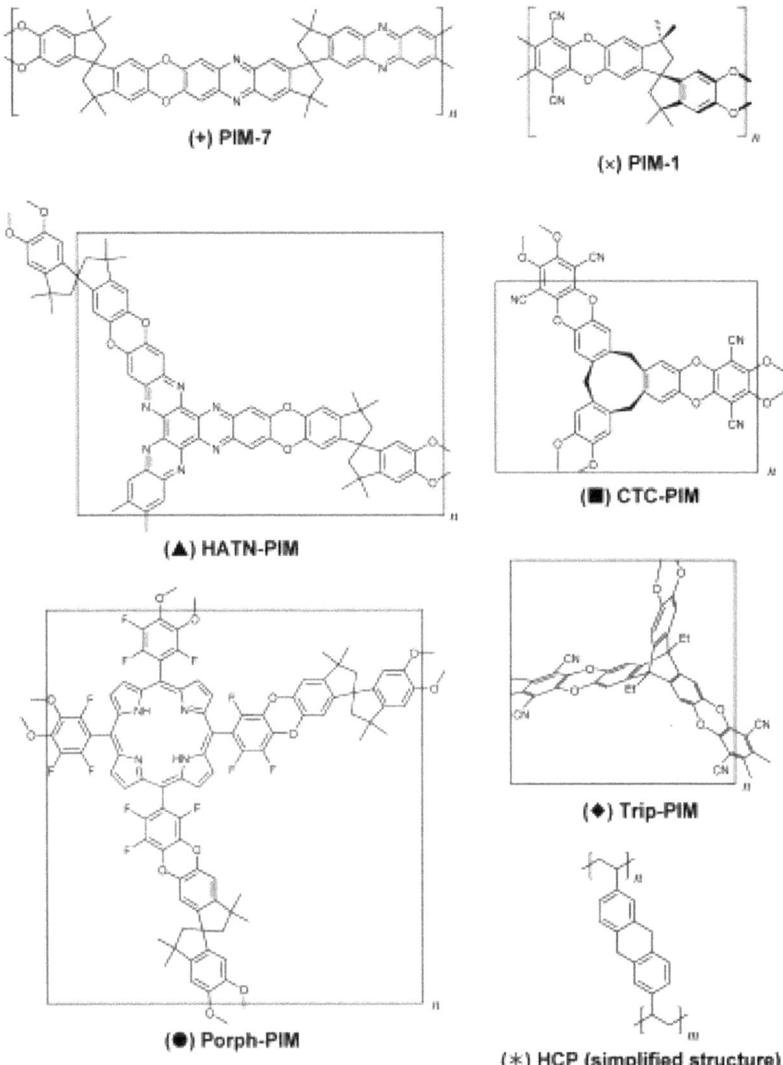

Figure 5.4 Molecular structures of PIMs and simplified structure of a hypercros-slinked polymer (HCP; in reality the HCP will incorporate a variety of crosslink structures).[13]
Reprinted with permission from N. B. McKeown, P. M. Budd and D. Book, *Macromol. Rapid Commun.*, 2007, **28**, 995. Copyright 2007 WILEY-VCH Verlag GmbH & Co. KGaA, Weinheim.

organic solvents. So, polymerisation of larger "iptycenes" has been shown to be a suitable method to obtain promising soluble porous polymers.

It is believed that PIMs could provide an alluring combination of characteristics, including low intrinsic density (as they are made of only light elements—C, H, N, O—which is a real advantage over other porous

Figure 5.5   Free-standing film of PIM-1 in tetrahydrofuran (THF).
Reproduced from Ref. 1 with permission from The Royal Society of Chemistry.

materials),[15] chemical homogeneity (an advantage over carbons), chemical and thermal stability, and synthetic reproducibility. It is extremely interesting and noteworthy that the PIM's synthetic strategy offers the potential to tailor the structure of the micropore by varying the monomer precursors; for instance, by using monomers that contain pre-formed cavities it is possible to provide sites for small gases, such as hydrogen, making the formed networks extremely apt for the storage of hydrogen. For example, the monomer with a bowl-shaped receptor (CTC) has been incorporated into a PIM-network by a benzodioxane-forming reaction between tetra-fluoroterephthalonitrile and CTC.[15]

5.4   Porosity of PIMs

A study by McKeown elucidated different alternated PIMs with high surface areas, for example, HATN-PIM, (BET surface area $= 680$ m^2 g^{-1}) and CTC-PIM (770 m^2 g^{-1}). At very low relative pressure, Porph-PIM (960 m^2 g^{-1}) showed high uptake, which is considered to be a characteristic of microporous materials (Figure 5.7, Table 5.2).[13] PIMs in general show long hysteresis curves extending down to low relative pressures. This is due to the swelling nature or the complex microporous structure of PIMs, where cavities are formed owing to the complex nature of the interconnected free volume.[13]

PIM-1 has a BET surface area of 720 m^2 g^{-1}, while its derivatives with CO1 in different ratios[14] show a slight increment or similar surface area values.

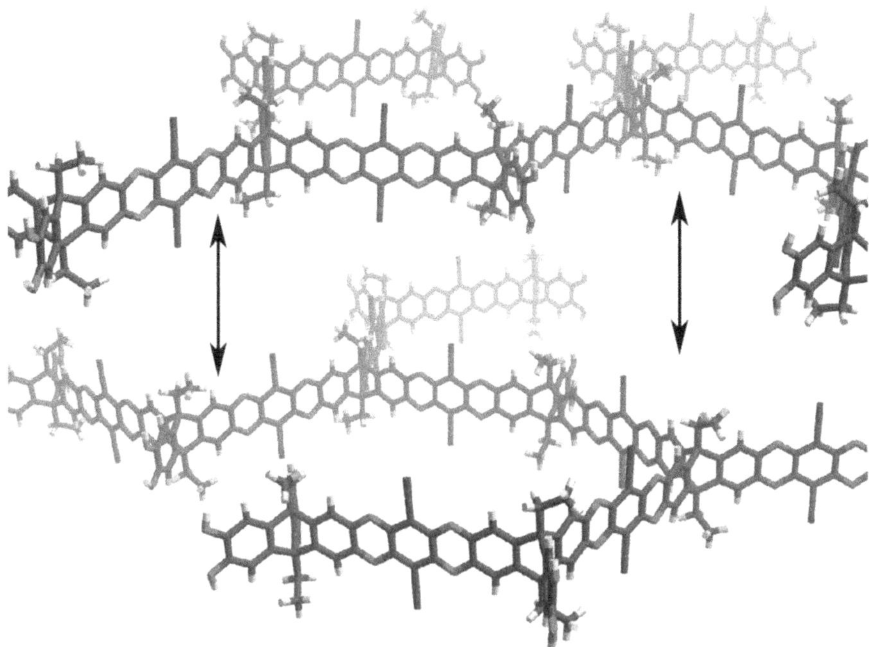

Figure 5.6 Representation of two ideal fragments of the network of Trip(Et)-PIM, showing how the shape of each macromolecule, as dictated by the architecture of the triptycene units, prevents close intermolecular interactions between the planar "struts". The arrows indicate the loose layered structure, which provides space.[11]
Reprinted with permission from B. S. Ghanem, M. Hashem, K. D. M. Harris, K. J. Msayib, M. Xu, P. M. Budd, N. Chaukura, D. Book, S. Tedds, A. Walton and N. B. McKeown, *Macromolecules*, 2010, **43**, 5287. Copyright 2010, American Chemical Society.

Both PIM1-CO1-40 and PIM demonstrate high N_2 uptake at extremely low pressure with hysteresis, which may be an indication of the microporous nature. Trip(R)-PIMs[11] exert surface areas in the range of 618 m^2 g^{-1} (when R = octyl) to 1760 m^2 g^{-1} (when R = methyl). Flexible, linear side chains occupy larger volume than branched ones. That is why higher surface area is observed with branched alkyl-derived PIMs than their linear chain derivatives. Polyimide-linked PIMs have BET surface areas in the range of 471–683 m^2 g^{-1}.[3]

^{129}Xe nuclear magnetic resonance (NMR) experiments are a useful tool to explore the micropores lying on the frameworks of porous polymers. Emmler *et al.* studied Xe adsorption in PIMs (0–3 bar) to explore their microporosity.[14] High free volume copolymers, which form films, were synthesised when the proportion of the PIM-1 spiro-units were exchanged by ethanoanthracene (CO1) units. PIM1-CO1-40 has a 60 : 40 ratio of spiro-units to CO1 units and is used as a comparison with PIM-1. A systematic investigation was performed to compare the N_2 sorption, Xe sorption, positron

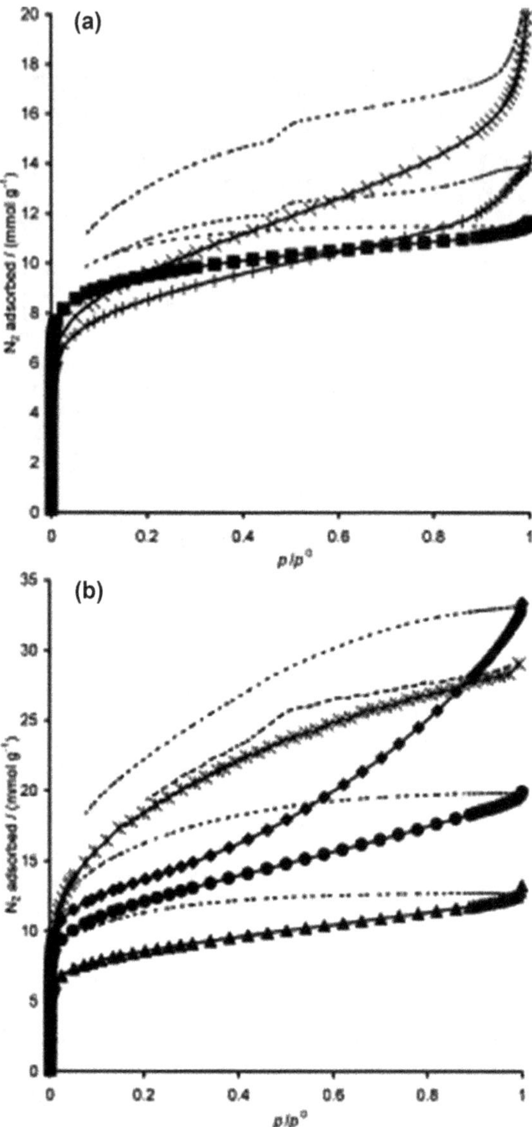

Figure 5.7 Nitrogen adsorption (solid lines with symbols) and desorption (dashed lines) isotherms at 77 K for (a) PIM-7 (+), PIM-1(x), CTC-PIM (■) and (b) HATN-PIM (▲), Porph-PIM (●), Trip-PIM (◆).[13]
Reprinted with permission from N. B. McKeown, P. M. Budd and D. Book, *Macromol. Rapid Commun.*, 2007, **28**, 995. Copyright 2007 WILEY-VCH Verlag GmbH & Co. KGaA, Weinheim.

annihilation lifetime spectroscopy (PALS) and ^{129}Xe NMR, and all confirmed that these polymers have free volume holes or pores on the nanometre length scale (*i.e.*, microporosity as defined by IUPAC). Compared with PIM1-CO1-40, PIM-1 shows smaller micropores. Both PIM-1 and

Table 5.2 Apparent surface areas from N_2 and H_2 sorption, and gravimetric hydrogen uptake at 77 K.[13] Reprinted with permission from N. B. McKeown, P. M. Budd and D. Book, *Macromol. Rapid Commun.*, 2007, **28**, 995. Copyright 2007 WILEY-VCH Verlag GmbH & Co. KGaA, Weinheim.

Sample	N_2 BET surface area (m^2 g^{-1})	H_2 Langmuir surface area (m^2 g^{-1})	H_2 uptake, 1 bar (% mass)	H_2 uptake, 10 bar (% mass)
PIM-1	750	540	0.95	1.45
PIM-7	680	530	1.00	1.35
HATN-PIM	680	590	1.12	1.56
CTC-PIM	770	630	1.35	1.70
Porph-PIM	960	760	1.20	1.95
Trip-PIM	1050	1100	1.63	2.71
HCP	1466	1250	1.27	2.75

PIM1-CO1-40 show outstanding sorption capacities for xenon, even greater than for the highly permeable polymer poly[1-(trimethylsilyl)-1-propyne] (PTMSP; Figure 5.8).[14]

5.4.1 Understanding the Porosity of Porous Polymers

To date, many types of porous solids have been revealed such as small molecules of clathrate,[16,17] rigid oligomers containing triptycenes,[3] polymers of intrinsic porosity (PIPs),[18] crystalline covalent organic frameworks (COFs),[19] amorphous conjugated microporous polymers,[20] porous aromatic frameworks (PAFs),[21] *etc.* These porous solids show high porosity or high surface areas originating from their high "international molecular free volumes" (IMFVs). To understand the mathematic nature of IMFVs is very important to discover or design novel porous materials with specific functional pores. Budd and McKeown provided the convex and concave model to explain the origin of IMFV.[3] Convex shapes and concave shapes are always used as mathematical three-dimensional (3D) packing models to explain the origin of unpacked space or the porosity of a certain solids. The packing efficiency coefficient of convex shapes, which are always used to describe an ordered pack of spheres, is at least 0.74. Disordered convex packing will decrease the packing efficiency coefficient to lower than 0.64. For random packing, the most inefficient packing geometric model is concave faces, which can be explained by forming incomplete interpenetration of concavities.[22,23] This give the chance to produce large IMFVs, and thus leads to high porosity.[3]

Figure 5.9 illustrates the model of recent reports regarding porous solids to give a more clear understanding of the porosity. The empty part of Figure 5.9 (A) is the so-called IMFV, which comes from the concave shape model. Figure 5.9 (B) gives the model of the crystalline concave. This model can be used to explain crystalline clathrates, which show permanent pores after the removal of solvents.[16,17,24–28] Usually, after removal of the

Table. Order of Xe Sorption Measurements

no.	PIM-1 (°C)	PIM1-CO1-40 (°C)
1	85	85
2	25	25
3	100	100
4	25	25
5	25	-

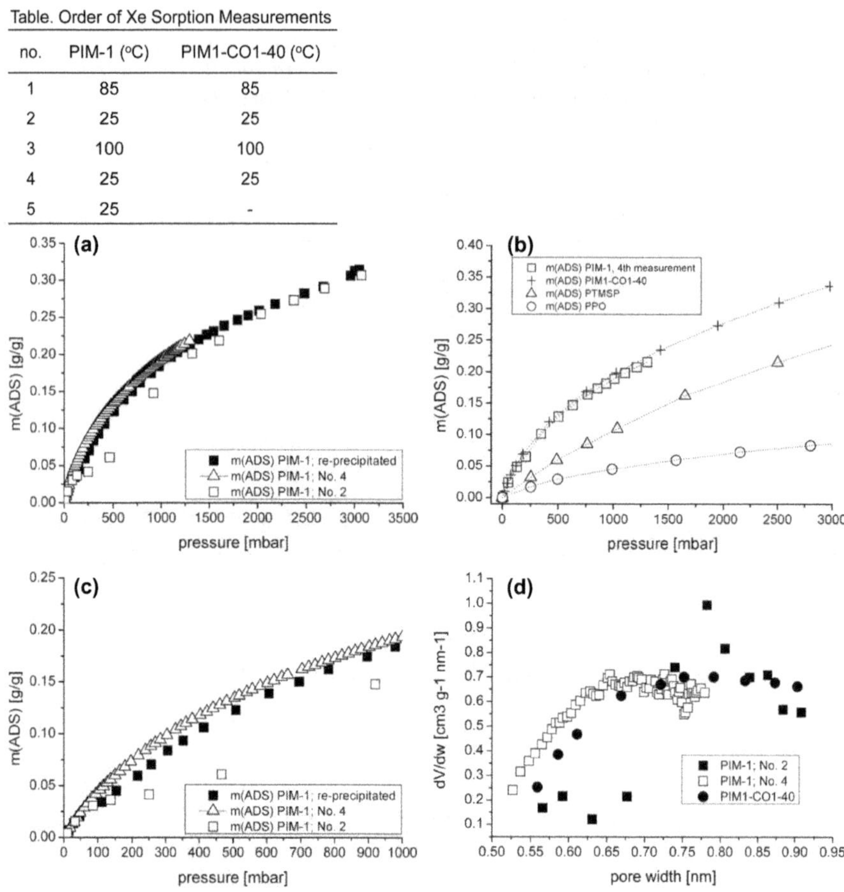

Figure 5.8 (a) Sorption of Xe on PIM-1 at the second and forth sorption cycles and after redissolving in CHCl$_3$ and reprecipition in MeOH at 25 °C for the complete curve. (b) Comparison of Xe sorption curves of PIM-1, PIM1-CO1-40 (both after 100 °C Xe sorption), PTMSP and poly(phenylene oxide) (PPO) at 25 °C. (c) Detail up to 1000 mbar. (d) Pore width distributions determined by the Horvath–Kawazoe method from Xe adsorption at 25 °C for PIM1-CO1-40 and for PIM-1 batch CT/02/07 before and after Xe adsorption at 100 °C.[14]
Reprinted with permission from T. Emmler, K. Heinrich, D. Fritsch, P. M. Budd, N. Chaukura, D. Ehlers, K. Ratzke and F. Faupel, *Macromolecules*, 2010, **43**, 6075. Copyright 2010, American Chemical Society.

residual solvent molecules, the clathrate crystal collapses, as described in Figure 5.9 (C). Such amorphous solids always provide more free volume than the crystal. Figure 5.9 (D) represents the packing of rigid, non-porous oligomers. Though no obvious porosity can be found in the oligomer, the packing model shows the assembly also to show the high free volume. For example, organic molecules of intrinsic microporosity (OMIMs) reported by Budd and McKeown show surface areas of 600 m^2 g^{-1}. Connecting these

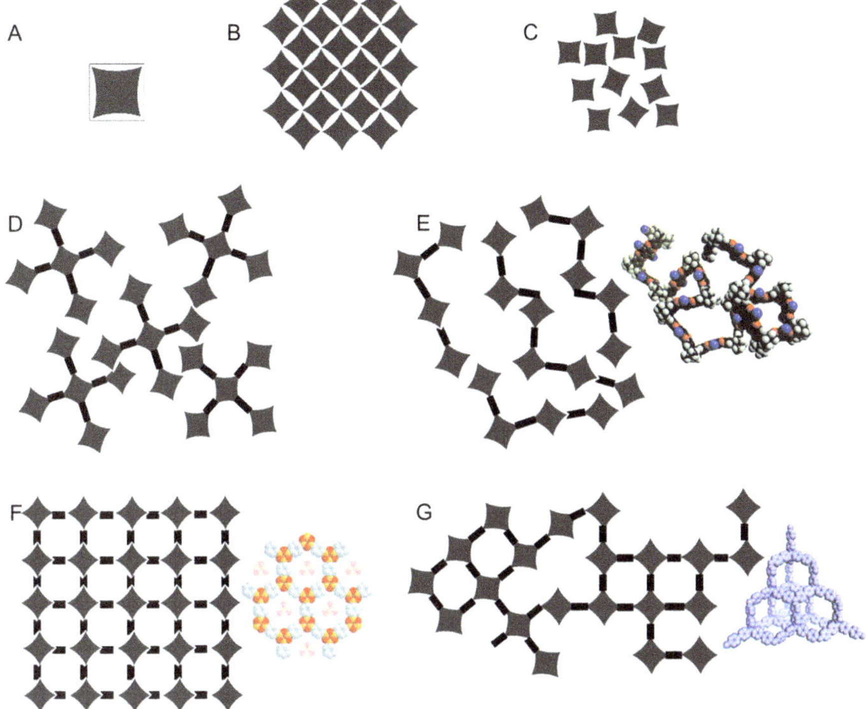

Figure 5.9 (A) The concave shape model, which can give IMFV values. (B) The model of the crystalline concave. (C) The collapsed structure after removal of solvents in the crystalline concave. (D) Packing of rigid, non-porous oligomers. (E) Connecting these kinds of rigid, concave-shaped oligomers into polymer chains will give a polymer of intrinsic porosity. Reproduced from Ref. 1 with permission from The Royal Society of Chemistry. (F) Crystalline concave molecule network.[19] Reprinted with permission from A. P. Côté, A. I. Benin, N. W. Ockwig, M. O' Keeffe, A. J. Matzger and O. M. Yaghi, *Science*, 2005, **310**, 1166. Copyright 2005, American Association for the Advancement of Science. (G) Non-crystalline concave molecule network. Reproduced from Ref. 42 with permission from The Royal Society of Chemistry.

kinds of rigid, concave-shaped oligomers into polymer chains will give a polymer of intrinsic porosity [Figure 5.9 (E)]. All the above models are for the packing of single molecules, oligomers and non-network polymers. For network polymers, they can be illustrated by Figure 5.9 (F) and (G), where Figure 5.9 (F) represents a crystalline concave molecule network, such as covalent organic frameworks, while Figure 5.9 (G) describes a non-crystalline concave molecule network such as conjugated microporosity polymers and porous aromatic frameworks. In conclusion, inefficient packing of concave shape molecules, such as single molecules, rigid oligomers, non-network polymers, crystalline network polymers and non-crystalline network polymers, can give large IMFVs, which lead to high porosity and surface area.

5.5 Applications

5.5.1 Gas Permeation Studies

The driving force of a membrane for gas separation is the pressure difference across the membrane. The yield of the separated gas can be expressed in terms of membrane permeance, which can be characterised by the amount of permeated gas that passes through a certain membrane area in a given time at a definite pressure difference. The values of permeability are often quoted in Barrer (1 Barrer $= 10^{-10}$ cm^3 s^{-1} cm^{-1} cm Hg$^{-1} = 3.35 \times 10^{-16}$ mol m m^{-2} s^{-1} Pa^{-1}; STP, standard temperature and pressure). Gas permeation phenomena can be described by a simple solution diffusion model, which involves (1) sorption or dissolution of the permeating gas in the membrane at the higher pressure side, (2) diffusion through the membrane and (3) desorption or dissolution at the lower pressure side. Thus, the permeability coefficient P can be determined by the product of the solubility coefficient S and the mutual diffusion coefficient D [eqn (5.1)]:

$$P = SD. \tag{5.1}$$

The selectivity (α) of an ideal mixture of two gases, A and B, is the difference between the permeability coefficients of the two gases and is equal to the product of the solubility selectivity (S_A/S_B) and the mobility selectivity (D_A/D_B):

$$\alpha_{A/B} = \frac{P_A}{P_B} = \frac{S_A \times D_A}{S_B \times D_B} \tag{5.2}$$

As discussed, a useful membrane must exhibit good selectivity for gas mixtures, combined with high permeability to minimise the membrane area required. Unfortunately, membranes with high selectivity always exhibit low permeability and *vice versa*. This can be described for a pair of gases by a double logarithmic plot of selectivity against the permeability of the fastest gases. As shown in Figure 5.10, Robeson showed an empirical upper bound in such plots in 1991[29] and updated upper bounds in 2008.[30]

Most polymers that are highly permeable have rigid, twisted polymer backbones and a huge amount of inter-chain voids or free volume. Examples include addition-type polynorbornene, substituted perfluoropolymers, polyacetylenes, PIMs and some polyimides.[15] Gas permeation experiments with PIM-1 membranes revealed that its permeability is exceeded by only very high free volume polymers, such as Teflon AF2400 and PTMSP, joined by selectivities above Robeson's[29] 1991 upper bound for several commercially important gas combinations, including O_2/N_2 and CH_4/CO_2.[31] Further studies have shown that higher permeability could be achieved simply by methanol treatment, which is beneficial to extract out residual casting solvent and enable chain relaxation.[32] This shows that PIMs are inherently different to many other polymers applied for gas permeation, including microporous ultra-permeable polymers.[1,33] PIM-1 films (P(oxygen) = 380 Barrer) and PIM-7 (P(oxygen) = 190 Barrer) show very high gas permeation and only polydimethylsiloxane (PDMS) and the "ultrapermeable" polymers,

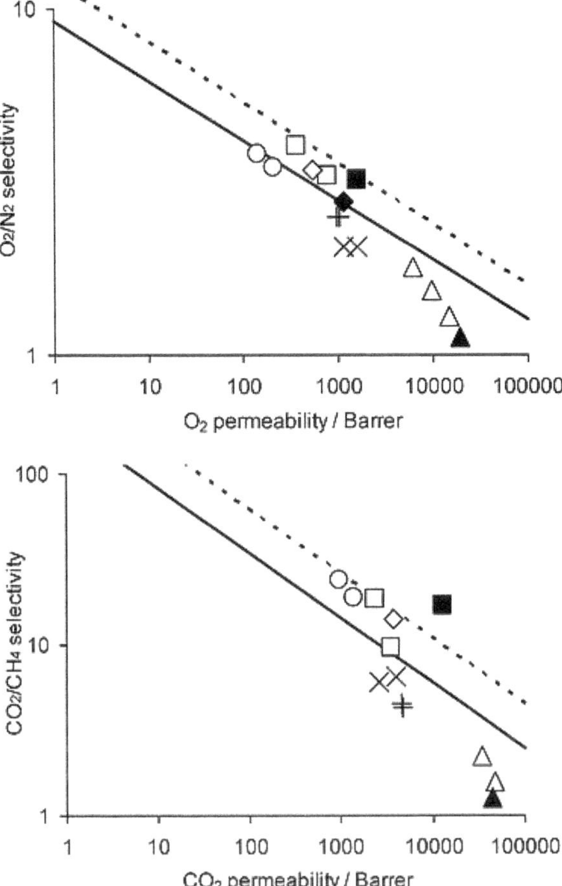

Figure 5.10 Double logarithmic plots of selectivity *versus* permeability for (top) O_2/N_2 and (bottom) CO_2/CH_4, showing (solid line) Robeson's 1991 upper bound and (dashed line) the 2008 upper bound, as well as data for ($\triangle$) poly(trimethylsilyl propyne) (PTMSP), ($\blacktriangle$) indan-based polyacetylene 2e, ($\times$) Teflon AF2400, (+) addition-type poly(trimethylsilyl norbornene), ($\square$) PIM-1, ($\blacksquare$) PIM-1 after methanol treatment, ($\bigcirc$) 6FDA-DMN polyimide, ($\diamondsuit$) PIM-PI-8 and ($\blacklozenge$) PIM-PI-8 after methanol treatment. Reproduced from Ref. 15 with permission from The Royal Society of Chemistry.

such as poly(1-trimethylsilyl-1-propyne) (PTMSP), demonstrate higher overall gas permeabilities.[2,34] Moreover, PIM-1 thin films exhibit much higher selectivity $(\alpha(O_2/N_2) = 3.0)$ than other polymer membranes of similar permeability and they show great progress across Robeson's upper bound plot for O_2/N_2. The properties of PIM-1 membranes can be tuned by post modification such as carboxylation.[8]

PIM-polyimides (PIM-PIs), with intrinsic microporosity, are found to be among the most permeable of all polyimides with high selectivities close to the upper bound for commercially important gas pairs[8] (Figure 5.11). It was

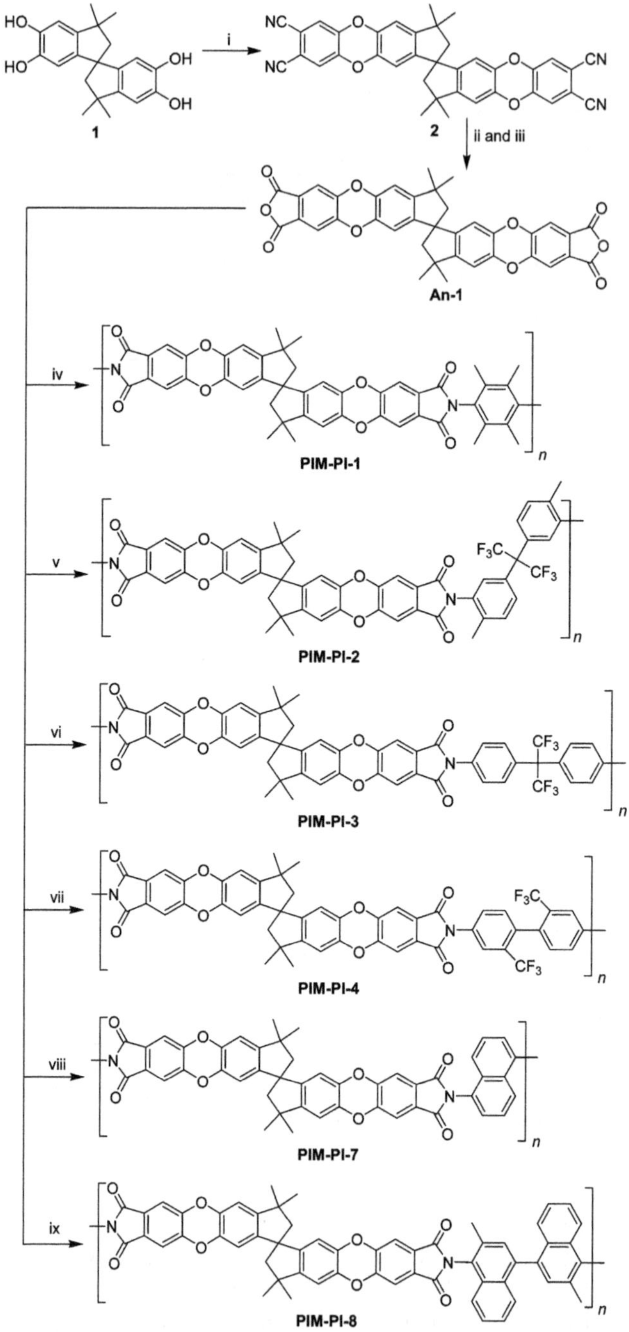

Figure 5.11 The preparation and structures of PIM-polyimides (PIM-PIs).[8] Reprinted with permission from B. S. Ghanem, N. B. McKeown, P. M. Budd, N. M. Al-Harbi, D. Fritsch, K. Heinrich, L. Starannikova, A. Tokarev and Y. Yampolskii, *Macromolecules*, 2009, **42**, 7881. Copyright 2009, American Chemical Society.

observed that the structure of the diamine used for the preparation of PIM-PIs plays an important role in the resulting permeability. The usual trade-off behaviour between permeability and permselectivity can also be observed for PIM-PIs. As we know, normal polyimides show different separation factors with $\alpha(N_2/CH_4) > 1$, while the α values are always less than 1 for most other polymers. Generally speaking, the diffusion separation of methane/nitrogen dominates over the solubility separation in many other normal polyimide membranes, while the case in PIM-PIs is to the contrary. PIM-PIs favour methane over nitrogen and solubility separation plays a very important role.

Extended immersion of the PIM-PI membrane in alcohol obviously affects the properties of permeation. After immersion in alcohol, the residual solvents can be removed and polymer chain relaxation may be achieved in the swollen state. It was observed that both ethanol and methanol treatment have similar effects on gas permeability. PIM-PI membranes treated with methanol or ethanol for several days show higher gas permeabilities than the films cast directly from chloroform, even though these films are subjected to removal of the chloroform in a vacuum at room temperature for a few days until a constant weight of the film is achieved (Table 5.3). These effects were also observed in PIM-1,[7] though PIM-1 has the higher permeability coefficient of the films. These phenomena indicate local interactions between low molecular weight alcohol and some binding sites on the PIM structure. It should be noted that an increase of permeability is always accompanied by a decrease in permselectivity.

The simple solution-diffusion model was often used to explain the gas permeation of a polymer membrane. The permeability coefficient (P) is the product of the solubility coefficient (S) and the diffusion coefficient (D). For a given gas molecule, the main factor for the diffusion coefficient of a certain

Table 5.3 Gas permeation parameters of PIM-PI films after different pre-treatments.[a,8]

Polymer	Type of pre-treatment	$P(O_2)$ (Barrer)	$P(N_2)$ (Barrer)	$\alpha(O_2/N_2)$
PIM-PI-1	Cast from $CHCl_3$	276	90.7	3.0
	Cast from $CHCl_3$, immersed in MeOH for 1 day	733	216	3.4
PIM-PI-8ii	Cast from $CHCl_3$	$488/290^b$	$142/89^b$	$3.4/3.3^b$
	Cast from $CHCl_3$, immersed in MeOH for 5 days	1150	416	2.8
	Cast from $CHCl_3$, immersed in EtOH for 5 days	1071	377	2.8
PIM-1	Cast from $CHCl_3$	580	180	3.2
	Cast from $CHCl_3$, immersed in MeOH for 5 days	1610	500	3.2

[a]The films after primary removal of the solvent ($CHCl_3$) were kept in a vacuum at room temperature until constant weight was achieved (3–5 days). The films after contact with alcohols were kept in an ambient atmosphere for 1 day. No traces of alcohols were present as indicated by gas chromatography analysis.
[b]The first value is after heating in He at 75 °C; the second value is after vacuum treatment at room temperature.

polymer is the amount of free volume in the amorphous polymer backbone. The solubility coefficient is related to the binding strength between the polymer chain and the gas molecules. Reports reveal that all porous polymers above or close to Robeson's upper bound line have a rigid structure unit. For example, methyl- and trifluoromethyl-substituted polyimides (*e.g.*, 2,2-bis(3,4-dicarboxylphenyl) hexafluoropropane dianhydride–diamino tetramethyl benzene, 6FDA-4MDAB) and polypyrolones (*e.g.*, 6FDA–2,4,6-triaminopyrimidine, 6FDA-TAP) show enhanced rigidity because the rotation of the polymer main chains is hindered by those substituents (Table 5.4). Another example comes from PIM-1 and PIM-7. Their unique spirocyclic fused-ring structures not only provide high rigidity but also prohibit molecular chain rotation. To put it another way, these examples express a targeted design concept of PIMs. In the case of gas selectivity, micropore size distribution is a key point for porous polymers. The kinetic diameter of oxygen is 0.346 nm, while it is larger for nitrogen at 0.364 nm.[35,36] It is anticipated that smaller micropores favour the separation of gas molecules such as O_2/N_2. It seems that permeability and permselectivity are in contrast because the higher permeability requires fast diffusion or large pores and suitable binding energy, while the selectivity needs a comparable pore size with the gas molecules. It has been a challenge to date to find a suitable porous polymer with high porosity at 0.6 nm for high selectivity and suitable hierarchical pores for fast diffusion.

Table 5.4 Gas permeability data for a range of polymers of interest for gas separation membranes (TAB, tetraamino biphenyl).[2]

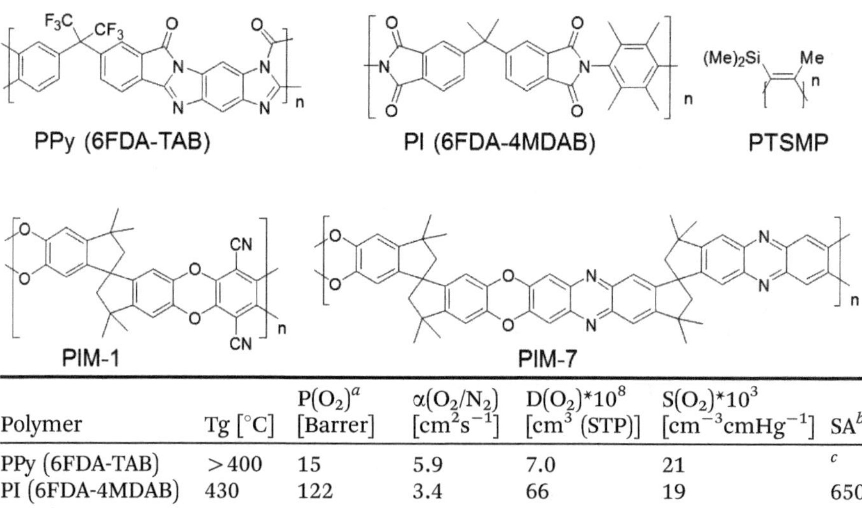

Polymer	Tg [°C]	$P(O_2)^a$ [Barrer]	$\alpha(O_2/N_2)$ [cm^{2}s^{-1}]	$D(O_2)*10^8$ [cm^3 (STP)]	$S(O_2)*10^3$ [cm^{-3}cmHg^{-1}]	SA^b
PPy (6FDA-TAB)	>400	15	5.9	7.0	21	c
PI (6FDA-4MDAB)	430	122	3.4	66	19	650
PTMSP	>200	9000	1.4	5200	15	950
PIM-1	>400	370	4.0	82	46	850
PIM-7	>400	190	4.5	62	30	750

^{a}Barrer = * 10^{-30} cm^3(STP).
^{b} = SA = BET surface area of poedered sample.
^{c}Not measured.

Generally speaking, gas separation by a normal polymer membrane is dominated by the mobility selectivity. In the case of PIM-1 and PIM-7, though the solubility selectivity is not the main factor for oxygen, the overall values of solubility are extremely high, even over Robeson's upper bound line, compared with the other polymers. This can be attributed to the excellent apparent solubilities or adsorption, which charge up the high permeability with selectivity maintenance.[2] High micropore distribution and high surface area are beneficial for a high gas uptake capacity in PIMs. Furthermore, introduction of polar groups in PIMs enhances the binding energy between the polymer backbone and gas molecules, which increase the gas sorption. For instance, nitrile groups lie on the lateral position of the main chain of PIM-1 and can provide a binding site for gas molecules and expand the free volume. So, the total adsorption of gas on PIM-1 is higher than that on PIM-7. PTMSP has the highest surface area among the polymers listed in Table 5.4 and exhibits very high diffusion coefficients.

PIM-1 and the zeolitic imidazolate framework (ZIF-8) have been applied to prepared mixed matrix membranes (MMMs) and their gas permeation and selectivity properties have also been explored. Free-standing MMMs with different ZIF-8 loadings have been investigated and permeability coefficients have been determined for gases such as hydrogen, oxygen, nitrogen, carbon dioxide, methane and helium. Increasing ZIF-8 loading in "as-cast" films leads to increased permeability and diffusion coefficients, as well as separation factors, $\alpha(H_2/N_2)$, $\alpha(H_2/CH_4)$, $\alpha(He/N_2)$, $\alpha(O_2/N_2)$ and $\alpha(CO_2/CH_4)$ (Table 5.5). It was also observed that these ZIF-8/PIM MMMs show higher permeability after treatment with ethanol. Data points on several Robeson diagrams are located above the 2008 upper bound (Figure 5.12). Results reveal that the insertion of ZIF-8 nano-crystals into PIM-1 can expand the free volume coming from highly porous ZIF-8 and the loosely packed polymer chains at the boundary between ZIF-8 and the PIM-1.[37]

Jansen *et al.* (*Angew. Chem. Int. Ed.* 2013, **52**, 1253–1256) incorporated a crystalline porous organic cage within PIM-1 and the resulting membrane had substantially enhanced permeability accompanied by high selectivity to CO_2/N_2 and CO_2/CH_4 (Figure 5.13). More interestingly, such MMMs show better resistance towards physical aging. The porous organic cage can be *in situ* crystallised from a single homogeneous solution with PIM-1 to

Table 5.5 Separation factors of PIM/ZIF-8 MMMs.[37]

[ZIF-8], vol%	H_2/N_2	H_2/CH_4	He/N	O_2/N_2	CO_2/CH_4
0 [63]	9.1 ± 0.9	5.3 ± 0.5	4.2 ± 0.4	3.2 ± 0.3	14.2 ± 1.4
11	10.2 ± 1.0	8.0 ± 0.8	5.2 ± 0.5	3.3 ± 0.3	15.0 ± 1.5
28	15.2 ± 1.5	13.0 ± 1.3	7.3 ± 0.7	4.5 ± 0.4	18.6 ± 1.9
36	15.2 ± 1.5	11.3 ± 1.1	7.7 ± 0.7	4.3 ± 0.5	13.4 ± 1.3
43	19.1 ± 1.9	15.5 ± 1.3	9.1 ± 0.9	4.8 ± 0.5	14.7 ± 1.4

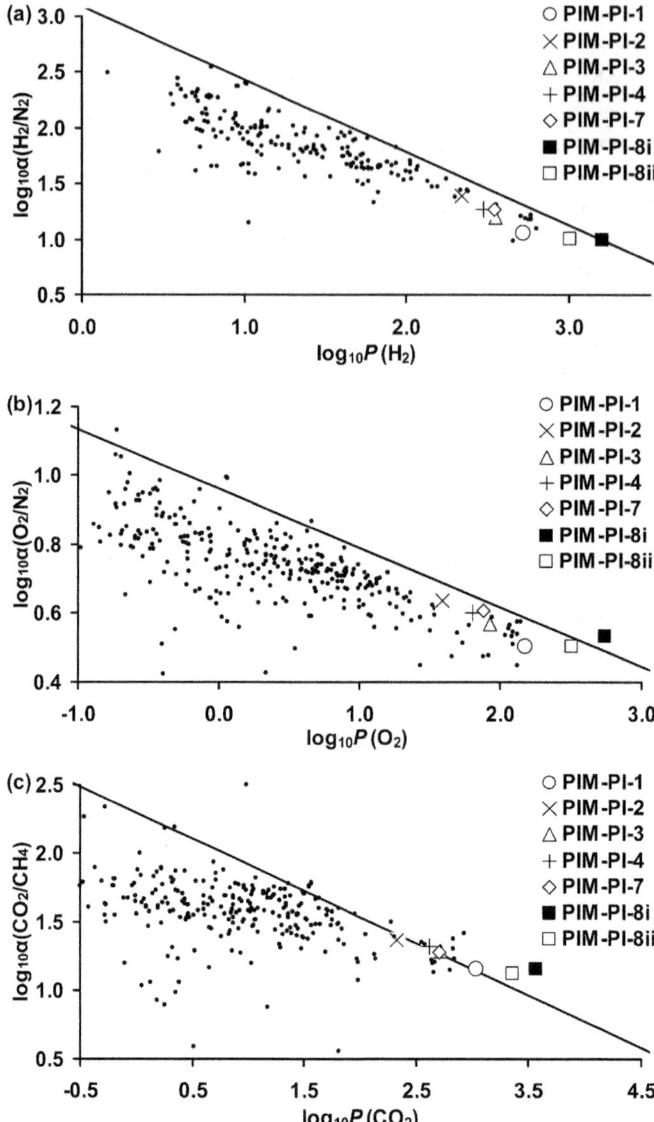

Figure 5.12 Robeson plots for PIM-PIs and PIM-PIs and the gas pairs (a) H_2/N_2, (b) O_2/N_2 and (c) CO_2/CH_4. The solid lines represent Robeson's 1991 upper bounds.[29]
Reprinted with permission from B. S. Ghanem, N. B. McKeown, P. M. Budd, N. M. Al-Harbi, D. Fritsch, K. Heinrich, L. Starannikova, A. Tokarev and Y. Yampolskii, *Macromolecules*, 2009, **42**, 7881. Copyright 2009, American Chemical Society.

generate MMMs. Such a convenient preparation method could be applied to other functional "porous" organic oligomers such as cucurbiturils, calixarenes and other rigid macrocyclic species.

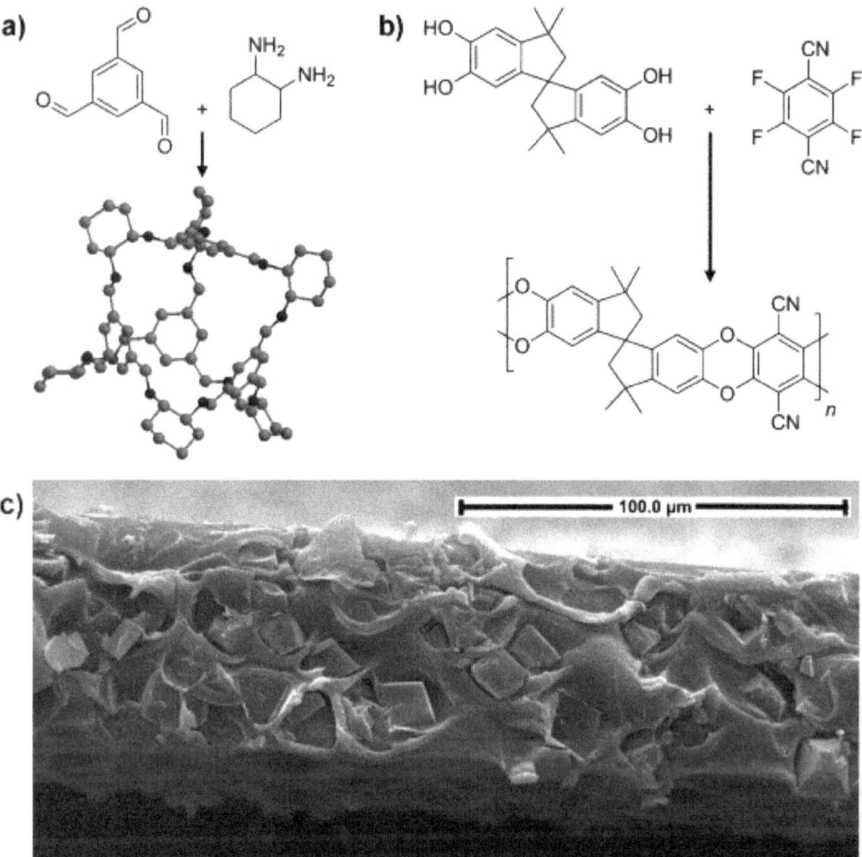

Figure 5.13 (a) Porous imine cage CC3 synthesised from 1,3,5-triformylbenzene and (*R*,*R*)-1,2-diaminocyclohexane by a condensation reaction. (b) PIM-1 synthesised from 5,5′,6,6′-tetrahydroxy-3,3,3′,3′-tetramethyl-1,1′-spirobisindane and 1,4-dicyanotetrafluorobenzene by a step polymerisation involving a double aromatic nucleophilic substitution. (c) SEM image of a cross-section of a PIM-1/CC3 composite membrane (weight ratio 10:2). Reprinted with permission from A. F. Bushell, P. M. Budd, M. P. Attfield, J. T. A. Jones, T. Hasell, A. I. Cooper, P. Bernardo, F. Bazzarelli, G. Clarizia and J. C. Jansen, *Angew. Chem. Int. Ed.*, 2013, **52**, 1253–1256; *Sci.*, 2013, **427**, 48. Copyright 2013 WILEY-VCH Verlag GmbH & Co. KGaA, Weinheim.

5.5.2 Storage

On investigation of many potential solutions for on-board hydrogen storage, porous polymers provide a fantastic microporous internal surface for reversible physical adsorption of hydrogen molecule. Till now, no report has shown a quantity of hydrogen that can be adsorbed onto any type of microporous material that meets the requirements of practical hydrogen storage. To date, this is still a great challenge and there is an urgent demand to discover a robust material for hydrogen adsorption. PIMs have a large amount of free volume and microporosity, and can be used as a good candidate for hydrogen adsorption. McKeown *et al.* systematically studied hydrogen storage in three PIMs (PIM-1, a CTC network and an HATN network) at 77 K by using volumetric and gravimetric techniques. The resulting adsorption isotherms indicated these PIMs can adsorb large amounts of hydrogen (*ca.* 1.7 wt%) at low relative pressure. They reach saturation before 10 bar and most of the hydrogen uptake is achieved at 1 bar[38] (Figure 5.14). The hydrogen uptake sequence at 1 bar and 77 K of these PIMs is in the order of CTC network > HATN network > PIM-1.

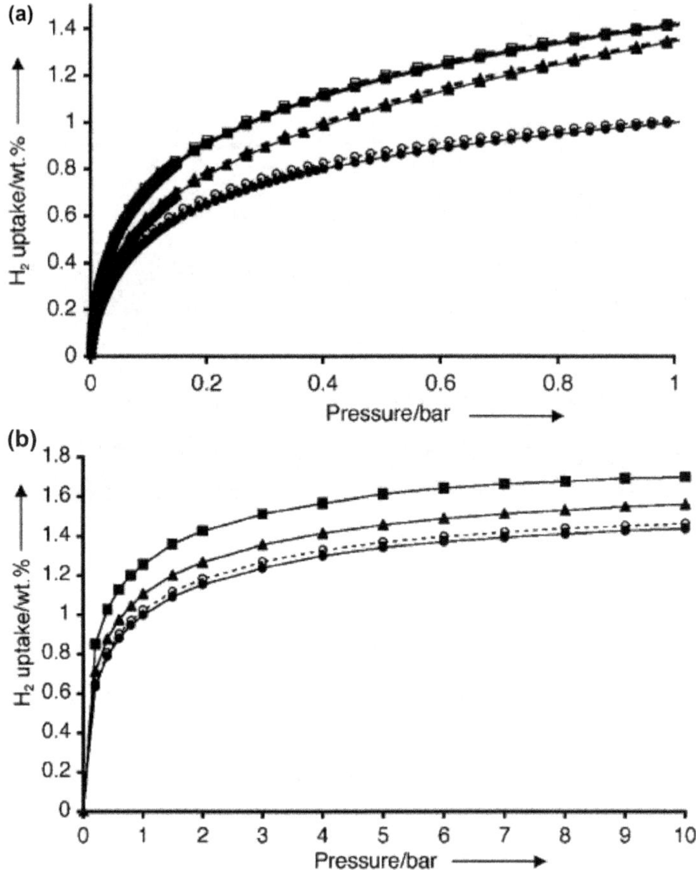

Figure 5.14 The H_2 adsorption (filled symbols) and desorption isotherms (open symbols) at 77 K for PIM-1 (●), HATN-network-PIM (▲) and CTC-network PIM (■), obtained using a) volumetric and b) gravimetric analysis.[30]
Reprinted with permission from N. B. McKeown, B. Gahnem, K. J. Msayib, P. M. Budd, C. E. Tattershall, K. Mahmood, S. Tan, D. Book, H. W. Langmi and A. Walton, *Angew. Chem., Int. Ed.*, 2006, **45**, 1804. Copyright 2006 WILEY-VCH Verlag GmbH & Co. KGaA, Weinheim.

Though the hydrogen uptake is lower than high surface area MOFs (carbonised porous polymers reported recently),[39–41] these values are comparable to those of zeolites, carbon and MOFs with similar surface areas.

5.5.3 Catalysis

As mentioned before in Section 5.2, transition metal ions can be introduced in phthalocyanine-, porphyrin- or hexaazatrinaphthylene-containing PIM networks to obtain catalytic activity towards a wide variety of reactions like hydrogen peroxide degradation or cyclohexene oxidation to 2-cyclohexene-1-one. Pd-incorporated PIM-7 is such a catalytic material. Pd^{2+} ions here act as a bridge among PIMs to form an extended network providing catalytic activity.[1]

The iron porphyrin-based PIMs can also act as excellent catalysts. Rigid and nonlinear linking units in phthalocyanines provide spiro-centres and nonlinearity to inhibit facial association and prevent solidification through π–π and other non-covalent interactions. This also improves the catalytic properties of phthalocyanines as it can catalyse H_2O_2 degradation more effectively than solidified low molar mass phthalocyanines.[4,5] PcCo-network-PIM shows a higher degradation rate of H_2O_2 compared to non-porous compounds (Figure 5.15). It is also important to note that the network-PIM is more efficient and selective than the cobalt-based homogeneous catalyst for the oxidation of cyclohexene. The synthesis method is also important for

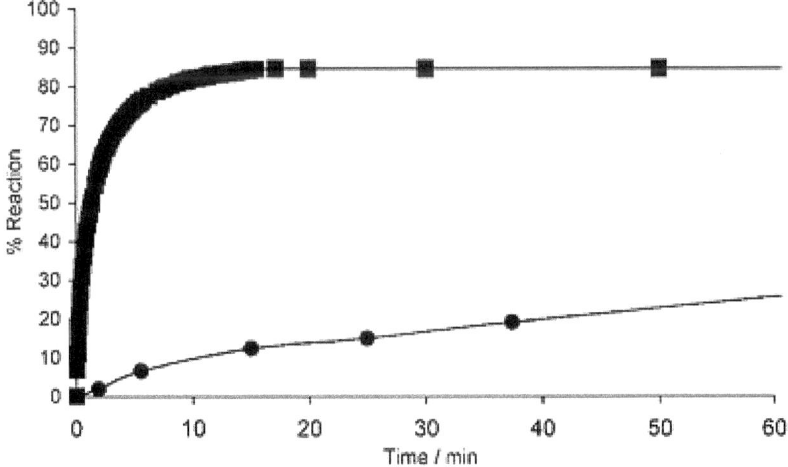

Figure 5.15 Dependence of the extent of reaction on time for the degradation of H_2O_2 (0.74 mol dm^{-3}, $T = 30\,^{\circ}C$) with ($\bullet$) low molar mass cobalt phthalocyanine (PcCo) and ($\blacksquare$) PcCo-network-PIM as the catalyst. Oxygen evolution measured with a gas burette.[1]
Reproduced from Ref. 1 with permission from The Royal Society of Chemistry.

catalytic efficiency, *e.g.*, the PcCo-network-PIM can be prepared in two ways, but the original phthalocyanine-forming reaction (surface area $= 500$ m^2 g^{-1}) proved to be a better method. HATN is a very good ligand, capable of forming a complex with up to three transition metal ions. The orange HATN network-PIM was exposed to a chloroform solution containing bis(benzonitrile) palladium(II) dichloride and gave a highly coloured compound with a mass loading of Pd of 26%. Similar palladium-containing compounds with higher specific surface areas could be derived from PIM-7 by means of its phenazine sub-units, which have the ability to act as ligands for coordination of a metal ion. Thus, the mixing of bis(benzonitrile) palladium(II) dichloride to a yellow solution of PIM-7 creates an immediate precipitation of a red solid, which is insoluble in all solvents of organic character. This material contains over 20% by mass Pd^{2+} and has a surface area of 650 m^2 g^{-1}. It can be noted that the Pd^{2+} ion was acting as a cross-link between the PIM macromolecules to give a network compound. Moreover, it was observed that a solvent cast-film of PIM-7, swollen in methanol, can also be cross-linked by Pd^{2+} ions to produce an insoluble network. This process has potential for reactive membrane fabrication.[1]

5.5.4   Adsorption of Organic Compounds

Network-PIMs can adsorb phenol from aqueous solution. Hence, they can be used for the separation of organic compounds from the aqueous phase. This process can also be utilised for the separation of phenol from waste water. One such example is HATN-network-PIM, which can adsorb up to 5 mmol g^{-1}. This material can also efficiently remove a low concentration of phenol from water. Another such example is PcCo-network-PIM. These network-PIMs behave like activated carbon. In contrast, PcCo-network-PIM cannot adsorb large dyes (*e.g.*, Napthol Green B) like activated carbons. This distinct size exclusion effect is due to its small pore size.[1]

References

1. N. B. McKeown and P. M. Budd, *Chem. Soc. Rev.*, 2006, **35**, 675.
2. N. B. McKeown, P. M. Budd, K. J. Msayib, B. S. Ghanem, H. J. Kingston, C. E. Tattershall, S. Makhseed, K. J. Reynolds and D. Fritsch, *Chem. – Eur. J.*, 2005, **11**, 2610.
3. N. B. McKeown and P. M. Budd, *Macromolecules*, 2010, **43**, 5163.
4. N. B. McKeown, S. Makhseed and P. M. Budd, *Chem. Commun.*, 2002, 2780.
5. N. B. McKeown, S. Hanif, K. Msayib, C. E. Tattershall and P. M. Budd, *Chem. Commun.*, 2002, 2782.
6. N. B. McKeown, *J. Mater. Chem.*, 2000, **10**, 1979.
7. P. M. Budd, B. Ghanem, K. Msayib, N. B. McKeown and C. Tattershall, *J. Mater. Chem.*, 2003, **13**, 2721.

8. B. S. Ghanem, N. B. McKeown, P. M. Budd, N. M. Al-Harbi, D. Fritsch, K. Heinrich, L. Starannikova, A. Tokarev and Y. Yampolskii, *Macromolecules*, 2009, **42**, 7881.

9. N. Ritter, M. Antonietti, A. Thomas, I. Senkovska, S. Kaskel and J. Weber, *Macromolecules*, 2009, **42**, 8017.

10. P. M. Budd, E. S. Elabas, B. S. Ghanem, S. Makhseed, N. B. McKeown, K. J. Msayib, C. E. Tattershall and D. Wang, *Adv. Mater.*, 2004, **16**, 456.

11. B. S. Ghanem, M. Hashem, K. D. M. Harris, K. J. Msayib, M. Xu, P. M. Budd, N. Chaukura, D. Book, S. Tedds, A. Walton and N. B. McKeown, *Macromolecules*, 2010, **43**, 5287.

12. J. Weber, O. Su, M. Antonietti and A. Thomas, *Macromol. Rapid Commun.*, 2007, **28**, 1871.

13. N. B. McKeown, P. M. Budd and D. Book, *Macromol. Rapid Commun.*, 2007, **28**, 995.

14. T. Emmler, K. Heinrich, D. Fritsch, P. M. Budd, N. Chaukura, D. Ehlers, K. Ratzke and F. Faupel, *Macromolecules*, 2010, **43**, 6075.

15. P. M. Budd and N. B. McKeown, *Polym. Chem.*, 2010, **1**, 63.

16. K. J. Msayib, D. Book, P. M. Budd, N. Chaukura, K. D. M. Harris, M. Helliwell, S. Tedds, A. Walton, J. E. Warren, M. C. Xu and N. B. McKeown, *Angew. Chem., Int. Ed.*, 2009, **48**, 3273.

17. C. G. Bezzu, M. Helliwell, D. R. Allan, J. E. Warren and N. B. McKeown, *Science*, 2010, **327**, 1627.

18. P. M. Budd, B. S. Ghanem, S. Makhseed, N. B. McKeown, K. J. Msayi and C. E. Tattershall, *Chem. Commun.*, 2004, 230.

19. A. P. Cote, A. I. Benin, N. W. Ockwig, M. O'Keeffe, A. J. Matzger and O. M. Yaghi, *Science*, 2005, **310**, 1166.

20. J. Jiang, F. Su, A. Trewin, C. D. Wood, N. L. Campbell, H. Niu, C. Dickinson, A. Y. Ganin, M. J. Rosseinsky, Y. Z. Khimyak and A. I. Cooper, *Angew. Chem., Int. Ed.*, 2007, **46**, 1.

21. T. Ben, H. Ren, S. Ma, D. Cao, J. Lan, X. Jing, W. Wang, J. Xu, F. Deng, J. Simmons, S. Qiu and G. Zhu, *Angew. Chem., Int. Ed.*, 2009, **48**, 9457.

22. Y. Jiao, F. H. Stillinger and S. Torquato, *Phys. Rev. E: Stat., Nonlinear, Soft Matter Phys.*, 2009, 79.

23. Y. Jiao, F. H. Stillinger and S. Torquato, *Phys. Rev. Lett.*, 2008, **100**, 4.

24. L. R. Nassimbeni, *Acc. Chem. Res.*, 2003, **36**, 631.

25. A. I. Cooper, *Adv. Mater.*, 2003, **15**, 1049.

26. S. J. Dalgarno, P. K. Thallapally, L. J. Barbour and J. L. Atwood, *Chem. Soc. Rev.*, 2007, **36**, 236.

27. L. A. Jones, T. Hasell, X. Wu, J. Bacsa, K. E. Jelfs, M. Schmidtmann, S. Y. Chong, D. J. Adams, A. Trewin, F. Schiffman, F. Cora, B. Slater, A. Steiner, G. M. Day and A. I. Cooper, *Nature*, 2011, **474**, 361.

28. J. L. Atwood, L. J. Barbour, A. Jerga and B. L. Schottel, *Science*, 2002, **298**, 1000.

29. L. M. Robeson, *J. Membr. Sci.*, 1991, **62**, 165.

30. L. M. Robeson, *J. Membr. Sci.*, 2008, **320**, 390.

31. P. M. Budd, K. J. Msayib, C. E. Tattershall, B. S. Ghanem, K. J. Reynolds, N. B. McKeown and D. Fritsch, *J. Membr. Sci.*, 2005, **251**, 263.
32. P. M. Budd, N. B. McKeown, B. S. Ghanem, K. J. Msayib, D. Fritsch, L. Starannikova, N. Belov, O. Sanfirova, Y. Yampolskii and V. Shantarovich, *J. Membr. Sci.*, 2008, **325**, 851.
33. Y. Aoyama, *Top. Curr. Chem.*, 1998, **198**, 131.
34. C. L. Mangun, Z. Yue, J. Economy, S. Maloney, P. Kemme and D. Cropek, *Chem. Mater.*, 2001, **13**, 2356.
35. D. R. Lide, *CRC Handbook of Chemistry and Physics*, CRC Press, Boca Raton, FL, 2004.
36. Y. Yamposkii, I. Pinnau and B. D. Freeman, *Materials Science of Membranes for Gas and Vapor Separation*, John Wiley & Sons Ltd, West Sussex, England, 2006.
37. A. F. Bushell, M. P. Attfield, C. R. Mason, P. M. Budd, Y. Yampolskii, L. Starannikova, A. Rebrov, F. Bazzarelli, P. Bernardo, J. C. Jansen, M. Lanč, K. Friess, V. Shantarovich, V. Gustov and V. Isaeva, *J. Membr. Sci.*, 2013, **427**, 48.
38. N. B. McKeown, B. Gahnem, K. J. Msayib, P. M. Budd, C. E. Tattershall, K. Mahmood, S. Tan, D. Book, H. W. Langmi and A. Walton, *Angew. Chem., Int. Ed.*, 2006, **45**, 1804.
39. D. Q. Yuan, D. Zhao, D. F. Sun and H. C. Zhou, *Angew. Chem., Int. Ed.*, 2010, **49**, 5357.
40. O. K. Farha, A. O. Yazaydın, I. Eryazici, C. D. Malliakas, B. G. Hauser, M. G. Kanatzidis, S. T. Nguyen, R. Q. Snurr and J. T. Hupp, *Nat. Chem.*, 2010, **2**, 944.
41. T. Ben, Y. Li, L. Zhu, D. Zhang, D. Cao, Z. a. Xiang, X. Yao and S. Qiu, *Energy Environ. Sci.*, 2012, **5**, 8370.
42. T. Ben and S. Qiu, *CrystEngComm*, 2013, **15**, 17.
43. G. Cheng, B. Bonillo, R. S. Sprick, D. J. Adams, T. Hasell and A. I. Cooper, *Adv. Mater.*, 2014, **24**, 5219.

Rational Design of Covalent Organic Frameworks for High Performance Gas Storage

HUI WANG, LING HUANG AND DAPENG CAO*

State Key Laboratory of Organic–Inorganic Composites, Beijing University of Chemical Technology, Beijing 100029, P.R. China
*Email: caodp@mail.buct.edu.cn

6.1   Introduction

Nanoporous materials have attracted extensive attention because of their outstanding performance in gas storage[1–3] and separation[4,5] *etc*. Scientists have made great efforts in designing various nanostructured materials, ranging from the disordered structure to regular configurations, by assembling the same and different building units according to different means. The designed materials contain zero-, one-, two-, and three-dimensional structures.

Recently, a new class of material, covalent organic frameworks (COF), without the presence of heavy metals was also synthesized by Yaghi and co-workers.[6,7] COFs are crystalline porous materials, and the formation of COFs is controlled by a dynamic covalent reaction, and it thus can form thermodynamically stable structures.[8] COFs consist of light elements (C, N, H, B, *etc*.), and are held together only by strong covalent bonds. These materials not only hold all the advantages of metal organic frameworks (MOFs), but also possess lower density and higher hydrothermal stability than MOFs. Therefore, this kind of material has attracted extensive attention from

Monographs in Supramolecular Chemistry No. 17
Porous Polymers: Design, Synthesis and Applications
By Shilun Qiu and Teng Ben

Published by the Royal Society of Chemistry, www.rsc.org

different fields, including organic and inorganic chemistry, physical and materials chemistry. The previously reported COF materials are either two- (2D) or three-dimension (3D) structures. Two-dimensional COFs induce a large electronic coupling between the π-orbitals of the eclipsed stacking sheets, which is beneficial to charge carrier transport in the pre-organized π-pathways.[9,10] It implies that 2D COFs will be suitable for semiconducting and photoconducting materials. In contrast, 3D COFs characteristically possess high specific surface areas (4210 m^2 g^{-1} for COF-103), and low densities (COF-108 is only 0.17 g cm^{-3}), and so 3D COFs are promising materials for gas storage.[11]

After Yaghi and co-workers successfully synthesized the COF materials by a solvothermal method,[6] many other researchers have proposed different synthetic methods.[9,12] Generally, COFs are synthesized mainly by the solvothermal method.[5,13] The method can grow highly crystalline COFs if a suitable temperature, pressure, solvent, reaction rate *etc.* are selected. However, the solvothermal method often takes a long time, which extends the exploring duration of COF materials. Thomas and co-workers reported an ionothermal method, which can also synthesize crystalline porous COFs.[14] However, it was difficult for the ionothermal method to control the crystallinity in the reversible cyclotrimerization reaction, which actually limited its applications. Enlightened by the synthesis of crystalline MOFs using the microwave method, Cooper and co-workers developed a rapid microwave-assisted method to prepare COF materials.[15] Microwave heating is quicker and cleaner than the solvothermal methods, suggesting its possibility for further applications. Biswal *et al.* reported a solvent-free mechanochemical method to synthesize COFs at room temperature.[16] Although the porosity and crystallinity of these COFs are moderate, it provides a synthetic strategy for the large-scale production of COFs. Medina *et al.* synthesized 2D COF films by vapor-assisted conversion at room temperature.[17] This is an excellent method that can precisely control the film thickness. Besides the aforementioned methods, COF films or monolayers have also been obtained by reactions on substrates.[18]

The characterization of COF materials is essentially important and complex. The main issues for characterizing COFs include the porosity, structural regularity, morphology and atomic connectivity style. X-Ray diffraction (XRD) data can provide the crystalline structures of 2D and 3D COFs.[19] Moreover, solid-state nuclear magnetic resonance (NMR) spectroscopy, infrared spectroscopy, elemental analysis, and X-ray photoelectron spectroscopy (XPS) can be used to evaluate the terminal groups, and linkage and chemical compositions of COFs. Generally, the specific surface areas and porosity of COFs are calculated by the argon, helium or nitrogen adsorption–desorption isotherms. The morphology of the COF material can be characterized by scanning electron microscopy (SEM). For the monocrystalline COFs, single-crystal XRD can be used to determine the structures.[20] However, for the COF materials with no single crystals, powder X-ray diffraction (PXRD) is used for structural analysis. Alternatively, we can also build the possible structures of the materials and simulate the PXRD using Materials Studio (Reflex module), and thus determine the exact periodical structure of the prepared COF by comparing the simulated PXRD with the experimental one.

Besides many efforts in experimental researches, theoretical methods also play an important role in rationally designing new materials and predicting the concerned properties, which can provide reference and guidance for practical experiments. This article first introduces the multiscale theoretical methods for designing COF materials and further evaluating the properties concerned. Then, we summarize the building blocks utilized for COF material design and propose the two rational strategies for the design of novel COF materials. Finally, we summarize the applications of COF materials in high performance gas adsorption.

6.2   Theoretical Methods

In fact, a lot of organic building blocks can yield immense COFs with diverse topology and functionalities. Experimentally exploring the properties of all the COFs of interest would be difficult. Fortunately, theoretical methods have been considered useful to assist experiments in the case of materials science, and can be an effective tool in explaining the mechanism of experimental phenomena, as well as in further designing novel materials and predicting the properties of interest.[21] The basic computational methods that can be used in molecular systems are quantum mechanics calculations and molecular simulations based on statistical mechanics and classical physics.

6.2.1   Quantum Chemistry

Quantum chemistry methods apply quantum mechanics to solve problems in chemistry. For physical chemists, quantum chemistry is employed to calculate the thermodynamic properties of gases, to interpret molecular spectra, to calculate properties of transition states in chemical reactions, to understand intermolecular forces, and to deal with bonding in solids. Organic chemists use quantum chemistry to evaluate the relative stabilities of molecules, to investigate the mechanisms of chemical reactions, and to analyze NMR spectra; while inorganic chemists use quantum chemistry together with ligand field theory to predict and explain the properties of transition–metal complexes.

A wave function-based method, second-order Møller–Plesset (MP2) perturbation theory, is usually considered to be the high-level approach, and could yield reliable results for the weak interactions. However, the MP2 method needs much more computational cost, and cannot be applied to large systems. Hence, an alternative method, density function theory (DFT), including PW91, B3LYP, and PBE, has been developed and widely used. In addition, the basis set superposition (BSSE) is critically necessary to describe the non-bonding interactions in the MP2 method, while it is unnecessary in the methods of PW91 and B3LYP. So far, the basis set of B3LYP has been proven to be sufficient to describe electronic behavior with high accuracy. Quantum chemistry methods are able to compute electronic, structural, optical, magnetic and also dynamic properties of molecular systems with good accuracy. However, quantum chemistry has still failed to deal with relatively large systems. The widely used software for quantum mechanics calculations are Gaussian 03[22] and the Vienna *ab initio* simulation package (VASP).[23,24] We can use DFT to evaluate the binding energy of the gas and material, and find the preferable adsorption sites. Therefore, we can understand the adsorption mechanism and design excellent materials for gas storage.

6.2.2 Molecular Simulations

6.2.2.1 Grand Canonical Monte Carlo Simulations

DFT uses quantum mechanics to describe the electrons and nuclei, while the Monte Carlo (MC) method uses statistical means to obtain the properties of interest in a system. The MC calculation can simulate stochastic events and use statistical methods for numerical integration by the following steps: (1) Translating the physical problems into a probabilistic or analogous model. (2) Solving the model by numerical sampling experiments. (3) Using statistical methods to analyze the results obtained. Various ensembles have been adopted in MC simulations to explore the different properties of the systems, among which grand canonical Monte Carlo (GCMC) is the most popular method for the study of adsorption, because the grand canonical ensemble is suitable for describing the open system, allowing mass exchange in the system.

In the grand canonical ensemble, the volume (V), temperature (T), and chemical potential (μ) are kept constant, while the number of particles (N) can vary. The chemical potential can be determined by the Widom test particle method[25] or an appropriate equation of state (EOS), such as Peng–Robinson EOS,[26] and it is closely related to a specific pressure at a fixed temperature. The fluctuation of the total molecular number in a simulation box is determined by three trial moves (*i.e.*, creating a particle at a random position, deleting a particle randomly, displacing and/or rotating a particle randomly). In the GCMC simulation, the most widely used model to describe the interactions between the adsorbate and framework is the Lennard-Jones (L-J) potential,[27] given by

$$U(r_{ij}) = 4\varepsilon_{ij}\left[\left(\frac{\sigma_{ij}}{r_{ij}}\right)^{12} - \left(\frac{\sigma_{ij}}{r_{ij}}\right)^{6}\right], \qquad (6.1)$$

where U is the potential energy between atoms i and j at a separating distance of r_{ij}; ε_{ij} and σ_{ij} are the potential well depth and van der Waals (vdW) radius, respectively. The potential parameters of the adsorbents can be obtained from the widely used Dreiding force field[28] or Universal force field,[29] while the parameters for the adsorbates can be taken from the transferable potentials for phase equilibrium united-atom (TraPPE-UA)[30] or optimized potential for liquid simulation (OPLS).[31] Another function describing the interaction between adsorbate and framework is the Morse potential, given by

$$U_{ij} = D_{ij}\left[e^{-\gamma_{ij}\left(\frac{r_{ij}}{r_{e,ij}}-1\right)} - 2e^{-\frac{\gamma_{ij}}{2}\left(\frac{r_{ij}}{r_{e,ij}}-1\right)} \right], \tag{6.2}$$

where D_{ij} is the well depth, γ_{ij} is the stiffness parameter that defines the well shape, and $r_{e,ij}$ is the equilibrium distance. The parameters can be obtained by fitting quantum chemistry calculation results to the above Morse potential.

The adsorption capacity calculated by the GCMC simulation is the absolute uptake N_{ab}, while the experiment data usually are the excess uptake N_{ex}. In order to conveniently compare the simulation results with the experiment data, N_{ab} is usually converted into N_{ex} by

$$N_{ex} = N_{ab} - \rho_b V_{pore}, \tag{6.3}$$

where ρ_b is the density of the bulk adsorbate, which can be calculated from the Peng–Robinson EOS, and V_{pore} is the pore volume of the adsorbent, which can be measured by experiment or calculated from the ideal gas law (eqn 6.4):[32,33]

$$V_{pore} = \frac{RN_m T}{pm_m}, \tag{6.4}$$

where R is the gas constant, p is the pressure, T is the temperature, and N_m is the total number of adsorbed molecules per molar mass m_m of the adsorbents. N_m can be calculated from the GCMC simulations of nonadsorbed helium $(\varepsilon_{He}/K_b = 10.22\ \text{K},\ \sigma_{He} = 2.58\ \text{Å};\ K_b$ is the Boltzmann constant) in COF materials at ambient temperature (298 K) and low pressure.[34]

The isosteric heat of adsorption q_{st} is an important thermodynamic quantity, because it can reflect the interaction strength of the adsorbates and the materials. It is the released heat for each molecule added to the adsorbed phase, given by[35]

$$q_{st} = RT - \left[\frac{\mathrm{d}(U_{total} - U_{intra})}{\mathrm{d}N_{total}} \right]_{T,V}, \tag{6.5}$$

where U_{total} is the total adsorption energy, which includes the interaction energy between the adsorbate and the adsorbent and also the interaction energy between adsorbate and adsorbate, and U_{intra} is the intramolecular energy of the adsorbates and is zero for rigid molecules. In experiments, q_{st} can be directly obtained by microcalorimetry measurements and also computed using the Clausius–Clapeyron equation.

The GCMC simulation can be performed by MUSIC code[36] and also the Sorption model of Materials Studio. Of course, the GCMC programs developed in-house are desired, because they will be more suitable for the specific systems. Therefore, we can get macroscopic adsorption properties with high accuracy in a statistical manner.

6.2.3 Multiscale Simulation

Recently, a novel multiscale theoretical method has been developed to investigate large size molecular systems with the desired accuracy, which combines molecular simulations and first-principles calculations (see Figure 6.1). Actually, as shown in Figure 6.1, the first-principles calculations can compute the interactions between two kinds of different atoms, and the interactions can be inputted into the molecular simulation to predict the physical chemical properties of a material. This method illustrates one of the primary challenges associated with hierarchical modeling of materials. The advantage of multiscale simulations lies in that the level of simulation details can vary in time and length, saving on computational cost without sacrificing the necessary detailed physical properties. The higher level model could be considered to be composed of many elements in the lower level model. Multiscale theory covers three scales, and there are different tools for treating different cases, among which the minimum parts (*i.e.*, the building blocks) in the molecular system are dealt with using accurate *ab initio* methods, such as MP, DFT, and Couple Cluster (CC). For the periodic structure or the larger molecular models that contain more than 50 atoms, mixed methods involving quantum mechanics/molecular mechanics (QM/ MM) or DFT are adopted. The above two aspects should predict all the necessary information to fit an *ab-initio*-based interatomic potential. Owing to the diversity of atom types in novel materials like COFs and MOFs, the site-to-site force field is commonly adopted to calculate the interaction between adsorbate and adsorbent. Comparing the classical potentials with the parameter from united atoms, the Morse-fitting and L-J-fitting potential based on first-principles calculations exhibits the advantage of the multiscale

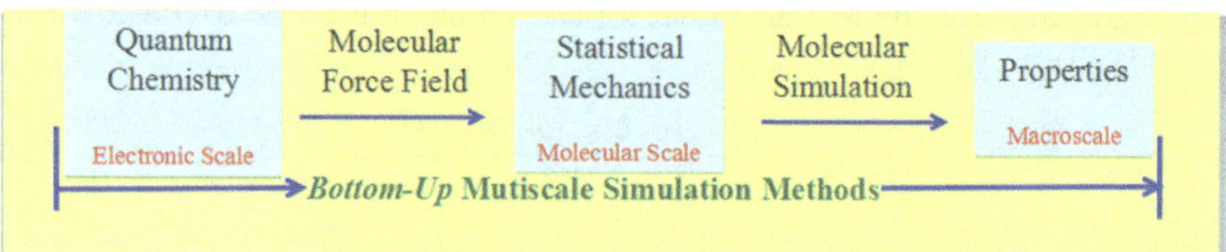

Figure 6.1 The scheme for the *bottom-up* multiscale simulation method, in which the quantum chemistry method, including first-principle calculations and DFT, was used to obtain the binding energy between gas molecules and COF materials. By fitting the binding energy into the molecular force fields and further inputting the force fields into a statistical mechanics-based molecular simulation, we can predict adsorption properties of COF materials. This bottom-up multiscale method spans three scales, including the electronic scale, the molecular scale, and the macroscale.

methods, because the first-principles calculations can accurately transfer the adsorbate–adsorbent interactions into the GCMC simulations. However, in the first-principles calculation, different methods may yield different binding energies, and by the same method the different basis sets also produce the different results. So, it is often important to calibrate the simulated results by using the experimental data.

In the calibration, the force field corresponding to different methods in the first-principles calculation is very important, because it is a bridge between the GCMC simulations and the first-principles calculation. A specific force field is needed to benchmark the calculated results and experimental data. In general, the classical L-J potential and Morse potential are used as the specific force field model. By inputting the specific force fields, the classical GCMC simulations can be performed to treat large systems containing thousands of atoms, and therefore obtain the physical chemical properties of the materials.

6.3 Building Blocks Utilized for COF Materials

As for the construction of COF materials, building blocks and synthetic strategies are the basic concern. The building blocks can be classified as two-dimensional (2D) or three-dimensional (3D) building blocks according to their dimensions. According to the rotational symmetry and coordination manner of the reactive groups, the building blocks can be classified as 2D-C_2, 2D-C_3, 2D-C_4, 3D-C_3, and 3D-T_d,[9] as shown in Figure 6.2, where C_x ($x = 2$, 3, 4) indicates that the ligands have rotational symmetry and x-coordination, and T_d indicates the tetrahedral coordination of the ligands. The self-condensation of 2D-C_2 (Figure 6.2a) or 2D-C_3 building blocks (Figure 6.2d) may produce hexagonal 2D COFs. The co-condensation of the 2D-C_2 building block with the 2D-C_3 building block can also produce hexagonal 2D COFs (Figure 6.2c). Apparently, these hexagonal 2D COFs have different pore sizes and surface areas due to the different building blocks. The co-condensation of 2D-C_2 building blocks with another 2D-C_2 building block may create a linear structure (Figure 6.2b). The co-condensation of 2D-C_2 building blocks with 2D-C_4 may yield tetragonal 2D COF structures (Figure 6.2e). On the contrary, 3D COFs can be fabricated by the self-condensation of tetrahedral nodes (3D-T_d) (Figure 6.2g) or the co-condensation of 3D-T_d with linear (2D-C_2; Figure 6.2f) or triangular (2D-C_3) building blocks. Furthermore, the co-condensation of 3D-T_d with 3D-C_3 can also produce the 3D structure (Figure 6.2h).

The design and synthesis routes of COF materials mentioned above have been widely used in practical applications, and a series of new materials have been developed. For example, the self-condensation reaction of 1,4-benzenediboronic acid (2D-C_2) forms the 2D COF-1 (Figure 6.2a). The co-condensation reaction between 1,3,5-benzene boronic acid (BTBA; 2D-C_3) and 2,3,6,7,10,11-hexahydroxytriphenylene (HHTP; 2D-C_3) produced 2D COF-6 (Figure 6.2d), and the co-condensation reaction between HHTP and

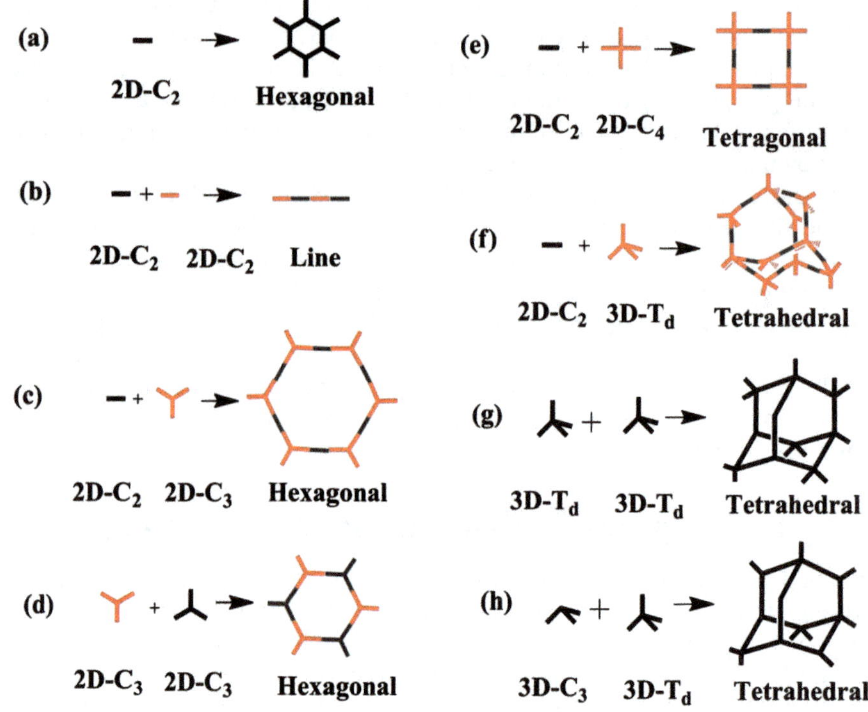

Figure 6.2   The geometry of the building blocks successfully utilized for the synthesis of COF materials. Two-dimensional and three-dimensional ligands are indicated by 2D and 3D, respectively. C_2, C_3, and C_4 indicate that the ligands have rotational symmetry and are two-coordination, three-coordination or four-coordination, and T_d indicates tetrahedral coordination of the ligands. (a)–(h) show the synthetic routes of COF materials with different structures.

4,4′-biphenyldiboronic acid (BPDA; 2D-C_2) produced COF-10 (Figure 6.2c). The self-condensation of tetrahedral tetra(4-dihydroxyborylphenyl)methane (TBPM) forms 3D COF-102 (Figure 6.2g). The co-condensation of TBPM with HHTP produced 3D COF-105. As presented in Figure 6.2e, tetragonal 2D-COFs need a 2D-C_4 building block. For example, 4,4′-biphenylboronic acid (2D-C_2) and octahydroxy phthalocyanine Co(II) (2D-C_4) synthesized the CoPc-BPDA COF.[37] Actually, the 2D-C_2 building blocks can be di-substituted units such as 1,4-benzenediboronic acid (BDBA) and BPDA. The 2D-C_3 building blocks can be tri-substituted derivatives, for example, BTBA and HHTP. 2D-C_4 building blocks are mainly produced by porphyrin and phthalocyanine derivatives. The 3D-C_3 building blocks can be *tert*-butylsilane triol, while 3D-T_d can be tetra(4-dihydroxyborylphenyl)-silane (TBPS) or TBPM. Besides, to construct functionalized COFs for particular applications, we can introduce functional groups into COFs, mainly including metal doping and functional group modification and topochemistry.

6.4 Strategies for the Design of New COF Materials

Yaghi and co-workers proposed a targeted design strategy to synthesize extended structures, which followed the principle known as reticular chemistry.[38] The concept of reticular chemistry is a key for the applicability of COF formation from well-defined building units. Accordingly, hypothetical COFs can be designed rationally and synthesized successfully. As shown in Figure 6.3, new materials can be designed and synthesized *via* either the guideline of reticular chemistry or modification on the existing backbones. On one hand, COFs are similar to MOFs constructed by combining diverse nodes and linkers.[39] Therefore, the design strategies for MOFs are also applicable for COFs, because a large number of types of nodes and linkers provide large possibilities in topology. In fact, the node and linker replacement strategy can also be divided into the linker replacement strategy and the node replacement strategy. On the other hand, introducing functional groups into the COF frameworks also contributes to the emergence of new materials without changing the net topology. The above strategies have been employed extensively in numerous researches.[2,38,40–42]

6.4.1 Linker Replacement Strategy

As mentioned in Section 6.4, the new materials can be derived by the combinations of diverse nodes and linkers. Creating new materials *via* linker replacement in the existing materials might be a good choice to develop highly efficient gas storage materials. For example, as shown in Figure 6.4,

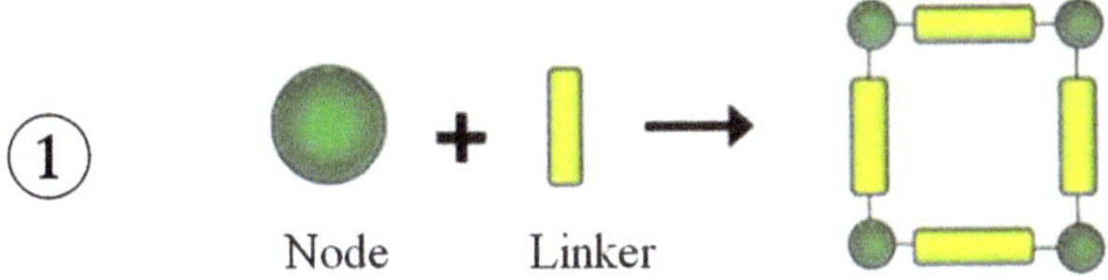

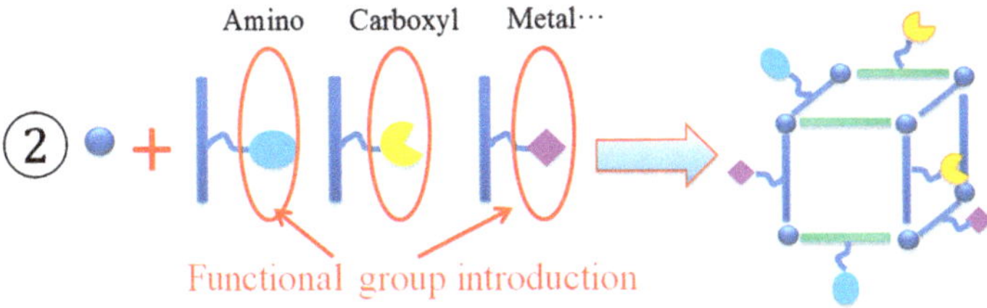

Figure 6.3 Two strategies for generating new COF materials. One is to use different ligand nodes and organic linkers to obtain the COF materials. The other is to adopt the functional groups to synthesize the functionalized COF materials.

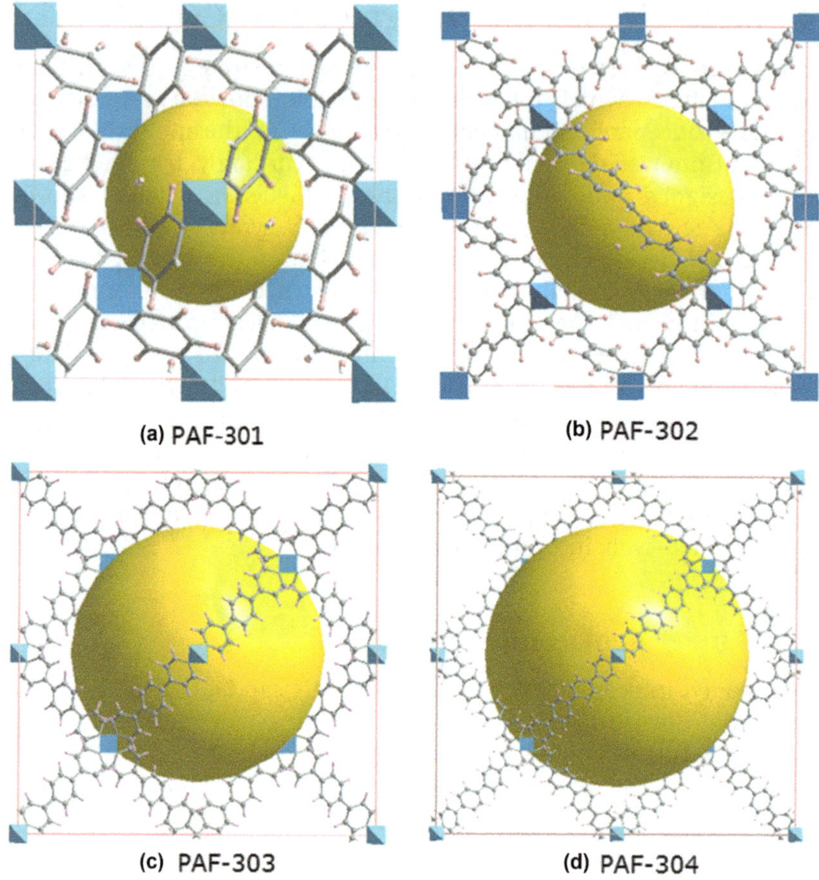

Figure 6.4 Unit cells of PAFs: (a) PAF-301, (b) PAF-302, (c) PAF-303, and (d) PAF-304. (Printed with permission from Lan *et al.*[43])

on the basis of the diamond-like structure, Lan *et al.* designed four similar structures [porous aromatic framework-(PAF-) 301, -302, -303, and -304] by replacing each C–C single bond in diamond with one, two, three, and four phenyl rings, respectively.[43] The free volumes of PAF-30X ($X = 1$, 2, 3, or 4) are 40.92%, 77.6%, 87.5%, and 92.1%, respectively, which indicates that if the linker is longer, the pore volume and the Brunauer–Emmett–Teller (BET) specific surface area are larger. This is a classical case of using the linker replacement strategy.

Another example is the substitution of phenylene in COF-102 with extended aromaticity (diphenyl, triphenyl, naphthalene, and pyrene molecules), as shown in Figure 6.5,[44] in which the new COF-102-X frameworks ($X = 2$, 3, 4, or 5) were designed. These new 3D structures combined triangular and tetrahedral building blocks, and were optimized by first principles calculations. As a result of GCMC predictions, the structures designed exhibited higher gravimetric uptake than the parent COF-102.

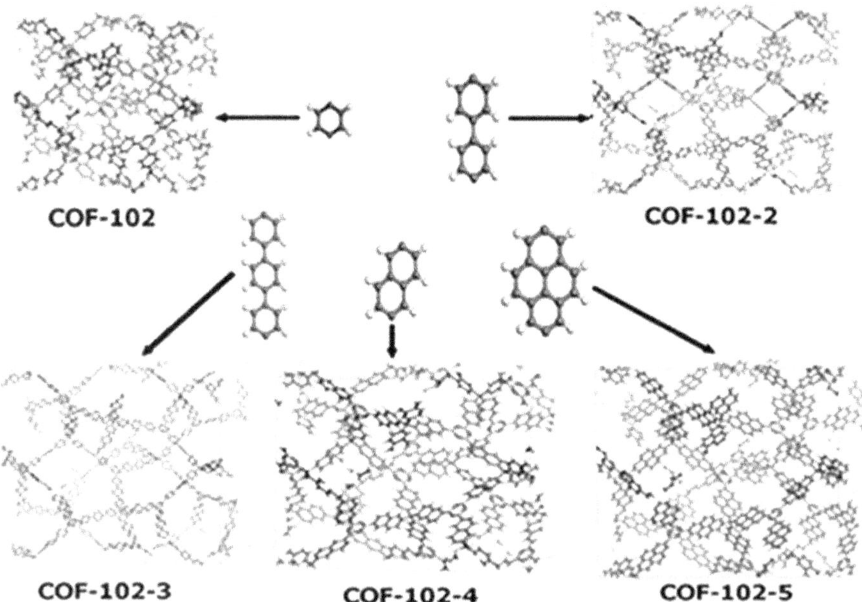

Figure 6.5   The substitution of phenylene moieties in COF-102 with extended aromaticity (diphenyl, triphenyl, naphthalene, and pyrene molecules). (Printed with permission from Klontzas *et al.*[44])

Yaghi and co-workers synthesized a series of 2D COFs (COF-5, COF-6, COF-8, and COF-10) using HHTP co-condensation with different building blocks, as shown in Figure 6.6.[6,39] The pore sizes of these materials vary from 0.9 nm (COF-6) to 3.2 nm (COF-10) due to the different lengths of the linkers. The BET surface areas also cover a wide range, from 750 m^2 g^{-1} (COF-6) to 1760 m^2 g^{-1} (COF-10), and the pore volumes of these materials vary from 0.32 cm^3 g^{-1} (COF-6) to 1.44 cm^3 g^{-1} (COF-10). Therefore, these differences in the structures will lead to great differences in the adsorption capacity. Jiang *et al.* used HHTP and pyrene-2,7-diboronic acid (PDBA), which is a longer linker than the ones used by Yaghi *et al.*, to synthesize a new COF, named TP-COF.[45] The TP-COF holds a pore size of 3.2 nm, and the pore volume and BET surface area are 0.79 cm^3 g^{-1} and 868 m^2 g^{-1}, respectively. TP-COF is a semiconducting and luminescent material due to its highly ordered π-conjugation system. Spitler *et al.* adopted an extremely long, rigid linker [4,4′-diphenylbutadiynebis(boronic acid); DPB] to react with HHTP to synthesize a 2D COF (HHTP-DPB COF). As a result, the extremely long, rigid linker led to the HHTP-DPB COF having an extremely large pore size of 4.7 nm, and also a high BET surface area of 2640 m^2 g^{-1}.[46] Actually, HHTP can co-condensate with various organic linkers to produce diverse COF materials with different properties. Inspired by these interesting materials, Choi *et al.* adopted the linker replacing strategy to design three new COFs [2D COF-05, 3D COF-05 (ctn), and 3D COF-05 (bor)].[47] 2D COF-05 uses HHTP

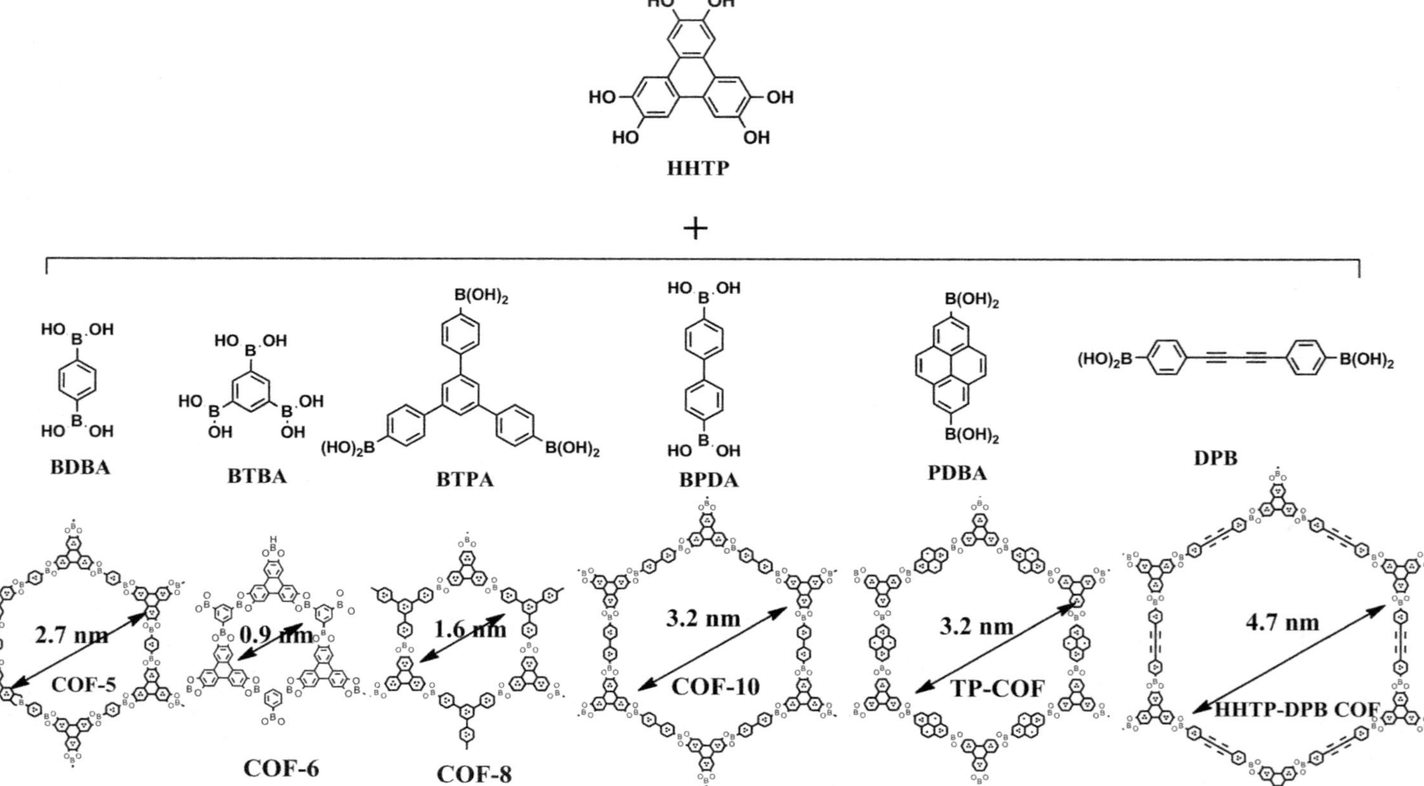

Figure 6.6 Synthesis of different hexagonal 2D COFs by co-condensation of HHTP with various linkers (BTPA, 1,3,5-benzenetris(4-phenylboronic acid)).

as the node and diphenylacetylene-4,4′-diboronic acid (DPABA) as the linker. 3D COF-05 uses HHTP as the node and tetra[4-(4-dihydroxybor-ylphenyl)ethynyl]phenyl methane (TBPEPM) as the linker. The accessible surface area and pore volume of 2D COF-05 are much larger than COF-10. Two 3D COF-05 exhibit around 30% larger accessible surface area and two times larger pore volumes than that of COF-108.

Porphyrin is also a fascinating building block, which is successfully used as a node to synthesize various tetragonal 2D COF materials[48–51] such as ZnP-COF, H_2-COF and CuP-COF. As an example, as shown in Figure 6.7, Yaghi *et al.* designed two COFs: COF-66 and COF-366, which have high charge carrier mobility.[50] COF-66 was synthesized *via* a condensation reaction between tetra(*p*-boronic acid-phenyl)porphyrin (TBPP) and 2,3,4,5-tetrahydroxy anthracene (THAn). COF-366 was obtained *via* a condensation reaction between tetra(*p*-amino-phenyl)porphyrin (TAPP) and ter-ephthaldehyde. Obviously, both COF-66 and COF-366 hold the similar por-phyrin node, but their different linkers lead to the difference of porosity. The BET surface areas for COF-66 and -366 were 360 and 735 $m^2\,g^{-1}$, respectively. The pore sizes of COF-66 and COF-366 were 2.3 and 2.0 nm, respectively. Total pore volumes for COF-66 and -366 were 0.2 and 0.32 $cm^3\,g^{-1}$,

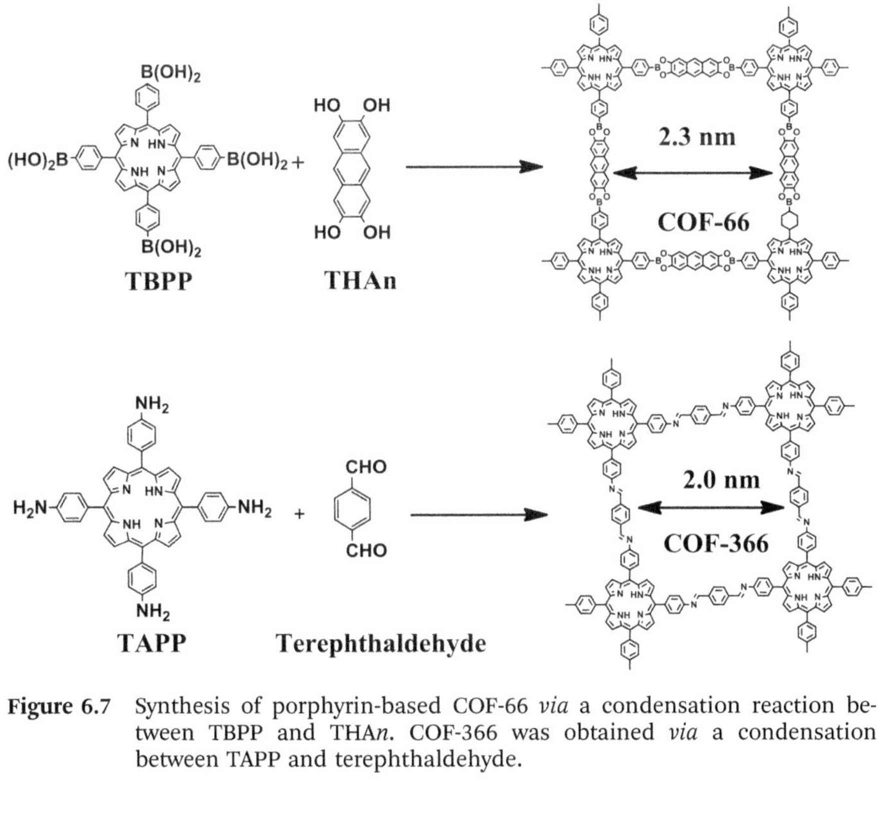

Figure 6.7   Synthesis of porphyrin-based COF-66 *via* a condensation reaction be-tween TBPP and THA*n*. COF-366 was obtained *via* a condensation between TAPP and terephthaldehyde.

respectively. Definitely, the properties of materials can be well tailored through the linker replacement strategy.

Phthalocyanine is also an important building block to construct tetragonal 2D COFs.[52–54] Spitler *et al.* synthesized Zn phthalocyanine (ZnPc) COF films (ZnPc-Py COF, ZnPc-NDI COF, ZnPc-DPB COF, and ZnPc-PPE COF) using the co-condensation of phthalocyanine with linkers of different lengths (Figure 6.8).[55] The diagonal pore widths of these COFs increased from 2.7 to 4.4 nm with the increase of linker length, and the BET surface areas ranged from 420 to 490 $m^2\ g^{-1}$. By the linker replacing strategy, we can tune the porosity and the BET surface area of the COFs while these COFs still maintain the desirable topology. As a result, Echegoyen *et al.* synthesized a new 2D CoPc-PorDBA COF by condensation of phthalocyanine with porphyrin (Por).[56] The CoPc-PorDBA COF showed a high BET surface area of 1315 $m^2\ g^{-1}$, the total pore volume was 0.88 $cm^3\ g^{-1}$, and the average pore size was about 3.5 nm.

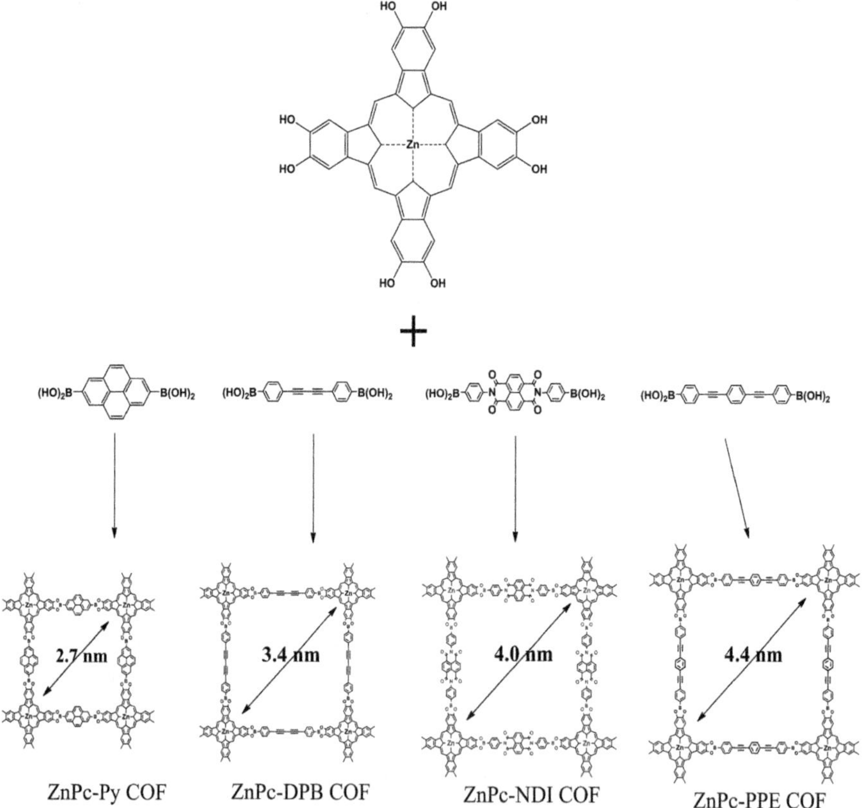

Figure 6.8 Synthesis of different tetragonal 2D COFs by co-condensation of phthalocyanine with various linkers [Py, pyrene; NDI, naphthalene diimide; PPE, poly(phenylene ethynylene)].

6.4.2 Node Replacement Strategy

Besides the linker replacement strategy, the node replacement strategy is also an alternative means to design new materials. Nowadays, the triphenylene scaffold and phenyl are the commonly used C_3-symmetric building units. In order to develop node units for new architectures and functions, Jiang *et al.* prepared a kind of star-shaped 2D COF, as shown in Figure 6.9. These structures were constructed *via* the condensation of one phenanthrene cyclotrimer-based C_3-symmetric building block with three C_2-symmetric units.[57] The star-COFs (star-COF-1, 2, and 3) possess inherent 1D channels, exhibiting BET surface areas of 2120, 1767, and 2129 m^2 g^{-1}, respectively. In addition, the intrinsic porosity mainly originates from the hexagonal cells, and the pore diameters range from 3.9 to 4.7 nm, showing the tunable possibility.

Yaghi *et al.* synthesized two new 2D COFs (COF-42 and COF-43) by the condensation of 2,5-diethoxyterephthalohydrazide with 1,3,5-triformylbenzene or 1,3,5-tris(4-formylphenyl)-benzene, as displayed in Figure 6.10.[58] The different node units lead to the different porosities of COF-42 and COF-43. COF-43, with the larger size node, exhibits a larger pore size (3.5 nm) and pore volume (0.36 cm^3 g^{-1}), compared to COF-42, while the pore size and pore volume of COF-42 are 2.8 nm and 0.31 cm^3 g^{-1}, respectively.

Generally, the linker replacement strategy may not change the structural topology, while with the node replacement strategy this is not always the case, as shown in Figure 6.11. In Figure 6.11a, condensation of terephthaldehyde with tetragonal TAPP produced COF-366. However, Zhou *et al.* synthesized a novel structure COF by condensation of terephthaldehyde with 4,4′,4″,4‴-(ethene-1,1,2,2-tetrayl)tetraaniline (ETTA). This structure has both hexagonal mesopores of 26.9 Å and triangle micropores of 7.1 Å.[13] This unique dual-pore feature may be useful for adsorption.

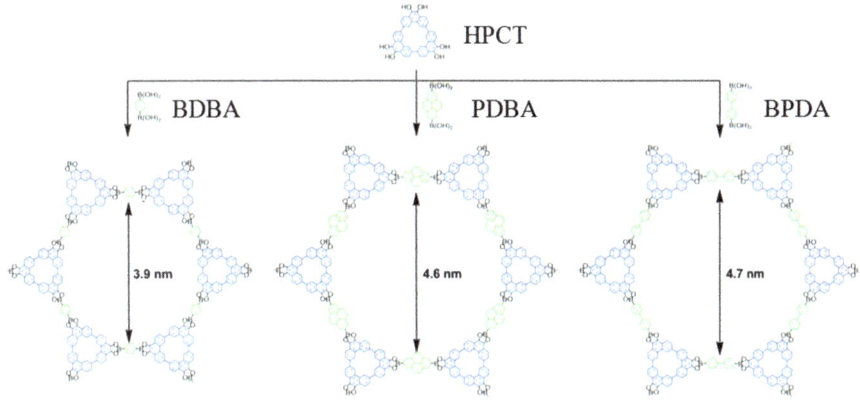

Figure 6.9 Synthesis of star-shaped 2D COFs by co-condensation of 9,10-hydroxyphenanthrene cyclotrimer (HPCT) with BDBA, PDBA, and BPDA. (Printed with permission from Feng *et al.*[57])

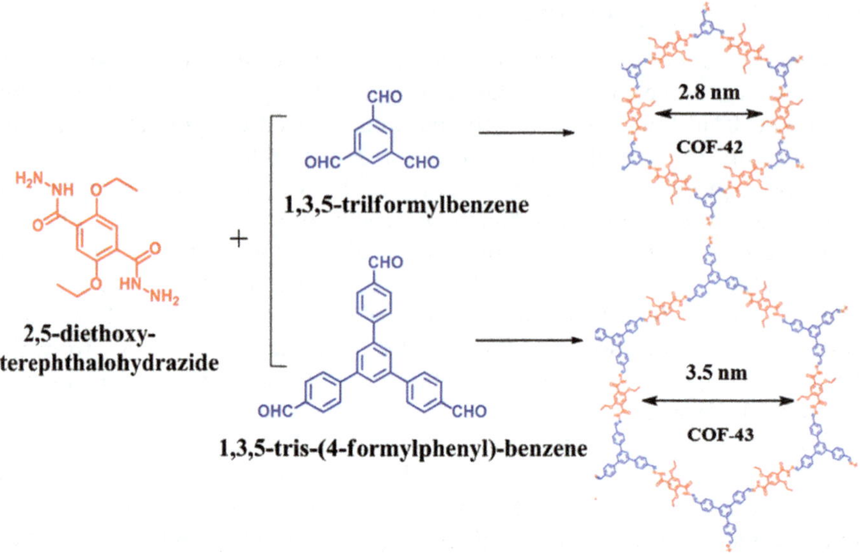

Figure 6.10 Synthesis of COF-42 and COF-43 by condensation of linear 2,5-diethoxy-terephthalohydrazide building blocks (red) with trigonal-planar 1,3,5-triformylbenzene (blue) or 1,3,5-tris(4-formylphenyl)-benzene (blue).

In addition, in Figure 6.11b, by using 1,4-diaminobenzene as the linker, and 1,3,5-triformylbenzene and 1,3,6,8-tetrakis(*p*-formylphenyl)pyrene as nodes, investigators synthesized COF-LZU1 and ILCOF-1, respectively.[9,59] Interestingly, the topology and porosity properties of the synthesized structures are entirely different. The pore size of COF-LZU1, with the small-size trigonal building unit node, is 1.8 nm, whereas the pore size of ILCOF-1, with a large-size tetragonal building unit node, is 2.3 nm. ILCOF-1 exhibits a tetragonally shaped channel, while COF-LZU1 shows a hexagonally shaped channel. The BET surface area of ILCOF-1 is 2723 $m^2\,g^{-1}$, which is much higher than other imine-linked COFs such as COF-LZU1 (410 $m^2\,g^{-1}$),[19] COF-366 (735 $m^2\,g^{-1}$), and COF-300 (1360 $m^2\,g^{-1}$).[60] The high BET surface area indicates that ILCOF-1 is a promising gas storage material.

6.4.3 Functionalization Strategy

In order to improve the performance of COFs, introduction of functional groups has been proven to be a good choice.[61,62] The properties of functionalized COFs can be significantly enhanced, especially for high performance gas storage. So, we will show various approaches to introduce functionalities into COFs in the following sections.

6.4.3.1 Metal Doping

To improve the storage capacity of hydrogen, methane, and carbon dioxide in COFs, the strategy of doping an electropositive metal into the

(a)

(b)

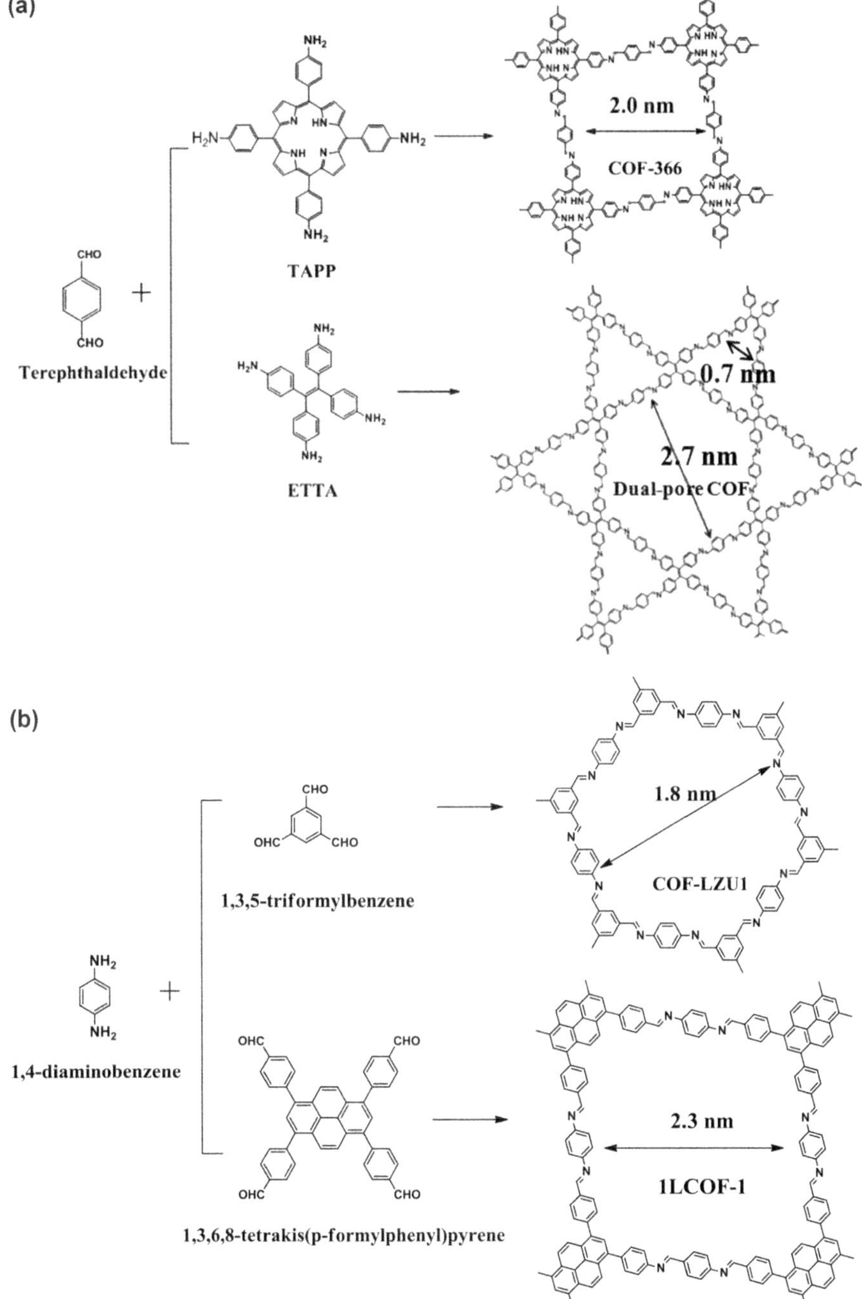

Figure 6.11   (a) Synthesis of amine-linked COFs by condensation of terephthalde-hyde with different building blocks. (b) Synthesis of amine-linked COFs by condensation of 1,4-diaminobenzene with different building blocks.

networks seems to be highly effective.[63,64] Cao *et al.* have studied the adsorption sites of Li atoms in COFs by multiscale simulation methods[65–67] since neutral Li atoms or anions have no contribution to the enhancement of storage capacity of COFs. To avoid unacceptable structural distortion induced by Li, the optimum adsorption sites for Li atoms are key issues. It is found that, for a single Li atom, COFs mainly exhibit five adsorption sites: (1) The hollow site in HHTP, (2) the open-hollow site in HHTP, (3) the oxygen site in the B_3O_3 or C_2O_2B ring, (4) the hollows site in TBPM or TBPS, and (5) the open-hollow site in TBPM or TBPS. The above configurations are presented in Figure 6.12. The calculated potential and binding energies are used to obtain the force field parameters for adsorbate molecules interacting with COFs, which were determined by fitting into the Morse potential. This method was widely used in force field fitting of the non-bond interaction.

In addition, Lan *et al.* studied the effects of the various metal dopants on CO_2 capture using multiscale simulations.[68] As a result, Mg, Be, and Ca failed to bind with COFs, while Li, Ti, and Sc can. However, Sc and Ti bind with COFs so strongly that structural deformation of the materials incurs. Hence, Lan *et al.* believed that Li shows the best potential for modification of COFs for CO_2 capture.

Besides lithium doping, a two-step strategy for chemical modification of COFs has been proven to be efficient for H_2 storage.[69] The first step was the boron doping in organic building blocks, then metal atoms (Sc, Ti, and Ca) were introduced to form trapping centers for H_2 molecules on B-doped COFs, as shown in Figure 6.13. Due to the presence of boron, the tendency of clustering of metal atoms was suppressed. The same phenomenon has also been found in hydrogen adsorption upon Li-doping the B-substituted phthalocyanine-1,4-phenylenebis(boronic acid) COF (Pc-PBBA COF).[70] The substitution of B can not only suppress the Li clustering, but can also improve the binding energy of H_2 to dope Li atoms. As a result, the positive charge on Li atoms enhances the affinity between hydrogen and the COF material, resulting in larger hydrogen storage capacity.

Li *et al.* proposed that substituting the bridging C_2O_2B rings with different metal-incorporating rings can not only promote the binding of H_2 on COFs, but can also avoid metal atom clustering. The binding energy of H_2 can be increased four-fold, and the storage capacity can be increased by a factor of two or three with suitable metal dopants.[71]

6.4.3.2 Other Functional Groups

Babarao *et al.* designed PAFs by introducing polar organic groups ($-OCH_3$, $-NH_2$, $-CH_2OCH_2-$) to the biphenyl unit, and investigated their separation properties for CO_2 using GCMC simulations. Ether-functionalized PAF-1 was found to have a high adsorption capacity for CO_2, reaching 10 mmol g^{-1} at 1 bar and 298 K, and also much higher selectivities for separating CO_2 from gas mixtures, compared to the amine functionality.[72]

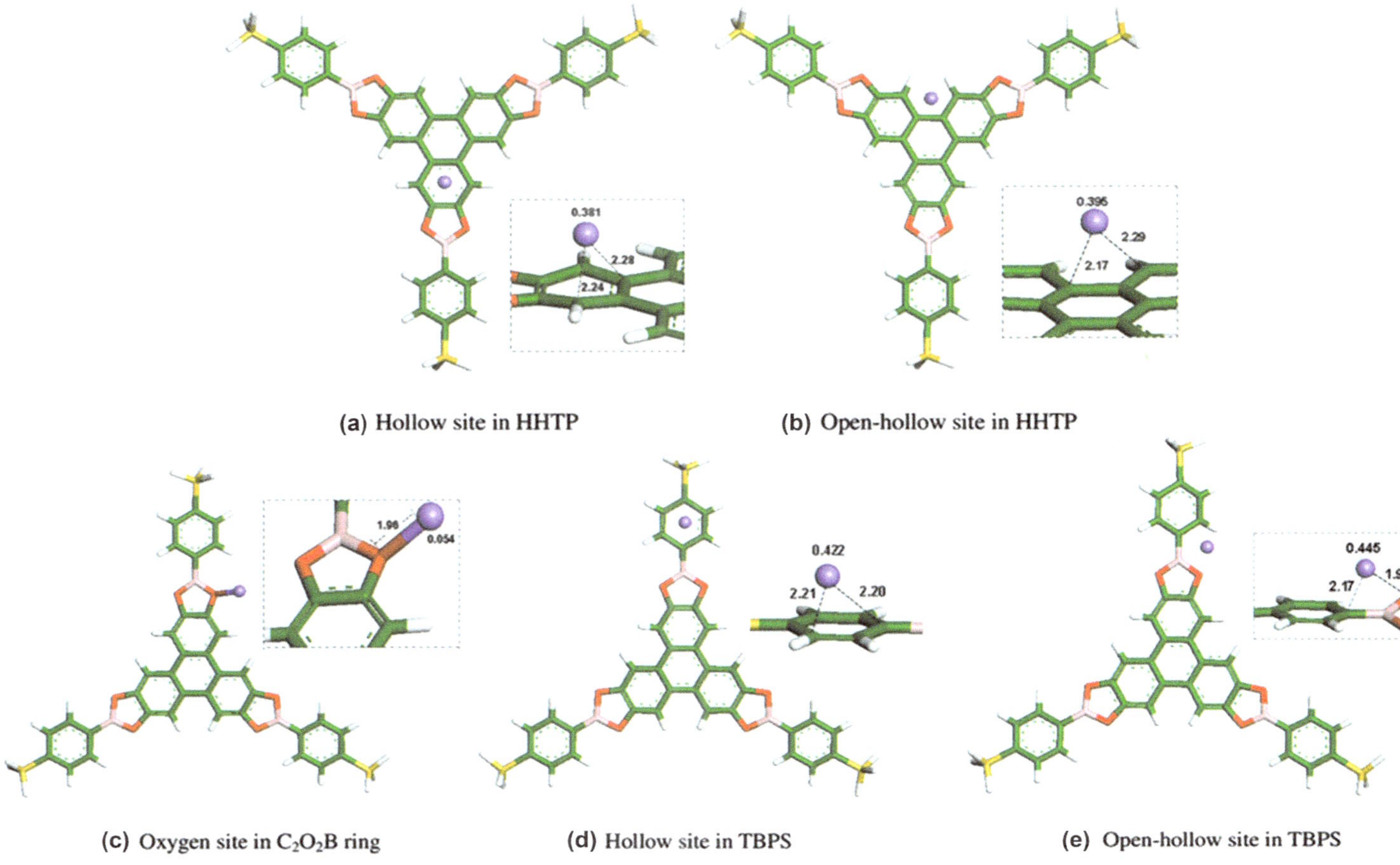

Figure 6.12 Adsorption sites of Li atoms on COF materials. (Printed with permission from Cao *et al.*[65])

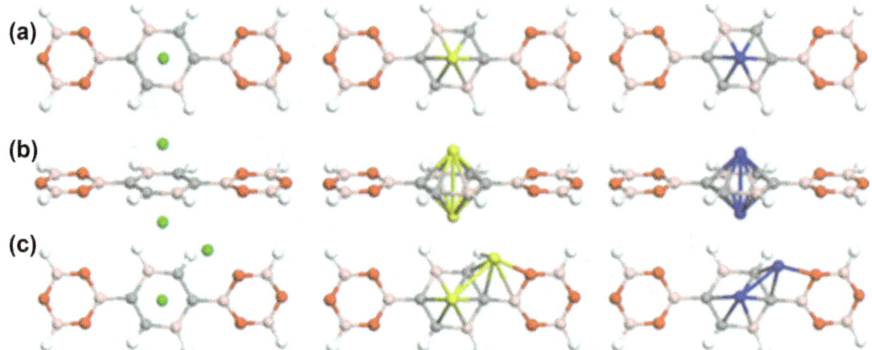

Figure 6.13 Optimized structures of B-COF-1 fragments with adsorbed metal atoms: (a) single-sided adsorption, (b) double-sided adsorption, and (c) dimer adsorption. Green, yellow, and blue spheres represent Ca, Sc, and Ti atoms, respectively.
(Printed with permission from Zou *et al.*[69])

Bunck and Dichtel employed a truncated mixed linker method to incorporate arbitrary functional groups into 3D COFs.[73,74] Condensing mixtures of **1** and **2** or allyl-functionalized monomer **3** under solvothermal conditions can obtain dodecyl-functionalized COF-102 (COF-102-C$_{12}$) and COF-102-allyl, as shown in Figure 6.14. The functionalized COFs maintained their crystallinity and large surface area, as well as the morphology. COF-102-allyl was subjected to the thiol–ene reaction, achieving full functionalization of the integrated allyl groups. The microporosity and crystallinity of COF-102-SPr remained intact, emphasizing the compatibility with the boroxine-linked framework under mild conditions.

Zhao *et al.* modified 3D COFs by substituting H atoms on benzene rings with other groups, including halogen groups, –CF$_3$, –NH$_2$, –CN, –OCH$_3$, and –CH$_3$.[75] They explored the effects of the substitutes on CH$_4$ storage using GCMC simulations, and the results indicated that halogen groups and –NH$_2$ are excellent substituent groups to increase CH$_4$ uptakes. Then, they further investigated the double-halogen-substituted COF-102 for methane storage. The GCMC simulation results showed that the methane delivery uptake can reach the Department of Energy (DOE) target of 180 V(STP)/V (STP, standard temperature and pressure),[76] suggesting that COF-102 with a double halogen substitution is beneficial for CH$_4$ uptake.

Tilford *et al.* synthesized alkyl-functionalized COF materials with benzene-1,3,5-triboronic acid reacting with 1,2,4,5-tetrahydroxybenzene substituted by dialkyl,[77] as shown in Figure 6.15. The BET surface area and the pore volume decrease when the alkyl groups become longer. This work provides a method to modify the pore interior of COFs in order to find promising materials for gas storage.

Jiang *et al.* introduced a strategy to functionalize the pore walls of COFs with various organic linkers (such as chromophoric moieties, alkyl chains, and aromatic units).[78] As shown in Figure 6.16a, they first synthesized COF-5

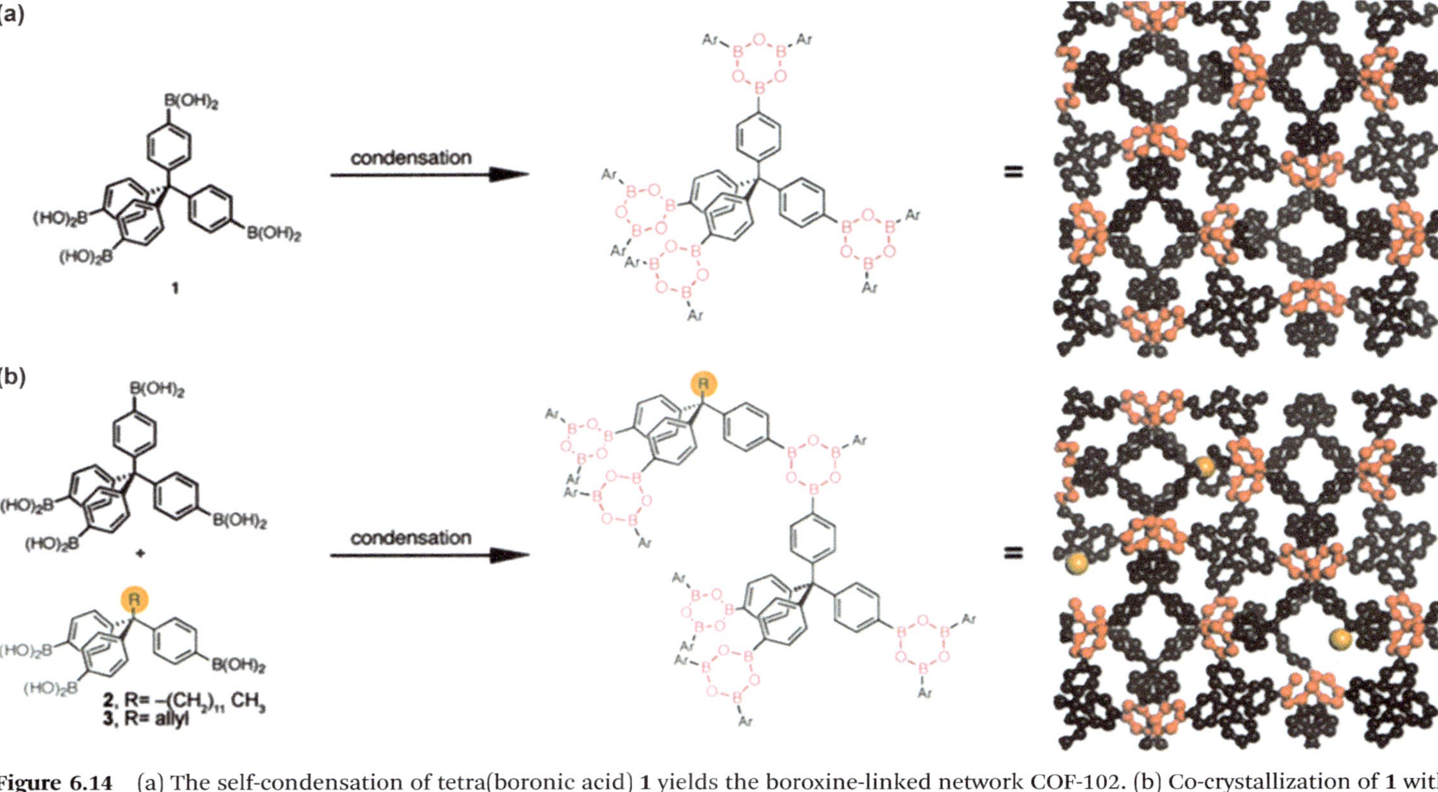

Figure 6.14   (a) The self-condensation of tetra(boronic acid) **1** yields the boroxine-linked network COF-102. (b) Co-crystallization of **1** with truncated monomer **2** or **3** produces internally functionalized 3D COFs. (Printed with permission from Bunck *et al.*[73])

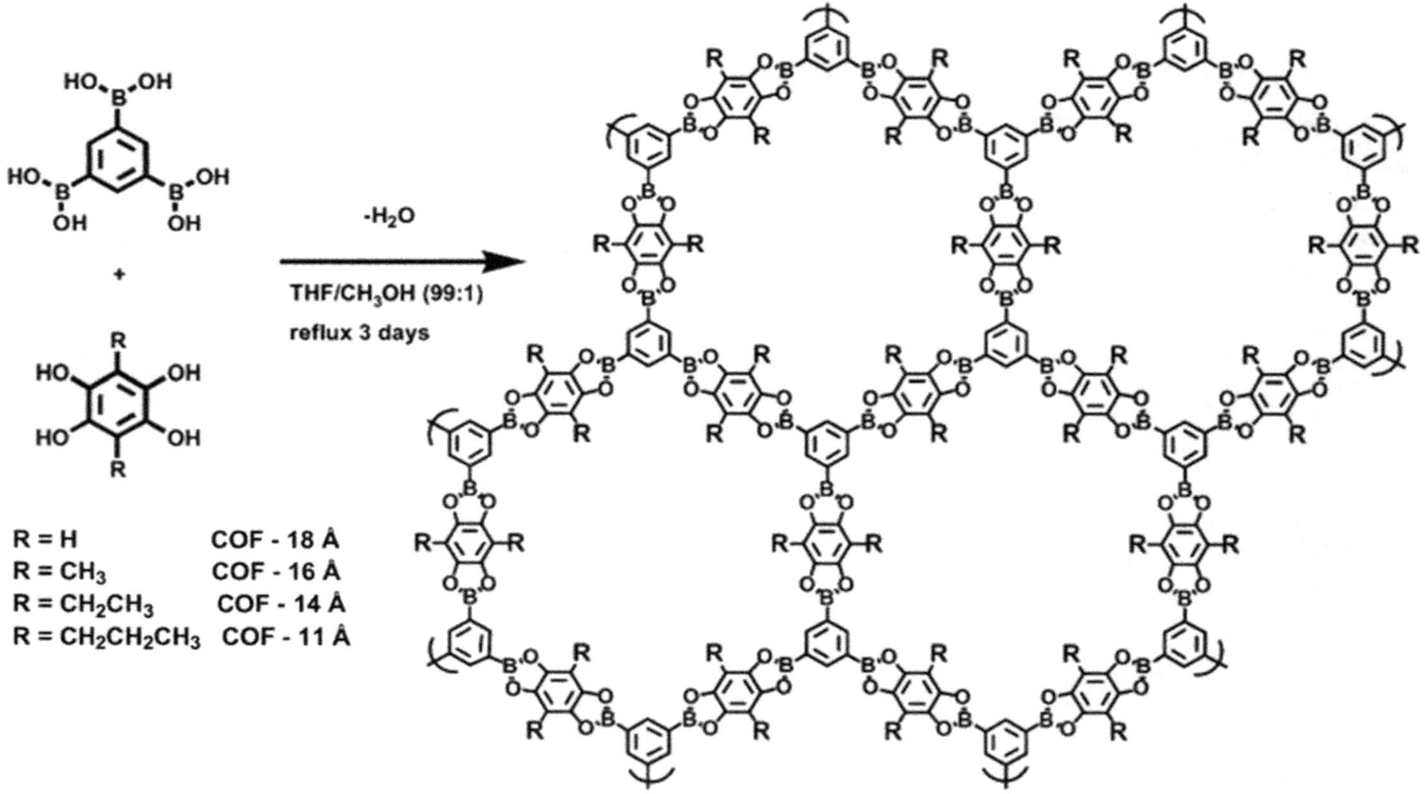

Figure 6.15 Synthesis of alkyl-functionalized COFs through condensation of benzene-1,3,5-triboronic acid and 2,6-disubstituted-1,2,4,5-tetrahydroxybenzene in tetrahydrofuran (THF) and methanol. (Printed with permission from Tilford *et al.*[77])

modified with a tunable content of azide units by replacing some BTBA with an azide-appended BTBA, and then the azide units on the COF walls reacted with alkynes to form triazole-linked groups on the wall surfaces. When the content of azide-appended BTBA increased from 5% to 100%, the BET surface area, pore size, and pore volume of AcTrz-COF-5 decreased from 2000 m^2 g^{-1} to 36 m^2 g^{-1}, 2.978 nm to 1.248 nm, and 1.481 m^3 g^{-1} to 0.0297 m^3 g^{-1}, respectively. Moreover, this surface engineering strategy was also applied to tetragonal COFs (Figure 6.16b). This pore surface engineering strategy of COFs provides a universal method to functionalize COFs with various organic groups, and it can change the pore size, pore volume, and BET surface area.[79] Thus, COFs with different functional groups will display an entirely different affinity for guest molecules.

Jiang *et al.* also reported a channel-wall functionalization to increase the carbon dioxide uptake of 2D COFs.[80] They used an imine-linked 2D COF as a scaffold, as shown in Figure 6.17, and further adopted succinic anhydride to react with the phenol group in $[HO]_{X\%}$-H_2P-COFs to introduce carboxylic acid groups into the channel walls. The BET surface area, pore size, and pore volume of $[HO_2C]_{X\%}$-H_2P-COFs decrease with an increase of the content of carboxylic groups. However, the CO_2 capacity of $[HO_2C]_{X\%}$-H_2P-COFs significantly increased at low pressure due to the functional group modification.

6.5   Applications

In recent decades, a large amount of CO_2 has been released into the air due to the burning of fossil fuels, which has led to the greenhouse effect, climate change, rising sea levels, and a series of environmental problems.[4] Thus, it is important to reduce the concentration of CO_2 and find alternative energies. One accessible method is using carbon capture and storage (CCS) technology to reduce the concentration of CO_2 in the air. The precombustion and postcombustion captures of CO_2 from flue gas streams and syngas are important factors of the CCS technology. Alternatively, developing renewable and clean energies will also eliminate CO_2 emission. Hydrogen has attracted extensive interest due to its high energy density and it is considered an environmentally friendly, clean energy. CH_4 is also an excellent candidate for substituting gasoline and diesel fuels in vehicles.[81] For commercial use of H_2 and CH_4 as clean energy, the main problem is to store the gases efficiently and safely. Compressed liquid ammonia is widely employed in industrial applications. However, it is difficult to treat the liquid ammonia because of its corrosiveness and toxicity. Thus, using an adsorbent to adsorb ammonia may be an alternative way to deal with this issue. COFs consist of light elements connected by strong covalent bonds, and they have high porosity with extremely large BET surface areas; they have therefore been deemed as ideal materials for gas adsorption. The storage capacities of COFs for hydrogen, carbon dioxide, methane, and ammonia have been widely explored. Generally, the gas storage capacity of COFs depends mainly on the

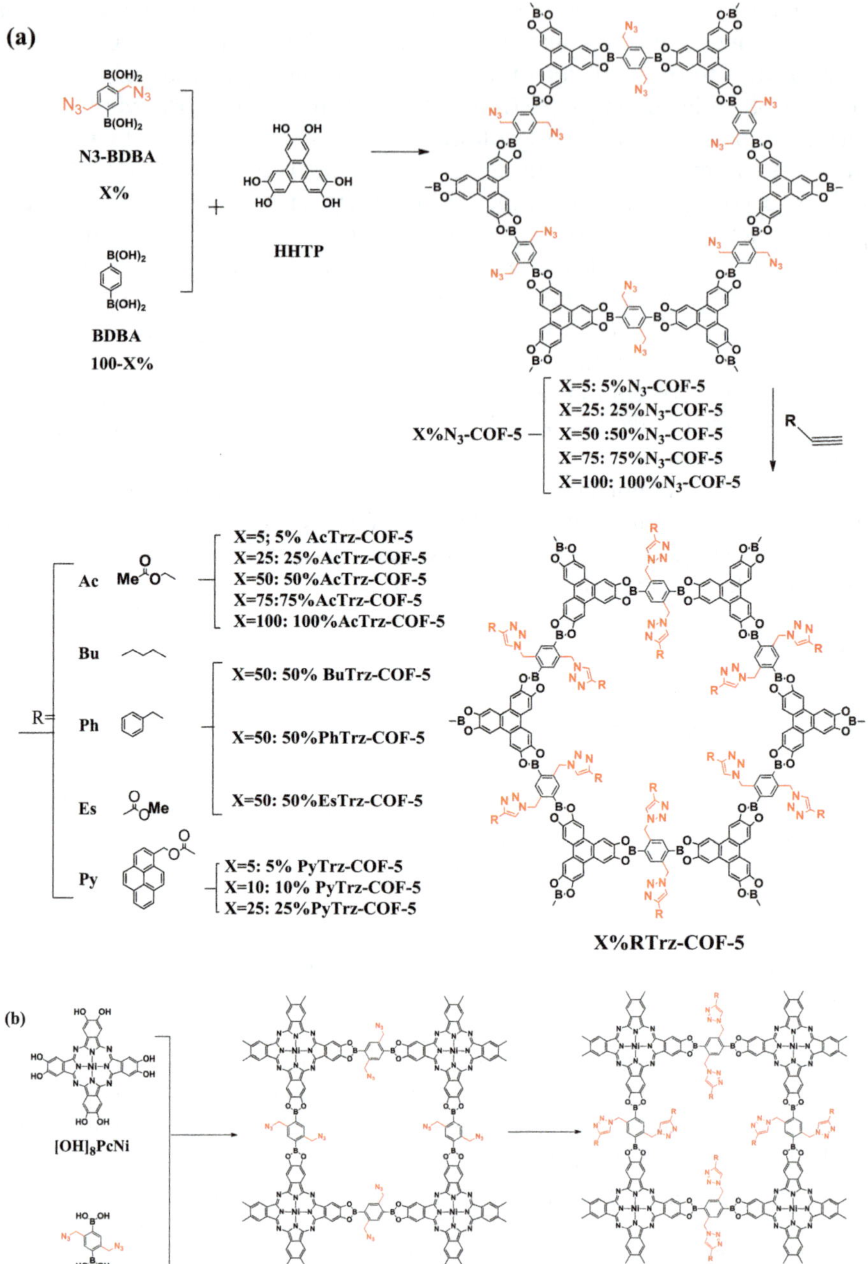
(a)
B(OH)2
N3
N3
B(OH)2
N3-BDBA
X%
B(OH)2
B(OH)2
BDBA
100-X%
HHTP
HO OH
HO OH
HO OH
X=5: 5%N3-COF-5
X=25: 25%N3-COF-5
X=50 :50%N3-COF-5
X=75: 75%N3-COF-5
X=100: 100%N3-COF-5
X%N3-COF-5
R
Ac
X=5; 5% AcTrz-COF-5
X=25: 25%AcTrz-COF-5
X=50: 50%AcTrz-COF-5
X=75:75%AcTrz-COF-5
X=100: 100%AcTrz-COF-5
Me O
Bu
Ph
Es
O Me
X=50: 50% BuTrz-COF-5
X=50: 50%PhTrz-COF-5
X=50: 50%EsTrz-COF-5
R=
Py
O
X=5: 5% PyTrz-COF-5
X=10: 10% PyTrz-COF-5
X=25: 25%PyTrz-COF-5
X%RTrz-COF-5
(b)
HO OH
HO OH
HO OH
HO OH
[OH]8PcNi
HO OH
HO OH
N3BDBA
100%N3-NiPc-COF
100%RTrz-NiPc-COF

surface areas and pore volume. As discussed above, COFs decorated with alkali metals or functionalized groups could improve the gas storage capacity.

6.5.1 Hydrogen Storage

Porous materials have been widely studied for hydrogen storage.[82–85] Recently, hydrogen storage on COF materials has also become a "hot" research topic owing to the large pore volumes and high BET surface areas of COFs. Furukawa and Yaghi found that COF-102 and COF-103 showed H_2 uptakes of 7.24 wt% and 7.05 wt%, respectively, while COF-10 exhibited the highest H_2 uptake of 3.92 wt% among the 2D COFs, compared to uptakes of 1.48, 3.58, 2.26, and 3.50 wt% for COF-1, 5, 6, and 8, respectively, at 77 K and 1 bar.[11] These high capacities show that COFs have the potential to store H_2. Although the hydrogen uptake of COFs at cryogenic temperature is considerable, their performance at ambient conditions is far from the DOE target.

Many strategies have been applied to improve the H_2 storage capacity, as illustrated above. Increasing the free volume of the materials is one method. As reported by Lan *et al.*, the hydrogen uptakes of PAF-304 (6.53 wt%) and PAF-303 (4.16 wt%) at 298 K and 100 bar are nearly tripled and doubled, respectively, compared to PAF-302 (2.21 wt%) under the same conditions. Excitingly, the highest hydrogen uptake of PAF-304 (6.53 wt%) exceeds the 2015 DOE target of 5.5 wt% at room temperature.[43] After substituting the phenylene of COF-102 with other aromatic moieties, keeping the net topology, all the designed structures exhibited higher gravimetric uptake compared with the matrix COF-102. In particular, the gravimetric uptake of one of these COFs overpassed 25 wt% at 77 K, and reached the DOE target of 6 wt% at ambient conditions.[44]

Metal doping is another efficient method for enhancing the hydrogen storage capacities due to the significant improvement of binding energy. Cao

---

Figure 6.16 (a) Surface engineering of COFs through the combination of a condensation reaction and click chemistry. In the first step, COFs bearing azide units on the walls are synthesized by condensation of HHTP with azide-appended benzene diboronic acid (N_3-BDBA) and benzene diboronic acid (BDBA) in a designated molar ratio ($X = 0$–100%). The content of N_3-appended wall units is tunable from 0 to 100%, depending on the molar ratio of N_3-BDBA to BDBA. Five members of X%N_3-COF-5 with different contents ($X = 5$, 25, 50, 75, and 100) were synthesized. The structure of 100%N_3-COF-5 is shown in the figure, with all of the walls occupied by the N_3-appended phenylene units. In the next step, the azide groups on the COF walls are clicked with alkynes to anchor various organic groups onto the walls of COFs (X%RTrz-COF-5). The density of R surface groups on the walls is determined by the azide content in X%N_3-COF-5. (b) Surface engineering of the typical tetragonal NiPc-COF. The NiPc-COF walls can be engineered, as demonstrated by 100%RTrz-COF (R = Es).

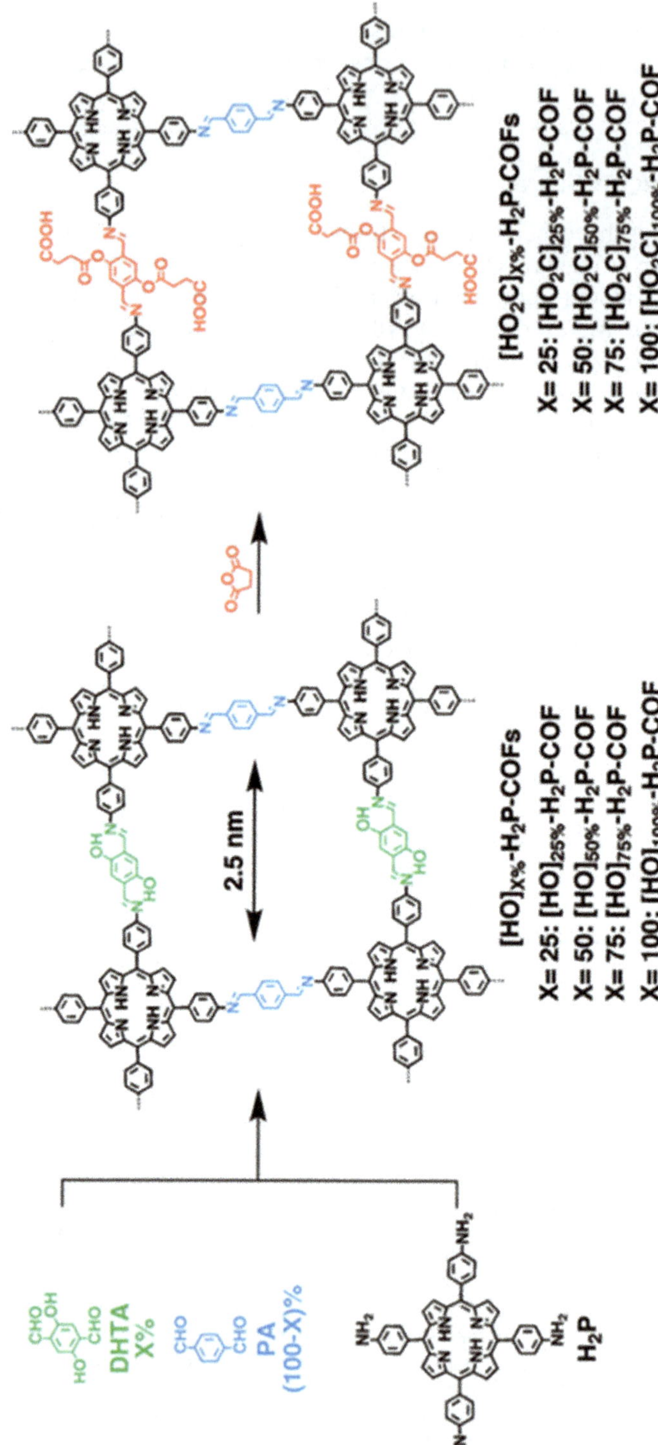

Figure 6.17 Synthesis of $[HO_2C]_{X\%}$-H_2P-COFs with channel walls functionalized with carboxylic acid groups through the ring opening reaction of $[OH]_{X\%}$-H_2P-COFs with succinic anhydride (DHTA, 2,5-dihydroxyterephthalic acid; PA, 1,4-phthalaldehyde). (Printed with permission from Huang *et al.*[80])

et al. found that Li-doped COF-108 and COF-105 showed extremely high hydrogen capacities of 6.73 and 6.84 wt% at 100 bar and 298 K, respectively.[65] In addition, Lan *et al.* reported that COF-202 possessed volumetric and gravimetric capacities of 8.08 g L^{-1} and 1.52 wt% at 100 bar and 298 K, respectively, while they reached 25.86 g L^{-1} and 4.39 wt%, respectively, after Li doping.[67] Klontzas *et al.* introduced lithium alkoxide into the 3D COF structures, and found that the hydrogen uptake of modified COF-105 exceeded the DOE target (6 wt%) for hydrogen applications.[86] All these observations indicate that metal doping does efficiently enhance hydrogen capacities.

In addition, the hydrogen storage capacities of Sc-, Ca-, and Ti-adsorbed B-COF-1 are estimated to be 6.9, 6.7, and 6.7 wt%, respectively, meeting the DOE target (6 wt%).[69] Guo *et al.* predicted that Li-doped Pc-PBBA COF has a volumetric H_2 and gravimetric uptake of 40.23 g L^{-1} and 4.70 wt%, respectively, at 250 K and 100 bar.[70] For Mg_2-doped COF-108, the storage capacity was enhanced up to 2.73 wt% by GCMC simulation at 298 K.[71]

By introducing undulated macrocyclic cyclotricatechylene (CTC) into 2D COFs, Zheng and co-workers found that the storage capacity of H_2 can be increased.[87] The obtained CTC-COF shows a higher H_2 uptake of 1.12 wt% at 1.05 bar than those of similar 2D COFs, and even close to those of 3D COFs. Mendoza-Cortes *et al.* found theoretically that the total H_2 uptake of COF-301-PdCl$_2$ reaches 60 g L^{-1} at 100 bar, which is much higher than the DOE 2015 target (40 g L^{-1}).[88]

6.5.2   Methane Storage

In addition to hydrogen, methane is also an appreciable energy form. The DOE set the target value for CH_4 storage as 180 V(STP)/V at 35 bar, 298 K. COF-103 and COF-102 were reported to possess total methane uptakes of 260 and 255 V(STP)/V at 100 bar, respectively, with excess uptakes of 234 and 229 V(STP)/V. The methane gravimetric capacities of COF-103 and -102 are 175 and 187 mg g^{-1}, respectively, at 35 bar and 298 K, while they are 40, 89, 65, 87, and 80 mg g^{-1} for COF-1, 5, 6, 8, and 10.[11] Moreover, lithium-doped COFs can double the CH_4 uptakes at pressures lower than 50 bar. The excess CH_4 gravimetric uptakes of Li-doped COF-103 and -102 reach 30.98 and 31.35 wt%, respectively, at 35 bar and 298 K, and the excess volumetric uptakes are 290 and 303 V(STP)/V.[66] Mendoza-Cortes *et al.* found that COF-103-Eth-*trans* and COF-102-Ant exceeded the DOE target for CH_4 [180 V(STP)/V] at 35 bar. Their performance can compare with PCN-14 and Ni-MOF-74, which are the best materials. The results show that vinyl bridging groups increase gas adsorption by the strong interaction between CH_4 and COFs at low pressure.[40]

Zhao and Yan studied the CH_4 adsorption capacities of modified COFs, and found that the excess methane uptakes of COF-102-X (X = I, Br, Cl, or NH$_2$) are 156, 153, 148, and 143 V(STP)/V, respectively. The adsorbents can be considered as promising adsorbents for methane uptake.[75] Rabbani *et al.*

investigated the gas storage performance of ILCOF-1 at low and high pressure. The CH_4 uptake of ILCOF-1 was 1.3 wt% at 273 K and 1 bar,[59] while the CH_4 uptake reached 11.2 mmol g^{-1} at 298 K and 40 bar, which is much higher than 2D COFs and comparable to 3D COFs.

6.5.3 Carbon Dioxide Storage

Using porous materials to capture CO_2 is a technically feasible and energetically efficient method. The CO_2 uptakes of COF-102, -103, -1, -5, -6, -8, and -10 achieve 27.3, 27.0, 5.2, 19.8, 7.0, 14.3, and 22.9 mmol g^{-1} at 55 bar and 298 K, respectively.[11] Lan *et al.* comprehensively studied COF doping with a series of metals and the effects of the doped metals on CO_2 capture using a multiscale simulation approach.[68] The results show among all the metals studied, Li is the best modifier for modification of COFs for CO_2 capture, because Li-doping modification enhances not only the gravimetric uptake, but also the volumetric capacity of CO_2 in COFs. Upon Li doping, the gravimetric uptake of CO_2 in Li-doped COF-105 and Li-doped COF-102 are about four-fold and eight-fold those of the non-doped counterparts at 298 K and 1 bar. Impressively, at 40 bar and 298 K, the uptake of CO_2 in Li-doped COF-105 reaches 2266 mg g^{-1}. Besides, at 20 bar and 298 K, the volumetric uptake of CO_2 in Li-doped COF-102 and Li-doped COF-105 reaches 313 and 139 V(STP)/V.

Choi *et al.* investigated the CO_2 adsorption capacity of 2D COFs (COF-5, COF-8, COF-10, and TP-COF), 3D COFs (COF-102 and COF-108), and three newly designed COFs [2D COF-05, 3D COF-05 (*ctn*), and 3D COF-05 (*bor*)] using the *ab initio* method and GCMC simulation.[47] The CO_2 uptakes of 3D COF-05 (*ctn*) and 3D COF-05 (*bor*) are 8582 mg g^{-1} and 9285 mg g^{-1} at 298 K and 55 bar, respectively, which is six times larger than that of COF-102.

Li *et al.* synthesized a new azine-linked 2D COFs (*i.e.*, ACOF-1), which had a BET surface area of 1176 m^2 g^{-1} and small pore size of 0.94 nm.[89] The CO_2 uptake of ACOF-1 is 177 mg g^{-1} at 273 K and 1 bar, which is higher than COF-5, COF-103, and TD-COF-5[90] under the same conditions. Rabbani *et al.* also investigated CO_2 storage performance of ILCOF-1 at low and high pressure.[59] The CO_2 uptake of ILCOF-1 is 6 wt% at 273 K and 1 bar, while the CO_2 uptake reaches 29.3 mmol g^{-1} at 298 K and 40 bar, which is higher than those of 2D and 3D COFs.

Jiang *et al.* reported that the CO_2 capacity of [HO$_2$C]$_{X\%}$-H$_2$P-COF significantly increased after functionalization with carboxylic acid groups.[80] The CO_2 adsorption uptake of [HO$_2$C]$_{100\%}$-H$_2$P-COF reaches 180 mg g^{-1} at 273 K and 1 bar, which is comparable to other kinds of materials [amine-PCN-58[91] (128 mg g^{-1}), PPN-6-CH$_2$DETA[92] (190 mg g^{-1}), PPN-6-SO$_3$Li[93] (187 mg g^{-1})]. Two-dimensional COFs may become an excellent kind of material for CO_2 capture by channel-wall functionalization. Figure 6.18 illustrates the relationship between the saturation CO_2 uptake *versus* the BET surface area and pore volume of the COFs. Obviously, the saturation CO_2 uptake shows an almost linear correlation with the pore volume and BET

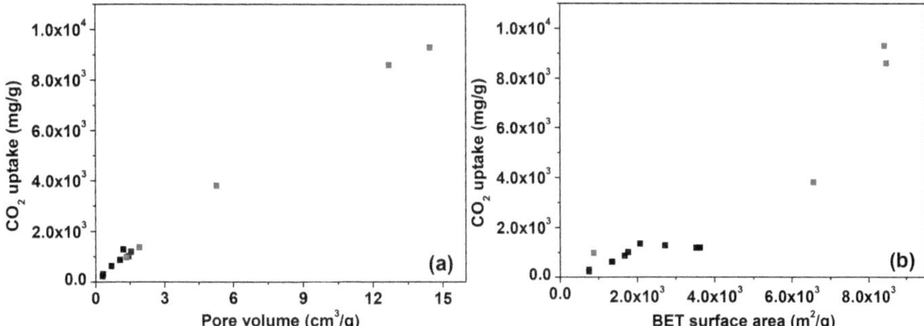

Figure 6.18   The saturated excess CO_2 storage capacity *versus* pore volume (a) and BET surface area (b) of selected COFs. Black points are taken from experimental data,[11,55] while red points are simulation results.[42]

surface area. This means that to develop high-capacity CO_2 storage materials, extremely high pore volume and BET surface area are desirable and beneficial.

6.5.4   Ammonia Storage

Storing ammonia by adsorbents would be a highly efficient and scalable method. COF-10 containing boron has been reported to show exceptionally high ammonia uptake (15 $mol\,kg^{-1}$) compared with other porous materials.[94] Moreover, COF-10 can release and re-adsorbed ammonia reversibly, and the total uptake is just slightly reduced (4.5%). The extremely high uptake of ammonia in COF-10 can be explained by the formation of a classical ammonia–borane coordination bond.

6.6   Summary and Perspectives

COFs have emerged as a new kind of crystalline porous material, which have the advantages of large surface areas, low density, functionality, and tunable properties due to the organic building blocks consisting of only light elements. Since Yaghi and co-workers reported the COF in 2005, COFs have been extensively employed in many applications, such as gas storage, luminescent probes, optoelectronics, and catalysis. Currently, they have attracted extensive interest from researchers with diverse expertise.

So far, the synthesis of functional COFs with highly regular and stable structures is still a great challenge. Generally, COFs do not have a highly regular single crystal structure. Alternatively, partial crystal COFs can only provide reasonable PXRD. Admittedly, the surface area and pore size distribution obtained by experiment are often deviated from those calculated by theory, because it is hard to control both kinetic and thermodynamic properties in the reversible reaction, leading to the imperfect formation of covalently bonded COF materials. Although a lot of theoretical investigations show excellent COF materials for gas storage, how to synthesize the designed COF materials is still a great challenge for experimental scientists.

Analysis of PXRD data associated with computational simulation can help determine the microscopic structure of COFs, particularly in the case of complex 3D COFs. Encouragingly, many researchers studying the computational modeling of COFs provide useful knowledge for the characterization of COFs. In fact, the theoretical calculations aim at guiding the experimental synthesis, and offer basic understanding of the microstructure and properties of COFs, while experiments are often used to verify the theory and assumptions, and to quantify the state variables. On the other hand, theoretical modeling (*e.g.*, the node and linker replacement and functionalization design strategies) can provide a material design strategy and molecular structure of new materials for experimental scientists. At the atomic and molecular scales, statistical mechanics-based theoretical methods can also predict physical chemical properties of newly designed COF materials, which will significantly reduce the costs of new material development with respect to experimental performance. In particular, researchers have built up multiscale theoretical approaches to predict the adsorption properties of COFs on the basis of the excellent agreement with the experimental results, and these methods can help towards the targeted design of novel COFs with high storage capacities. In addition, it is a key issue to enhance the complexity of COFs, and generate multifunctional COFs in the functional exploration.

In summary, even if there are still great challenges, COF materials have facilitated applications in various fields, indicating the promising future of these new porous materials. Numerous efforts should be made on further research of the design, synthesis, and characterization of COF materials. Of course, the two strategies (*i.e.*, the node and linker replacement strategy and the functionalization strategy) will help design more new COF materials and bring more functionalized COF materials into practical applications, including high-performance gas storage, luminescent probes, and catalysis *etc.*

Acknowledgements

This work was supported by the National NSF of China (91334203, 21274011), National 863 Program (2013AA031901) and Outstanding Talents Plans of BUCT.

References

1. A. R. Millward and O. M. Yaghi, *J. Am. Chem. Soc.*, 2005, **127**, 17998.
2. M. Eddaoudi, J. Kim, N. Rosi, D. Vodak, J. Wachter, M. O'Keeffe and O. M. Yaghi, *Science*, 2002, **295**, 469.
3. T. Ben, H. Ren, S. Ma, D. Cao, J. Lan, X. Jing, W. Wang, J. Xu, F. Deng, J. M. Simmons, S. Qiu and G. Zhu, *Angew. Chem., Int. Ed.*, 2009, **48**, 9457.
4. S. R. Venna and M. A. Carreon, *J. Am. Chem. Soc.*, 2010, **132**, 76.
5. J.-R. Li, J. Sculley and H.-C. Zhou, *Chem. Rev.*, 2011, **112**, 869.
6. A. P. Côté, A. I. Benin, N. W. Ockwig, M. O'Keeffe, A. J. Matzger and O. M. Yaghi, *Science*, 2005, **310**, 1166.

7. H. M. El-Kaderi, J. R. Hunt, J. L. Mendoza-Cortes, A. P. Cote, R. E. Taylor, M. O'Keeffe and O. M. Yaghi, *Science*, 2007, **316**, 268.

8. S. J. Rowan, S. J. Cantrill, G. R. L. Cousins, J. K. M. Sanders and J. F. Stoddart, *Angew. Chem., Int. Ed.*, 2002, **41**, 898.

9. X. Feng, X. Ding and D. Jiang, *Chem. Soc. Rev.*, 2012, **41**, 6010.

10. L. Chen, K. Furukawa, J. Gao, A. Nagai, T. Nakamura, Y. Dong and D. Jiang, *J. Am. Chem. Soc.*, 2014, **136**, 9806.

11. H. Furukawa and O. M. Yaghi, *J. Am. Chem. Soc.*, 2009, **131**, 8875.

12. S.-Y. Ding and W. Wang, *Chem. Soc. Rev.*, 2013, **42**, 548.

13. T.-Y. Zhou, S.-Q. Xu, Q. Wen, Z.-F. Pang and X. Zhao, *J. Am. Chem. Soc.*, 2014, **136**, 15885.

14. P. Kuhn, M. Antonietti and A. Thomas, *Angew. Chem., Int. Ed.*, 2008, **47**, 3450.

15. N. L. Campbell, R. Clowes, L. K. Ritchie and A. I. Cooper, *Chem. Mater.*, 2009, **21**, 204.

16. B. P. Biswal, S. Chandra, S. Kandambeth, B. Lukose, T. Heine and R. Banerjee, *J. Am. Chem. Soc.*, 2013, **135**, 5328.

17. D. D. Medina, J. M. Rotter, Y. Hu, M. Dogru, V. Werner, F. Auras, J. T. Markiewicz, P. Knochel and T. Bein, *J. Am. Chem. Soc.*, 2015, **137**, 1016.

18. J. W. Colson, A. R. Woll, A. Mukherjee, M. P. Levendorf, E. L. Spitler, V. B. Shields, M. G. Spencer, J. Park and W. R. Dichtel, *Science*, 2011, **332**, 228.

19. S.-Y. Ding, J. Gao, Q. Wang, Y. Zhang, W.-G. Song, C.-Y. Su and W. Wang, *J. Am. Chem. Soc.*, 2011, **133**, 19816.

20. D. Beaudoin, T. Maris and J. D. Wuest, *Nat. Chem.*, 2013, **5**, 830.

21. D. Frenkel and B. Smit, *Understanding Molecular Simulation: From Algorithms to Applications*, Academic press, 2001.

22. M. J. Frisch, G. W. Trucks, H. B. Schlegel, G. E. Scuseria, M. A. Robb, J. R. Cheeseman, J. A. Montgomery, Jr, T. Vreven, K. N. Kudin, J. C. Burant, J. M. Millam, S. S. Iyengar, J. Tomasi, V. Barone, B. Mennucci, M. Cossi, G. Scalmani, N. Rega, G. A. Petersson, H. Nakatsuji, M. Hada, M. Ehara, K. Toyota, R. Fukuda, J. Hasegawa, M. Ishida, T. Nakajima, Y. Honda, O. Kitao, H. Nakai, M. Klene, X. Li, J. E. Knox, H. P. Hratchian, J. B. Cross, V. Bakken, C. Adamo, J. Jaramillo, R. Gomperts, R. E. Stratmann, O. Yazyev, A. J. Austin, R. Cammi, C. Pomelli, J. W. Ochterski, P. Y. Ayala, K. Morokuma, G. A. Voth, P. Salvador, J. J. Dannenberg, V. G. Zakrzewski, S. Dapprich, A. D. Daniels, M. C. Strain, O. Farkas, D. K. Malick, A. D. Rabuck, K. Raghavachari, J. B. Foresman, J. V. Ortiz, Q. Cui, A. G. Baboul, S. Clifford, J. Cioslowski, B. B. Stefanov, G. Liu, A. Liashenko, P. Piskorz, I. Komaromi, R. L. Martin, D. J. Fox, T. Keith, M. A. Al-Laham, C. Y. Peng, A. Nanayakkara, M. Challacombe, P. M. W. Gill, B. Johnson, W. Chen, M. W. Wong, C. Gonzalez and J. A. Pople, Gaussian, Inc., Wallingford, CT, 2004.

23. G. Kresse and J. Furthmüller, *Phys. Rev. B: Condens. Matter Mater. Phys.*, 1996, **54**, 169.

24. L.-M. Yang and R. Pushpa, *J. Mater. Chem. C*, 2014, **2**, 2404.

25. B. Widom, *J. Chem. Phys.*, 1963, **39**, 2808.
26. D.-Y. Peng and D. B. Robinson, *Ind. Eng. Chem., Fundam.*, 1976, **15**, 59.
27. J. E. Lennard-Jones, *Proc. Phys. Soc.*, 1931, **43**, 461.
28. S. L. Mayo, B. D. Olafson and W. A. Goddard, III, *J. Phys. Chem.*, 1990, **94**, 8897.
29. A. K. Rappe, K. S. Casemit, W. A. Goddard, III and W. M. Skiff, *J. Am. Chem. Soc.*, 1992, **114**, 10024.
30. M. G. Martin and J. I. Siepmann, *J. Phy. Chem. B*, 1998, **102**, 2569.
31. W. L. Jorgensen, J. D. Madura and C. J. Swenson, *J. Am. Chem. Soc.*, 1984, **106**, 6638.
32. A. L. Myers and P. A. Monson, *Langmuir*, 2002, **18**, 10261.
33. O. Talu and A. L. Myers, *AIChE J.*, 2001, **47**, 1160.
34. S. E. Wenzel, M. Fischer, F. Hoffmann and M. Fröba, *Inorg. Chem.*, 2009, **48**, 6559.
35. J. Jiang and S. I. Sandler, *J. Am. Chem. Soc.*, 2005, **127**, 11989.
36. A. Gupta, S. Chempath, M. J. Sanborn, L. A. Clark and R. Q. Snurr, *Mol. Simul.*, 2003, **29**, 29.
37. V. S. P. K. Neti, X. Wu, M. Hosseini, R. A. Bernal, S. Deng and L. Echegoyen, *CrystEngComm*, 2013, **15**, 7157.
38. O. M. Yaghi, M. O'Keeffe, N. W. Ockwig, H. K. Chae, M. Eddaoudi and J. Kim, *Nature*, 2003, **423**, 705.
39. A. P. Cote, H. M. El-Kaderi, H. Furukawa, J. R. Hunt and O. M. Yaghi, *J. Am. Chem. Soc.*, 2007, **129**, 12914.
40. J. L. Mendoza-Cortes, T. A. Pascal and W. A. Goddard, *J. Phys. Chem. A*, 2011, **115**, 13852.
41. Z. Xiang, D. Cao, W. Wang, W. Yang, B. Han and J. Lu, *J. Phys. Chem. C*, 2012, **116**, 5974.
42. J. R. Hunt, C. J. Doonan, J. D. LeVangie, A. P. Cote and O. M. Yaghi, *J. Am. Chem. Soc.*, 2008, **130**, 11872.
43. J. Lan, D. Cao, W. Wang, T. Ben and G. Zhu, *J. Phys. Chem. Lett.*, 2010, **1**, 978.
44. E. Klontzas, E. Tylianakis and G. E. Froudakis, *Nano Lett.*, 2010, **10**, 452.
45. S. Wan, J. Guo, J. Kim, H. Ihee and D. Jiang, *Angew. Chem., Int. Ed.*, 2008, **47**, 8826.
46. E. L. Spitler, B. T. Koo, J. L. Novotney, J. W. Colson, F. J. Uribe-Romo, G. D. Gutierrez, P. Clancy and W. R. Dichtel, *J. Am. Chem. Soc.*, 2011, **133**, 19416.
47. Y. J. Choi, J. H. Choi, K. M. Choi and J. K. Kang, *J. Mater. Chem.*, 2011, **21**, 1073.
48. X. Feng, L. Chen, Y. Dong and D. Jiang, *Chem. Commun.*, 2011, **47**, 1979.
49. X. Feng, L. Liu, Y. Honsho, A. Saeki, S. Seki, S. Irle, Y. Dong, A. Nagai and D. Jiang, *Angew. Chem., Int. Ed.*, 2012, **51**, 2618.
50. S. Wan, F. Gándara, A. Asano, H. Furukawa, A. Saeki, S. K. Dey, L. Liao, M. W. Ambrogio, Y. Y. Botros, X. Duan, S. Seki, J. F. Stoddart and O. M. Yaghi, *Chem. Mater.*, 2011, **23**, 4094.

51. A. Nagai, X. Chen, X. Feng, X. Ding, Z. Guo and D. Jiang, *Angew. Chem., Int. Ed.*, 2013, **52**, 3770.
52. E. L. Spitler and W. R. Dichtel, *Nat. Chem.*, 2010, **2**, 672.
53. X. Ding, J. Guo, X. Feng, Y. Honsho, J. Guo, S. Seki, P. Maitarad, A. Saeki, S. Nagase and D. Jiang, *Angew. Chem., Int. Ed.*, 2011, **50**, 1289.
54. X. Ding, L. Chen, Y. Honsho, X. Feng, O. Saenpawang, J. Guo, A. Saeki, S. Seki, S. Irle, S. Nagase, V. Parasuk and D. Jiang, *J. Am. Chem. Soc.*, 2011, **133**, 14510.
55. E. L. Spitler, J. W. Colson, F. J. Uribe-Romo, A. R. Woll, M. R. Giovino, A. Saldivar and W. R. Dichtel, *Angew. Chem., Int. Ed.*, 2012, **51**, 2623.
56. V. S. P. K. Neti, X. Wu, S. Deng and L. Echegoyen, *CrystEngComm*, 2013, **15**, 6892.
57. X. Feng, Y. Dong and D. Jiang, *CrystEngComm*, 2013, **15**, 1508.
58. F. J. Uribe-Romo, C. J. Doonan, H. Furukawa, K. Oisaki and O. M. Yaghi, *J. Am. Chem. Soc.*, 2011, **133**, 11478.
59. M. G. Rabbani, A. K. Sekizkardes, Z. Kahveci, T. E. Reich, R. Ding and H. M. El-Kaderi, *Chem. – Eur. J.*, 2013, **19**, 3324.
60. F. J. Uribe-Romo, J. R. Hunt, H. Furukawa, C. Klock, M. O'Keeffe and O. M. Yaghi, *J. Am. Chem. Soc.*, 2009, **131**, 4570.
61. K. C. Kim, D. Yu and R. Q. Snurr, *Langmuir*, 2013, **29**, 1446.
62. M. Zhang, M. Bosch, T. Gentle, III and H.-C. Zhou, *CrystEngComm*, 2014, **16**, 4069.
63. T.-F. Gao and H. Zhang, *Struct. Chem.*, 2014, **25**, 503.
64. P. Srepusharawoot, E. Swatsitang, V. Amornkitbamrung, U. Pinsook and R. Ahuja, *Int. J. Hydrogen Energy*, 2013, **38**, 14276.
65. D. Cao, J. Lan, W. Wang and B. Smit, *Angew. Chem., Int. Ed.*, 2009, **48**, 4730.
66. J. Lan, D. Cao and W. Wang, *Langmuir*, 2010, **26**, 220.
67. J. Lan, D. Cao and W. Wang, *J. Phys. Chem. C*, 2010, **114**, 3108.
68. J. Lan, D. Cao, W. Wang and B. Smit, *ACS Nano*, 2010, **4**, 4225.
69. X. Zou, G. Zhou, W. Duan, K. Choi and J. Ihm, *J. Phys. Chem. C*, 2010, **114**, 13402.
70. J.-H. Guo, H. Zhang, Z.-P. Liu and X.-L. Cheng, *J. Phys. Chem. C*, 2012, **116**, 15908.
71. F. Li, J. Zhao, B. Johansson and L. Sun, *Int. J. Hydrogen Energy*, 2010, **35**, 266.
72. R. Babarao, S. Dai and D.-E. Jiang, *Langmuir*, 2011, **27**, 3451.
73. D. N. Bunck and W. R. Dichtel, *Angew. Chem., Int. Ed.*, 2012, **51**, 1885.
74. D. N. Bunck and W. R. Dichtel, *Chem. Commun.*, 2013, **49**, 2457.
75. J. Zhao and T. Yan, *RSC Adv.*, 2014, **4**, 15542.
76. J. Hu, J. Zhao and T. Yan, *J. Phys. Chem. C*, 2015, **119**, 2010.
77. R. W. Tilford, S. J. Mugavero, P. J. Pellechia and J. J. Lavigne, *Adv. Mater.*, 2008, **20**, 2741.
78. A. Nagai, Z. Guo, X. Feng, S. Jin, X. Chen, X. Ding and D. Jiang, *Nat. Commun.*, 2011, **2**, 536.

79. H. Xu, X. Chen, J. Gao, J. Lin, M. Addicoat, S. Irle and D. Jiang, *Chem. Commun.*, 2014, **50**, 1292.

80. N. Huang, X. Chen, R. Krishna and D. Jiang, *Angew. Chem., Int. Ed.*, 2015, **54**, 2986.

81. X. Peng, X. Cheng and D. P. Cao, *J. Mater. Chem.*, 2011, **21**, 11259.

82. E. I. Volkova, A. V. Vakhrushey and M. Suyetin, *Int. J. Hydrogen Energy*, 2014, **39**, 8347.

83. L. J. Murray, M. Dinca and J. R. Long, *Chem. Soc. Rev.*, 2009, **38**, 1294.

84. H. Wu, W. Zhou and T. Yildirim, *J. Am. Chem. Soc.*, 2007, **129**, 5314.

85. M. Latroche, S. Surble, C. Serre, C. Mellot-Draznieks, P. L. Llewellyn, J.-H. Lee, J.-S. Chang, S. H. Jhung and G. Ferey, *Angew. Chem., Int. Ed.*, 2006, **45**, 8227.

86. E. Klontzas, E. Tylianakis and G. E. Froudakis, *J. Phys. Chem. C*, 2009, **113**, 21253.

87. J.-T. Yu, Z. Chen, J. Sun, Z.-T. Huang and Q.-Y. Zheng, *J. Mater. Chem.*, 2012, **22**, 5369.

88. J. L. Mendoza-Cortes, W. A. Goddard, H. Furukawa and O. M. Yaghi, *J. Phys. Chem. Lett.*, 2012, **3**, 2671.

89. Z. Li, X. Feng, Y. Zou, Y. Zhang, H. Xia, X. Liu and Y. Mu, *Chem. Commun.*, 2014, **50**, 13825.

90. Z. Kahveci, T. Islamoglu, G. A. Shar, R. Ding and H. M. El-Kaderi, *CrystEngComm*, 2013, **15**, 1524.

91. H.-L. Jiang, D. Feng, T.-F. Liu, J.-R. Li and H.-C. Zhou, *J. Am. Chem. Soc.*, 2012, **134**, 14690.

92. W. Lu, J. P. Sculley, D. Yuan, R. Krishna, Z. Wei and H.-C. Zhou, *Angew. Chem., Int. Ed.*, 2012, **51**, 7480.

93. W. Lu, D. Yuan, J. Sculley, D. Zhao, R. Krishna and H.-C. Zhou, *J. Am. Chem. Soc.*, 2011, **133**, 18126.

94. C. J. Doonan, D. J. Tranchemontagne, T. G. Glover, J. R. Hunt and O. M. Yaghi, *Nat. Chem.*, 2010, **2**, 235.

Conjugated Microporous Polymers

ROBERT DAWSON[a] AND ABBIE TREWIN*[b]

[a] Department of Chemistry, University of Bath, Claverton Down, Bath BA2 7AY, UK; [b] Department of Chemistry, Lancaster University, Lancaster LA1 4YB, UK
*Email: a.trewin@lancaster.ac.uk

7.1 Introduction

Microporous materials have important potential applications for many fields of science, including gas storage, heterogeneous catalysis and chemical separations.[1] Examples of microporous networked materials include metal organic frameworks (MOFs),[1a,2] covalent organic frameworks (COFs),[3] zeolites[4] and microporous organic polymers (MOPs).[5]

MOPs exhibit high microporosity, are constructed from lightweight elements and feature strong covalent linkages between the repeating building blocks. MOPs exhibit a range of synthetic diversity resulting in a number of different MOPs being developed. These include hyper-crosslinked polymers (HCPs),[5] polymers of intrinsic microporosity (PIMs),[6] porous aromatic frameworks (PAFs)[7] and conjugated microporous polymers (CMPs).[8]

CMPs were first reported in 2007 within a publication by Cooper *et al.*[8a] In this seminal paper, the synthesis of CMP-1, -2, -3 and -4 was described and their resultant porosity properties reported. The potential that CMP materials offer for structure control and their relevance to organic electronics was immediately apparent. Since then, several hundred CMP

Monographs in Supramolecular Chemistry No. 17
Porous Polymers: Design, Synthesis and Applications
By Shilun Qiu and Teng Ben
© The Royal Society of Chemistry 2016
Published by the Royal Society of Chemistry, www.rsc.org

materials have been synthesised with surface areas of more than 1000 m^2 g^{-1} and increasingly complex chemical functionality.

For CMPs, the material building blocks are linked in a manner that allows π-conjugation and form three-dimensional (3-D) networks. There are a large number of synthetic strategies for forming linkages that allow π-conjugation, resulting in a great range of synthetically and functionally diverse CMP materials. The CMP classification is restricted to those materials that cannot be classified as crystalline (these can be classified as COFs, *e.g.*, those that are imine-linked) and exhibit microporosity.

The synthetic diversity available results in a wide range of materials with extremely diverse functionality that can be placed within the CMP classification. Some materials can be placed into multiple classifications. For example, PAF-5, a porous aromatic framework constructed from 3-D linkages between phenyl rings, has π-conjugation that extends throughout the framework, is amorphous and is microporous, therefore allowing it to be considered a CMP. For some materials with the exact same network topology and chemistry, the classification depends upon the synthetic route. For example, covalent triazine networks (CTN) synthesised under microwave, ionothermal and acid-catalysed conditions can exhibit varying degrees of order within the framework packing [covalent triazine framework-1 (CTF-1),[9] P1 and P1M].[10] Arguably, only the amorphous analogue can be considered as a CMP, with the crystalline version being classified as a COF.

Detailed structural information can be obtained from experimental data for crystalline materials, often with exact knowledge of respective atomic positions within the material framework. For CMPs, which are mostly highly amorphous, we have little experimental data that can aid in the construction of a representative atomistic model. Even when there are some features of order, such as peaks in powder X-ray diffraction (PXRD) plots, it is very challenging to utilise this information to rationalise the structure in a meaningful way. As a result, very little is known about the atomistic structure and packing of the CMP framework.

MOPs, including CMPs, have been shown to be very robust with good physicochemical stabilities; for example, PAF-1 can be boiled in acid and still retains its structural properties, showing exceptional chemical stability.[7,11] In contrast, some studies suggest that MOFs and COFs do not exhibit physicochemical stability; for example, chemical decomposition of COF-1 can occur in air.

Ultra high surface areas [Brunauer–Emmett–Teller (BET) surface areas (SA_{BET}) over 6000 m^2 g^{-1}] have been exhibited by crystalline materials, including MOFs and COFs.[12] PAF-1 has the highest surface area of any MOP at 5600 m^2 g^{-1},[7] but as yet, MOPs do not generally reach the ultra high surface areas routinely achieved by the highly crystalline MOFs and COFs. CMPs do not reach these ultra high surface areas as yet, although surface areas between 2000 and 3000 m^2 g^{-1} have been achieved. For example, CTF-1 has been synthesised with surface areas up to 3000 m^2 g^{-1}, but harsh conditions and long reaction times are required.[9,13]

The unique structural properties that CMPs possess of high surface area combined with extended π-conjugation result in a range of applications and great potential for exploring novel functionality. CMPs exhibit excellent performance in gas adsorption and heterogeneous catalysis, and their extended π-conjugation results in very interesting electronic structure properties opening up potential for light manipulation, for example, light emitting or light harvesting, and electric storage devices, for example, as supercapacitors.

Here, we review CMPs and assess the synthetic strategies, structural properties and design concepts that have thus far been adopted. The future direction and applications of CMPs will also be discussed and potential new strategies will be explored to incorporate new functionality and improve our ability to control the structural properties obtained.

7.2   Synthesis

The synthesis of conjugated microporous polymers requires the use of chemistries that extend the π-systems of their building blocks. Many of these chemistries include the use of group 10 transition metal catalysed carbon–carbon coupling reactions. Such reactions include palladium catalysed Heck,[14] Sonogashira–Hagihara[15] and Suzuki[16] reactions, and the nickel catalysed Yamamoto[17] reaction. These chemistries are amongst the most popular routes to synthesise MOPs due to their versatility and functional group tolerance. These synthetic strategies result in materials that include some of the highest reported surface areas of any porous material.[17b,18] Another popular route to CMPs is the use of trimerisation reactions, such as the cobalt $(Co_2(CO)_8)$ catalysed cyclotrimerisation of alkyne groups[19] and the acid catalysed cyclotrimerisations of nitriles[20] or carbonyls.[21] Schiff-base chemistry has also been used to produce imine-linked conjugated networks.[22] Oxidative coupling using iron(III) chloride has also been used to produce CMP networks from thiophenes[23] and carbazoles,[24] as well as mixtures of the two.[25] Figure 7.1 shows these chemistries and the resulting networks that are formed *via* each route.

The majority of CMPs are synthesised using conventional heating of the monomers within a solvent. However, a number of other routes to CMPs have been reported, for example, using solvothermal techniques, microwave heating and electropolymerisation.

In fact, the same nominal materials can be synthesised *via* a number of different approaches. For example, the cyclotrimerisation of nitrile groups (such as 1,4-dicyanobenzene to produce CTF-1) can be achieved *via* three different synthetic methods. The first report of this material used an ionothermal synthesis in molten zinc chloride. At temperatures of 400 °C in one equivalent of $ZnCl_2$ the material produced was shown to have some degree of order resembling a stacked AAA eclipsed formation and a surface area of 791 m^2 g^{-1}.[20a] However, by changing the synthesis conditions to higher temperatures and longer reaction times an even higher surface area

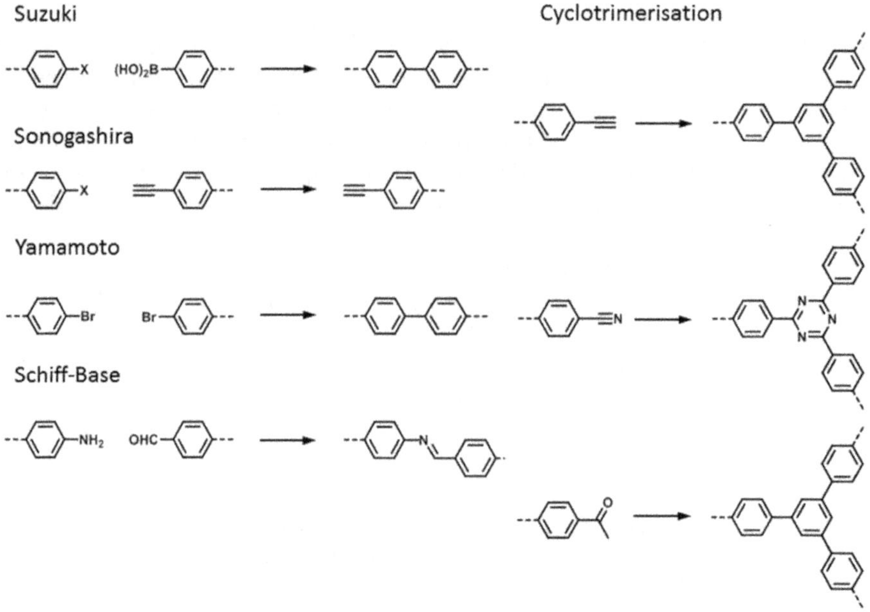

Figure 7.1 Schematic showing the synthetic reactions for CMPs.

material could be obtained (up to 3270 m^2 g^{-1}) at the expense of short range order.[26] The same reaction can also be carried our using trifluoromethanesulfonic acid at room temperature to yield a non-porous, non-crystalline material, while using trifluoromethanesulfonic acid and microwave heating yields an ordered material with no porosity.[20b] The trifluoromethanesulfonic acid with microwave approach has, however, yielded ordered porous materials from other monomers, which are amorphous when synthesised using ionothermal conditions.[20b]

CMPs, for example, synthesised from carbazoles, can also be polymerised using electrochemical methods. The thickness of the films deposited onto the electrodes can be varied simply by changing the number of scan cycles.[27] This method of network formation is particularly useful for the exploitation of the conjugated structures of these materials, hence their use in electronic devices. Indeed, solar cells containing CMPs have been reported,[28] as has the use of such materials for electrochemical sensing.[29]

The morphology of CMPs can be directed using a number of techniques. For example, the use of oil-in-water emulsions can be used to produce CMP spheres when the oil phase is polymerised.[30] Hollow spheres can also been produced by polymerisation around silica particles and etching the silica.[31] Similarly, bulk polymerisation around silica nanoparticles and their subsequent etching produces a network containing a hierarchical pore structure of larger pores from the silica and the micropore structure of the bulk CMP.[32] Polymerisation around a silica aerogel also produces a network containing both meso- and micro-pores.[33]

7.3 Structure

7.3.1 Basic Structure

CMPs are constructed from monomeric building blocks, directly analogous to secondary build units of crystalline framework materials, including MOFs and COFs. The building blocks can be classified into C_n types, where n describes the number of reactive groups that can form polymer links. Self-condensation requires building blocks that have a minimum of three reactive groups, *i.e.*, type C_3, to ensure a 3-D framework is formed. Cross-coupling can mean that building blocks with two reactive groups, *i.e.*, type C_2, can be used when in combination with building blocks with three or more reactive groups. Figure 7.2 shows an example of a C_2, C_3, C_4 and C_6 CMP building block.

Once the building blocks have reacted together to form a network, the structure can be defined in terms of network nodes and struts. The nodes are the connection points within the network. They can be an integral part of the building block, for example, the phenyl ring of the C_3 building block example in Figure 7.2, or be formed as part of the network formation process, for example, a triazine ring. Linking the nodes together are the struts. Struts can be an integral part of the building block, for example, in CTF-1, or be formed as part of the network formation process, for example, in a Yamamoto reaction with the C_3 building block in Figure 7.2.

The 3-D nature of the node–strut topology can be described by their respective nodal dimensionality. The nodal dimensionality is defined as a measure of how close the structure is to being 2-D (all reactive groups of the node lie within a two-dimensional plane) or 3-D (the reactive groups do not lie within a two-dimensional plane but, at its extreme, extend to three-dimensional space). The dimensionality is the average of the two torsional angles between two linking nodes. The ideal two-dimensionality is $0°$ and three-dimensionality is $90°$. Figure 7.3 shows pictorially the ideal 2-D and

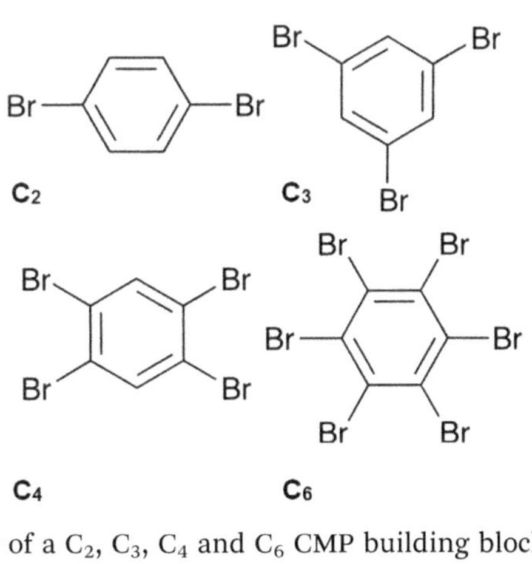

Figure 7.2   Example of a C_2, C_3, C_4 and C_6 CMP building block.

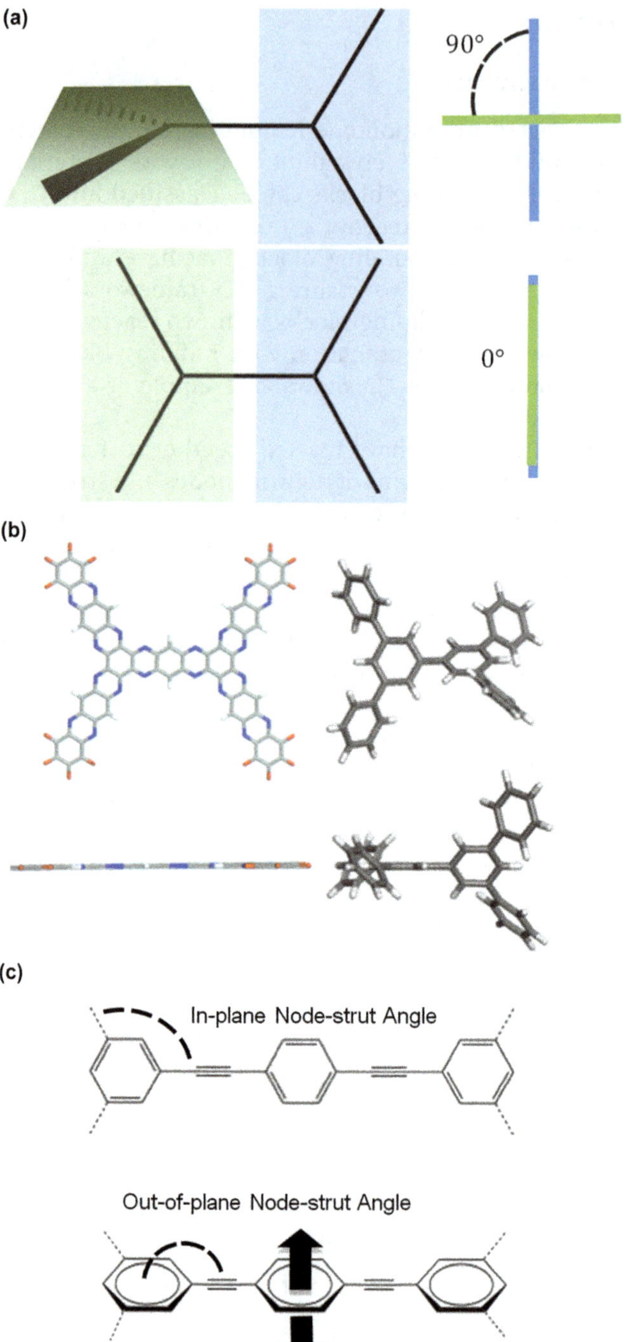
(a)
90°
0°
(b)
(c)
In-plane Node-strut Angle
Out-of-plane Node-strut Angle
Strut bend

ideal 3-D nodal dimensionality for a C_3–C_3 node–strut topology. The nodal dimensionality can be assessed by analysis of fragment node–strut–node models that have had their structure optimised through computational calculation at an appropriate level of theory. It must be noted that the resultant structures are a thermodynamic product and not necessarily representative of the range of structures within an amorphous kinetic product. Figure 7.3 shows the node–strut topology and the nodal dimensionality for aza-CMP, with nodal dimensionality close to $0°$, and for POP-3, with nodal dimensionality close to $90°$. It is difficult to assess the nodal dimensionality of the node–strut topology without atomistic models being built that have had the geometry of the respective node and strut optimised.

7.3.2 Structural Order

CMP frameworks are kinetic products and as such it is unlikely that the thermodynamic product will be achieved for each node–strut within the framework. A kinetic product is likely to be disordered due to the inherent flexibility of the CMP nodes and struts leading to a large number of possible configurations available within the framework structure. Each unique combination of node and strut will have an inherent flexibility; different struts can result in different inherent flexibilities when combined with differing nodes. This flexibility arises from the ability to deviate from the thermodynamically favourable node–strut structure. This occurs through deviation of the node–strut angles, out-of-nodal-plane bending, the nodal dimensionality and from bending of the struts themselves. These are shown in Figure 7.3(c).

The more intrinsic flexibilities the node and strut have, the more randomness that is introduced into the framework. Very stiff resultant node–struts that do not have intrinsic flexibility can introduce some degree of order within the network due to the strong structure-directing effect of the stiff building units.

Exploring the flexibility that is available to a CMP node–strut is challenging and involves sampling of the different potential structures that are available to each CMP node–strut. This can be achieved through cluster building, molecular dynamic (MD) simulation and Monte Carlo simulation.

7.3.3 Structural Packing

CMPs pack to maximise van der Waals interactions between the polymer chains. This packing can be frustrated by topological considerations

---

Figure 7.3   (a) Nodal dimensionality of a C_3–C_3 node–strut topology. Top: an idealised 3-D nodal dimensionality. Bottom: an idealised 2-D nodal dimensionality. (b) Examples of node and strut topologies. Left: aza-CMP with nodal dimensionality close to $0°$, top and side view. Right: POP-3 with nodal dimensionality close to $90°$, top and side view. (c) Schematic showing the degrees of freedom available to a CMP through the in-plane and out-of-plane node–strut angle and strut bend.

imposed by the node–strut topology and by the degree of network inter-penetration available to the framework.

Within a singular framework, stiff networks are less able to respond to the local environment by adapting their molecular configuration than more flexible polymers. This means that, for the individual framework, a stiff polymer is more likely to be open and the nodes well dispersed. A flexible polymer will be able to pack more efficiently, even if no network inter-penetration is taken into account.

Network interpenetration is energetically favourable as it increases the van der Waals interactions between frameworks and therefore allows space to be packed more efficiently. Frameworks where the node-to-node distance is comparatively large compared to the width of the strut will have space for network interpenetration to occur. Those that have comparatively short struts may inhibit network interpenetration simply due to geometric space considerations.

The flexibility of the network can also directly influence the degree of network interpenetration. CMPs with long flexible node–struts are able to respond to the local environment through bending of the strut and twisting around the node to allow individual frameworks to find space to pack efficiently in to.

7.4   Structure and Porosity Analysis

For crystalline materials that have a repeating, defined unit cell, we can pin point the exact atomistic positions within the crystal lattice with Å-scale precision using techniques such as single crystal X-ray diffraction. Deter-mining the absolute atomistic structure of an amorphous material in this way is not possible. Furthermore, structural studies are hindered by the insolubility of the CMP material. There is very little experimental data that can aid in determining the atomistic structure and packing, but we can unambiguously characterise the chemical and electronic structure through analytical and property characterisation techniques. This can result in an in-depth understanding of the chemical structure of the local environment within the framework. We can use this information to build representative atomistic models of the amorphous structure to rationalise the structure and packing on a larger scale. Here, we describe the main techniques used to elucidate the structural, chemical, electronic and porosity properties of CMPs and relate to examples.

7.4.1   Structure Analysis

Basic analysis, including infrared spectroscopy, elemental analysis and X-ray photoelectron spectroscopy, can determine the composition, bonding and percentage of unreacted end groups. Solid-state ^{1}H and ^{13}C nuclear magnetic resonance (NMR) spectroscopy is an increasingly important technique, which, when combined with X-ray diffraction and scattering analysis, can result in clear rationalisation of the network structure.

Solid-state NMR (ssNMR) is a useful and widely used tool to determine the structures of CMPs. Due to the amorphous nature and insolubility of most CMP networks, the precise structure of CMPs is difficult to determine. Solid-state NMR can give information about the bonding within the networks and can be useful to ascertain the incorporated functional groups. Due the large amount of carbon and hydrogen atoms in the networks, and the relatively quick collection times, ^{1}H–^{13}C cross-polarisation magic angle spinning (CP/MAS) NMR is widely employed. More quantitative data can, however, be obtained through the use of proton decoupled carbon NMR (^{13}C[^{1}H]) in order to determine the relative abundance of different carbon environments. Further to ^{13}C NMR, if other NMR active nuclei (*e.g.*, ^{14}N) are present in the network these can also be used to investigate their environments.[34]

The majority of CMPs have a large number of aromatic carbons and these can generally be seen in the ssNMR spectra at around 120–140 ppm. ssNMR is particularly useful in the determination of functional groups, for example, the alkyne bonds of Sonogashira synthesised CMPs can clearly be observed at around 91 ppm. The high sensitivity of the technique has also been used in order to gain further insight into the formation of such networks.[35] The different resonances of the reacted and unreacted alkyne groups have been followed by ssNMR, and showed that further crosslinking continues after the initial gelation of the network, explaining the increase in SA$_{\mathrm{BET}}$ over longer reaction times. Integration of the different terminal and internal group resonances can give insights into the degree of crosslinking within a network, with low crosslinking densities generally leading to low surface area materials.[35]

The incorporation of functionalities, such as catalytic centres, into the backbone of CMPs can be observed due to their characteristic resonances. For example, the incorporation of Tröger's base into a CMP network can easily be determined by the –CH$_2$– resonances at 59 and 67 ppm.[36] This technique is particularly useful in networks in which one monomer can polymerise with itself (homocoupling) without the incorporation of the functional monomer, *e.g.*, in the synthesis of networks *via* Sonogashira coupling, where possible alkyne–alkyne coupling can happen. Successful incorporation of both monomers into the network can therefore be determined using ssNMR.

X-Ray diffraction and scattering analysis, including PXRD, small angle X-ray scattering (SAXS) and wide angle X-ray scattering (WAXS), can give valuable information about the local environment within the framework and highlight any degree of order that may be present. The PXRD patterns of highly crystalline materials, including MOFs, exhibit sharp Bragg reflection peaks. Many COFs approximate to an ideal crystalline form, but imperfections, such as layer slippage, break the crystalline order and therefore exhibit peak broadening of the Bragg diffraction peaks that would be

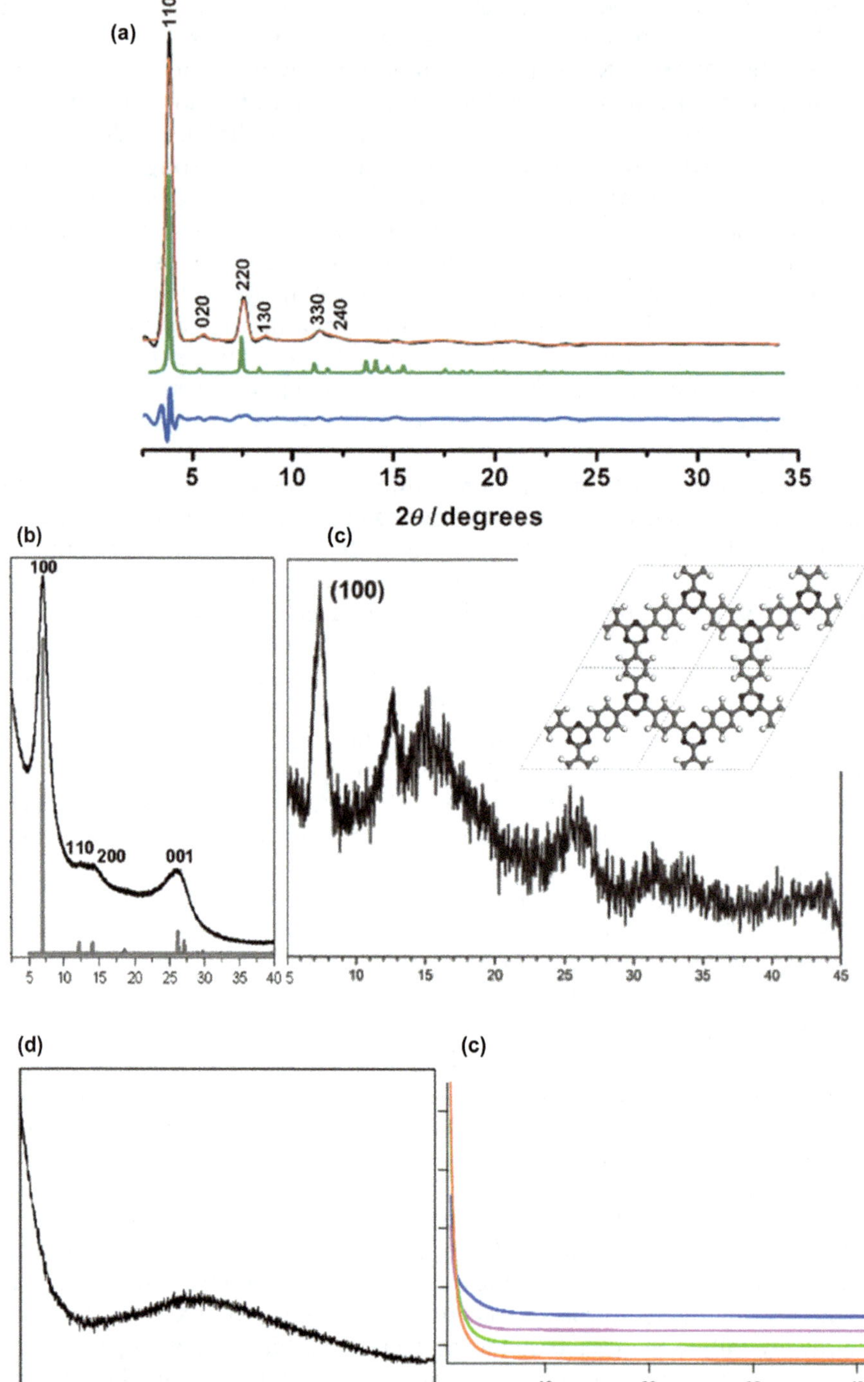

observed for an idealised structure. For truly amorphous materials, we see no peaks within the diffraction pattern. If some degree of local order or defined repeating motif is present within the structure, an amorphous halo that can be described as a broad hump in the PXRD pattern is observed. This halo does not centre on any Bragg diffraction peaks of any possible idealised crystalline form available, but is instead related to repeating molecular or void distances within the framework. These distances, obtained from the *d*-spacing, can sometimes be correlated to the node–node distance, molecular widths or to pore voids. However, these peaks can be difficult to distinguish from Bragg diffraction peaks if the idealised unit cell has the same dimensions as the molecular repeat units, as is often the case for materials with idealised layered structures, for example, CTF-1 and P1M.

Figure 7.4 shows the PXRD for a crystalline material that shows Bragg peak broadening (ILCOF-1),[37] a material with some degree of long range order (CTF-1),[9] a material with the same framework structure as CTF-1 but synthesised *via* a different route with a much reduced degree of order (P1M),[10] an amorphous material with a halo that can be related to the molecular and void structure (PAF-5)[38] and a truly amorphous material with no sign of any long or short range order (TPE-CMP).[39] For ILCOF-1, Figure 7.4(a), the peaks are well defined and clearly identifiable to the idealised pattern, shown below it. The peaks in the pattern for CTF-1, Figure 7.4(b), are broadened but are still clearly identifiable to the idealised pattern. For P1M, Figure 7.4(c), one peak is relatively sharp in comparison the rest of the pattern, identified as the (100) peak from the idealised pattern, but the remaining peaks are not clear and are not easily identifiable. It is worth noting that for CTF-1 and P1M, these peaks can be directly related to molecular distances, including the node–node distance (~ 0.8 nm), second neighbour distances (~ 1.4 nm),

---

Figure 7.4 PXRD patterns for materials with varying degrees of order. (a) ILCOF-1: the experimental pattern (top) compared to the idealised pattern (bottom). The experimental pattern shows sharp peaks that are directly related to Bragg diffraction peaks. Reprinted with permission from M. G. Rabbani, A. K. Sekizkardes, Z. Kahveci, T. E. Reich, R. Ding, H. M. El-Kaderi, *Chem. – A Eur. J.*, 2013, **19**, 3324–3328. (b) CTF-1: broadened peaks that centre on Bragg reflection peaks of an idealised structure. Reprinted with permission from P. Kuhn, M. Antonietti, A. Thomas, *Angew. Chem. Int. Ed.*, 2008, **47**, 3450–3453. (c) P1: very broad ill-defined peaks that are roughly aligned to Bragg reflection peaks. Inset shows the structure. Reprinted with permission from S. Ren, M. J. Bojdys, R. Dawson, A. Laybourn, Y. Z. Khimyak, D. J. Adams, A. I. Cooper, *Adv. Mater.*, 2012, **24**, 2357–2361. (d) PAF-5: large amorphous hump that roughly correlates to molecular distances within the net. Reprinted with permission from H. Ren, T. Ben, F. Sun, M. Guo, X. Jing, H. Ma, K. Cai, S. Qiu, G. Zhu, *J. Mater. Chem.*, 2011, **21**(28), 10348–10353. (e) TPE-CMP: truly amorphous, no peaks present in the PXRD pattern. Reprinted with permission from Y. Xu, L. Chen, Z. Guo, A. Nagai, D. Jiang, *J. Am. Chem. Soc.*, 2011, **133**, 17622–17625. Copyright 2014 American Chemical Society.

third neighbour distances (~ 1.6 nm) and π–π stacking (~ 0.35 nm), which could all be present in an amorphously organised framework with very well defined repeating units similar to the continuous random network model described by Bates *et al.*[40] The PXRD for PAF-5, Figure 7.4(d), shows a clear, very broad amorphous halo centred on a *d*-spacing of ~ 4.5 Å and ranging between 9.0 Å and 2.5 Å. This could be related to a combination of molecular widths, network distances and imperfections, π–π stacking, and void structure within the framework. The PXRD for TPE-CMP, Figure 7.4(e), shows no peaks or amorphous halos, indicating a truly amorphous material. Similar analysis of SAXS and WAXS patterns can be used to elucidate structural information. SAXS can be used to probe smaller scattering angles and WAXS wider scattering angles.

7.4.2 Porosity Analysis

Due to the inefficient packing of the rigid struts, CMPs have an open structure even after removal of all solvents. This open structure leads to high surface areas and pore volumes. Surface areas of CMPs are most often reported as Brunauer–Emmett–Teller (BET) surface areas, which takes into account the adsorption of multiple adsorbate layers.

CMPs display a range of BET surface areas, most often in the range of 500 to 1500 m^2 g^{-1}. For example, the Sonogashira linked CMP-1 network has a surface area of around 830 m^2 g^{-1}.[15] Higher surface area CMPs tend to have more contorted or 3-D monomers, such as tetrahedral monomers;[41] however, such monomers can impact on the conjugation length of the networks. Another synthetic strategy to yield higher surface area networks is the choice of coupling chemistry used. For example, the synthesis of networks from tetraphenylmethane building blocks yields networks with a wide range of surface areas. Here, Yamamoto coupling chemistry leads to the highest surface area network (*e.g.*, PAF-1; SA_{BET} 5640 m^2 g^{-1}),[17b] while Sonogashira coupling of tetrakis(4-iodophenyl)methane and tetrakis(4-ethynylphenyl)methane leads to a network with a surface area of 1917 m^2 g^{-1}.[41] Click chemistry yields networks with between 1100 and 1440 m^2 g^{-1}.[41,42] These are shown in Figure 7.5.

Unlike crystalline materials, such as MOFs and COFs, which can have unimodal pore sizes, CMPs often exhibit a range of pore sizes due the their amorphous structures. The range of pore sizes means that the pores are not only in the micropore region (*i.e.*, less than 2 nm) but commonly spill over into the mesopore classification (2–50 nm). However, to still be classed as CMPs the main pore sizes have to be in the micropore range, that is that the largest pore is usually below 2 nm or that the pores are centred in the micropore region. The (nonlocal) density functional theory [(NL)DFT] method is the most common method for calculating the pore sizes; however, there are currently no commercially available DFT models developed solely to describe CMPs or MOPs in general. Instead the model chosen should be that which fits the isotherm best.

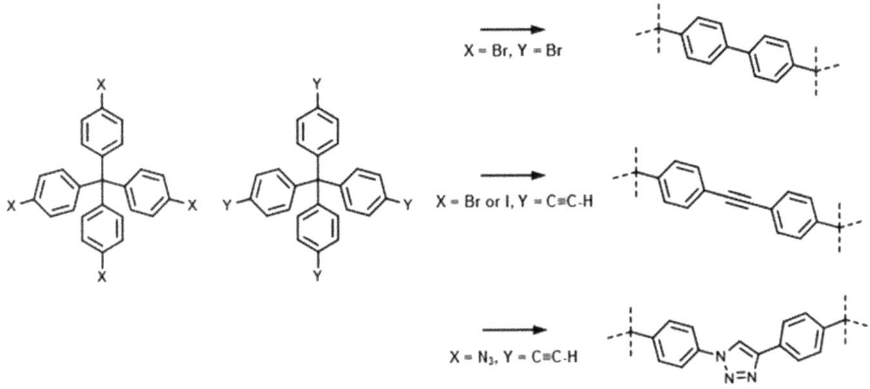

Figure 7.5 Tetrahedral monomers linked using Yamamoto (top), Sonogashira (middle) and click chemistry (bottom).

The wide range of pore sizes within CMPs leads to a number of different isotherm shapes, ranging from Type I isotherms for materials with more monodisperse pore sizes to isotherms with large hystereses. The hystereses may be either related to larger meso- and macro-pores or due to swelling effects. After synthesis the solvent in the material's pores is removed, inducing strain into the structures. This strain can be released either by the addition of another solvent, or by gases condensing into the pores under cryogenic conditions. This is shown as a large increase in the uptake of nitrogen gas at higher relative pressures.[43]

7.5   Structure Control and Functionality

7.5.1   Structure Control

Efficient packing of amorphous materials results in denser materials with reduced void volume that is not connected and hence occluded. Similarly, inefficient packing results in a more open structure with void space that is well connected in 3-D. Hence, control of the packing of CMP materials can lead to control of porosity properties. The packing of CMPs can be directly related to structural features, including the node, strut and the resulting node–strut topology. By systematically changing the node and strut combination, we can assess the influence on the packing of the material by assessing the porosity.

For MOF and COF materials, increasing the strut length results in an increase in surface area and pore volume as it results in the nodes being further apart in three dimensions and hence a more open structure. For CMP materials, a similar approach has been attempted for the series of CMP materials, CMP-0, -1, -2, -3 and -5, with increasing respective strut lengths that are composed of phenyl and ethynyl groups but with the same trisubstituted phenyl node. The longer the length of the strut, the more inherent flexibility the strut has. Figure 7.6(a) shows the node and strut

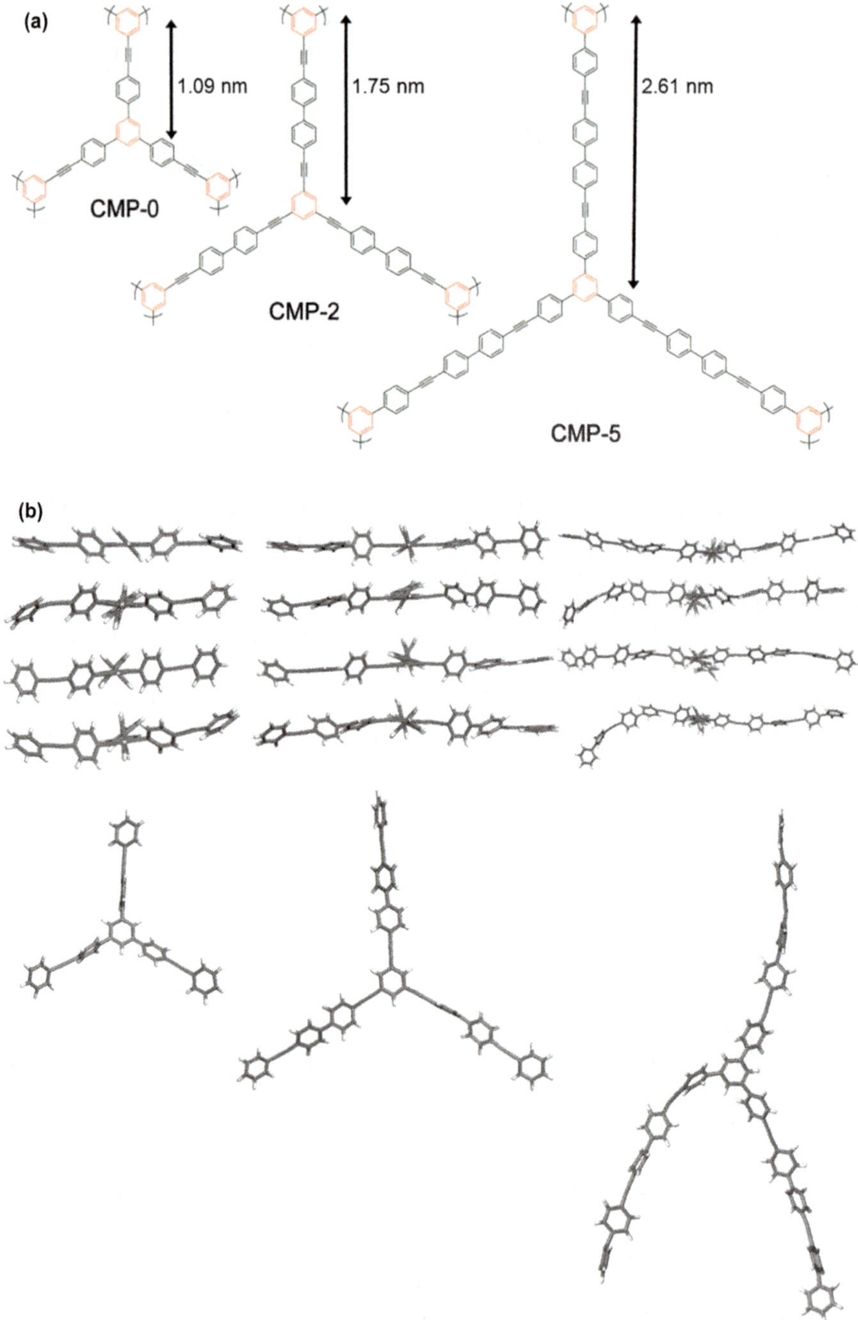

Figure 7.6   (a) Node and strut topology for CMP-0, CMP-2 and CMP-5. The node is highlighted in red and the node-to-node distance for each CMP is shown. (b) Side view of the node and strut topology of a fragment of CMP-0, CMP-2 and CMP-5 at the beginning of an MD simulation (top) and during the simulation. The top view was taken at the end of the MD simulation (bottom).

topology for CMP-0, -2 and -5. CMP-0 has short rigid struts, whereas CMP-5 has relatively long and flexible struts. This flexibility can be observed by assessing the molecular conformations that are available to the node and strut during an MD simulation. Figure 7.6(b) shows molecular models of the node and strut for CMP-0, -2 and -5 at the start, where the geometry of the structure is fully optimised, and snapshots during an MD simulation. The structure of CMP-0 does not change significantly as there is only limited conformational freedom. CMP-2 shows a greater degree of conformational freedom in the in-plane and out-of-plane angles and the strut bend, but the structure does not differ greatly to the optimised structure. For CMP-5, the final structure differs greatly to the optimised structure showing the greater degree of conformational flexibility available to it. The more flexible the strut is, the more able it is to respond to local structure and hence pack space efficiently to increase van der Waals interactions. Therefore, CMP-0 has the highest surface area and pore volume and CMP-5 has the lowest.

This effect was also observed when a mix of struts was used for statistical co-polymers for a series of co-polymer network (CPN) materials. Here, the average strut length, obtained from the ratio of the strut and their respective lengths, determines the packing and hence the porosity. This trend was also extended to include nitrogen-containing CMPs (NCMPs) that have nitrogen nodes and the same set of struts. The node-to-node distance was plotted against the porosity properties, showing a clear structural trend towards lower surface area and porosity properties for longer node-to-node distances, shown in Figure 7.7(a). We can further extend this grouping to include other CMP materials of the $C_3 + C_3$ type, including homocoupled CMP networks (HCMPs), covalent organic polymers (COPs), aza-CMP and covalent triazine-based porous polymer frameworks (CTFs). Figure 7.7(b) shows the node-to-node distance plotted against the BET surface area and shows a clear trend for lower surface areas for materials with a longer node-to-node distance.

Further structure control can be obtained through consideration of the nodal geometry. A series of polyphenylene CMPs, where the linkages for the strut to the nodal phenyl systematically adopted the *meta-*, *para-* and *ortho-* positions, have been studied.[44] It was found that this simple change in nodal geometry had profound effects upon the resulting macro structure, porosity and electronic structure properties. The *meta-* structure was found to be more open and therefore exhibited a higher surface area and greater uptake capacity, whereas the *ortho-* and *para-* structures gave stronger interactions with the adsorbed gas. The linkage geometry was also found to control the extended conjugation of the network, with the *para-* structure being the most efficient and further facilitated exciton migration.

7.5.2 Synthetic Conditions

The Sonogashira CMP networks (CMP-1 to -4) were first synthesised in a toluene and triethylamine mixture for 3 days to yield networks with surface

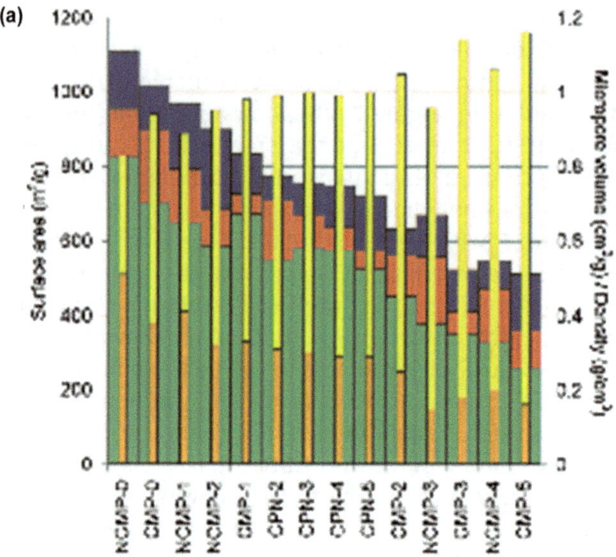

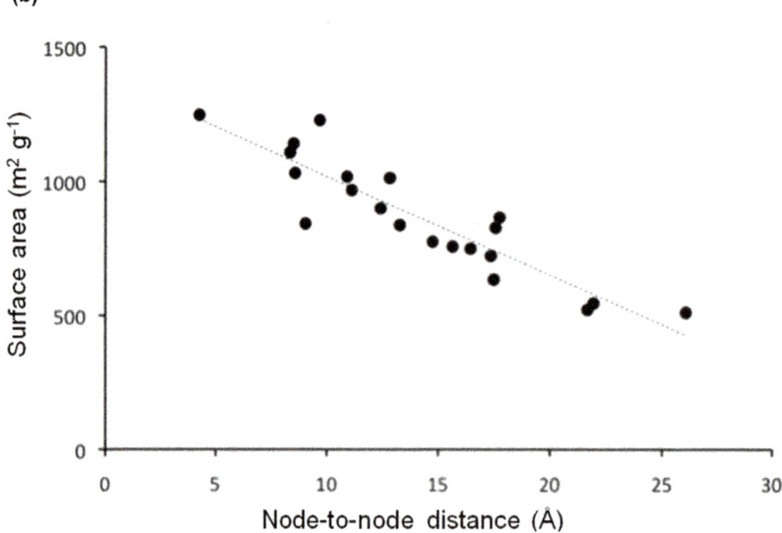

Figure 7.7   (a) Porosity properties of a series of CMP, NCMP and CPN materials plotted in order of strut length. Reprinted with permission from J.-X. Jiang, A. Trewin, F. Su, C. D. Wood, H. Niu, J. T. A. Jones, Y. Z. Khimyak, A. I. Cooper, *Macromolecules* **2009**, 42, 2658–2666. Copyright 2014 American Chemical Society. (b) Surface area *vs.* node-to-node distance for CMP materials with C_3 nodes and varying struts.

areas up to 830 m^2 g^{-1}. It was later shown that these networks could be synthesised in a range of different solvents. Most importantly, it was shown that the choice of solvent has a great effect on the porosity of the network, particularly when incorporating functional monomers. For example, a nitro-functionalised dibromobenzene yielded a low surface area ($247\ \mathrm{m^2\ g^{-1}}$) network synthesised in toluene. However, when the solvent was switched to dimethylformamide (DMF) the measured surface area was almost quadruple ($967\ \mathrm{m^2\ g^{-1}}$) that in toluene.[45] Of the four solvents in this study [dioxane, DMF, tetrahydrofuran (THF) and toluene], toluene produced on average the lowest surface area networks with the highest proportion of mesopores. In contrast, DMF showed the highest average surface areas and highest proportion of micropores (*i.e.*, isotherms with more Type I character). Further to this, the addition of catalyst was also found to be important. When the catalyst is added at the reaction temperature, more consistent and reproducible networks can be synthesised.[1a,45,46]

Temperature and reaction time have also been shown to have an important effect on the surface areas of CMPs. The ionothermal synthesis of triazine networks shows varying surface areas under different reaction temperatures. For example, CTF-1 has a surface area of $791\ \mathrm{m^2\ g^{-1}}$ at 400 °C and a reaction time of 10 hours, extending the reaction was shown to increase the surface area to $920\ \mathrm{m^2\ g^{-1}}$.[26] The surface area further increased with increasing reaction temperature to $1600\ \mathrm{m^2\ g^{-1}}$ at 500 °C, $1750\ \mathrm{m^2\ g^{-1}}$ at 600 °C and $2530\ \mathrm{m^2\ g^{-1}}$ at 700 °C. A maximum surface area of $3270\ \mathrm{m^2\ g^{-1}}$ was obtained using a two-step heating procedure. This strategy of high reaction temperatures has been used for a range of triazine networks, including those containing fluorine,[47] bipyridine[48] and tetraphenylethylene.[49]

Oxidation reactions have also been used to synthesise CMP networks. These have been carried out using iron(III) chloride linking together either thiophenes[23,50] or carbazole groups[24a,24c–e,25] to yield conjugated networks.

7.5.3 Chemical Functionality

The versatility of the different synthesis routes, as well as the mild reaction conditions of many CMPs, has led to the incorporation of a number of different functionalities for a range of potential applications. The first reports of functionalised CMPs showed the versatility of the Sonogashira reaction and showed that a number of simple functionalities could be incorporated into the backbone of CMPs.[45,46] By changing the functionalities in the backbone of the polymer the properties of the network could also be changed. One simple demonstration of the functionality of these CMPs was their hydrophobicity/hydrophilicity. Hydrophobic networks, such as those containing fluoro functionalities, have large contact angles with water and hence float on water, whereas hydrophilic hydroxyl-functionalised CMPs have small contact angles and are water dispersible.[46] Simple functionalities were also shown to have an effect on the gas uptakes of the networks. Compared to CMP-1 hydroxy, amine and carboxylic acids increased the heats of adsorption for CO_2 gas.[51]

Simple functionalities have also been used to perform catalytic transformations. Triazine networks have been used to perform cycloadditions to epoxides either using just the basic triazine units or *via* the incorporation of pyridines.[52]

Leading on from more simple functional groups, more targeted functionalities have been incorporated into a range of CMPs. One example is the incorporation of Tröger's base network synthesised *via* Sonogashira coupling. The Tröger's base was functionalised with iodo groups that were then able to react with 1,3,5-triethynylbenzene to form a network. This network had a surface area of 750 m^2 g^{-1}, similar to other Sonogashira networks, and could be used to catalyse the addition of diethyl zinc to an aldehyde.[36] Porphyrin moieties have also been incorporated into CMP networks *via* a number of routes, including Sonogashira,[53] Suzuki,[54] Yamamoto[55] and oxidative coupling[24e,56] reactions, particularly for use in oxidation catalysis,[24e,54a] as well as gas storage.[53b,56b]

A further route to incorporate different functionalities into the networks is *via* post-synthetic modification. Here, a network is synthesised that can be modified *via* a further synthetic step to change the surface of the network. A simple example of the post-synthetic modification of a CMP would be the thiol-yne reaction, where a CMP network containing alkyne groups is reacted after its formation with a thiol-containing small molecule. Urakami *et al.* first synthesised a benzothiadiazole CMP network using the Sonogashira reaction to yield a network with alkyne groups.[57] A thiol-yne click reaction was then performed on the alkynes in this network to produce a network with pendant carboxylic acid groups. Another CMP network containing amine groups has also been used to perform post-synthetic modifications.[58] Here, the amine groups were modified with anhydrides of varying chain lengths. Alternatively, the reverse reaction can also be performed using an aldehyde-containing network, which is then reacted with an amine.[59] It is to be noted that the post-synthetic modification strategy often reduces the surface areas and pore volumes of the resulting modified networks in comparison to the starting networks. The inclusion of metal nanoparticles is also another method of post-synthetic modification such as the incorporation of palladium[60] or gold[61] nanoparticles.

The post-synthetic methodology has also been used to incorporate catalytic centres. The incorporation of chiral 1,1′-bi-2-naphthols (BINOLs) into CMPs has been used to perform catalytic reactions using the alcohol groups to bind to catalytically active centres such as phosphoric acids for asymmetric transfer hydrogenations[23b,50] or titanium for diethyl-zinc additions.[62] Other metal-containing functionalities have also been incorporated into CMP networks. The coordination of metals to pyridine or phenylpyridine *via* Sonogashira reactions has also been reported for catalytic transformations such as reductive amination[63] or aza-Henry reactions, α-arylations and oxyaminations.[64]

7.5.4 Electronic and Optical Properties

Due to their extended π-conjugation throughout the framework, CMP materials have interesting electronic and optical properties.[65] CMPs are known to be luminescent and fluorescent, giving them potential applications as chemosensors, light harvesting devices and in polymer light emitting diodes (PLEDs). They have also shown great potential as super capacitance devices due to their appropriate pore sizes and ability to conduct electrons.

The emission of CMPs can be tuned by varying the constituent monomers of the networks.[66] For example, tetraphenylethylene can be co-polymerised with 1,4-benzenediboronic acid to yield a CMP network *via* Suzuki coupling that fluoresces at 560 nm. When a tetraphenylmethane monomer is introduced into the network, the emission is blue shifted, which is ascribed to a shortening of the conjugation length. The introduction of electron-rich thiophene groups redshifts the emissions. Tuning of the emissions and band gaps of pyrene-containing networks has also been reported.[66b] A network consisting of only pyrene units has an emission at 620 nm. When benzene or biphenyl co-monomers are introduced into the structure, the emissions can be shifted to shorter wavelengths. It was also shown that an alternating structure of benzene and pyrene synthesised *via* Suzuki coupling has a narrower emission than that produced by the more random polymerisation resulting from Yamamoto coupling. The use of polyphenylene-based networks was one of the first reports of the use of the conjugated structure of CMPs for light harvesting by transferring the energy from the network to coumarin 6 confined within the pores.[67] The emission of the network was quenched while intensity of the coumarin 6 emission was 21 times that when the dye itself was excited. Energy transfers in CMP networks with benzothiodiazole,[68] pyrene[69] and hexabenzocoronene[70] moieties have also be reported after the inclusion of C_{60} by observing the quenching of the C_{60} emissions.

The use of the fluorescence quenching in CMPs has also been used to detect organic vapours such as the nitro-containing compounds used in explosives. Networks synthesised using the oxidative coupling of carbazoles have been used to detect nitrotoluenes and nitrobenzene.[24d]

7.6 Applications

The applications of CMPs are built upon three key intrinsic properties: their microporosity, extended π-conjugation and synthetic versatility. The combination of these key properties leads to a wide range of applications available to CMPs with great potential for future applications yet to be discovered, particularly for energy applications.[71] A review by Xu *et al.* details the current applications of CMPs;[72] here, we summarise the main applications.

Gas sorption and storage is an area where CMPs have obvious applications. Their high microporosity coupled with synthetic versatility and ability to incorporate functionality within the pore walls gives potential strategies to increase gas uptake. At present, CMPs show promising gas uptake for hydrogen and carbon dioxide,[73] with potential for other gases, including methane and other organic molecules. POP-3 is able to take up

3.07 wt% of hydrogen.[19] The uptake of CO_2 is highly dependent upon the overall capacity of the material, which is related to the pore volume and, significantly, to the chemical functionality incorporated within the network. CMPs with electron-withdrawing triazine nodes, *e.g.*, TNCMP-2, exhibit high uptakes as CO_2 can interact strongly with the triazine and N nodes.[74] Furthermore, the enthalpy of adsorption ΔH_{ads} can also be increased by incorporation of pendent functional groups, including carboxylic acids, amines and hydroxyl groups.[75] Post-synthetic modification, for example, with Li ions, can also increase the uptake of CO_2.[76]

The wide range of potential strategies for including chemical functionality, coupled with the microporosity, lead to CMPs being ideal frameworks for which catalytically active sites can be incorporated, *e.g.*, incorporation of benzothiadiazole.[77] Key to this is the ability to incorporate organic coordination sites for metal ions directly into the CMP network. These sites include porphyrins, which can act as nodal building blocks, or aza groups within the conjugated network.[63,78] Additionally, they can act as a framework for which nanoparticles, such as Pd, can be dispersed and act as heterogeneous catalysis active sites. This also increases the volume of H_2 uptake at room temperature.[79]

CMPs are known to be luminescent[3e] and fluorescent,[8c] giving them potential applications as chemosensors,[80] light harvesting devices[81] and in PLEDs.[27a] They have also shown great potential as super capacitance devices due to their appropriate pore sizes and ability to conduct electrons.[82] In particular, the conjugated structure and the ability to incorporate different functionalities into CMPs lends them to their use as photocatalysts. Examples of photocatalysis include reductive halogenation,[83] oxidations,[84] photodegredation,[85] oxidative hydroxylation,[83a] aza-Henry reactions[64,86] and hydrogen evolution.[87]

7.7 Limitations

Compared to other microporous polymers, CMPs are some of the most versatile since there is no requirement for reversible reactions to form crystalline networks, as for COFs, or the need for contorted monomers required for PIMs. There are, however, some limitations of CMPs: for example, the common use of transition metal catalysts, such as palladium or nickel, limits their use in large scale applications (*e.g.*, carbon capture) due to the high cost. Further to this, many of the monomers, particularly those that are tetrahedral and lead to high surface areas, are also relatively high cost compared to the monomers often used for the synthesis of HCPs.

While there are now a number of reports of soluble and dispersible CMPs,[90] they are mainly produced as insoluble networks and therefore the processability of CMPs is poor in contrast to some PIMs. Until a general method to produce a wide vary of processable CMPs is developed the use of this class of porous polymer may be limited.

7.8 Future Directions

Exploring the potential for CMPs is an exciting area that is rapidly developing. This is mainly due to the exceptional ability to incorporate functionality by either pre-synthetic modification of building block monomers or by post-synthetic modification of the network.

Post-synthetic modification of CMPs opens up a wide range of new functionality that can be incorporated. A strategy recently adopted by Ratvijitvech *et al.* is to design in functionality specifically for post-synthetic modification. A CMP building block with a pendant amine group was synthesised, which was subsequently used for reaction with anhydrides to produce a series of amide-functionalised networks.[58]

This ability for functionalisation is particularly important for gases with weak interactions, for example, in the case of hydrogen storage, where chemical interactions will be an important tool in reaching the Department of Energy (DOE) 2015 target for hydrogen storage of 9 wt%, 81 kg m^{-3}, with a temperature of 253 to 323 K at a pressure of 100 atm. Potential areas for exploration include hybrid materials, where materials that utilise chemical storage of hydrogen, for example, in hydrides, is combined with physisorptive materials to potentially take advantage of the benefits of both approaches.

CMPs are yet to reach the exceptional surface areas and pore volumes of ultra porous materials. However, a similar material, PAF-1, did achieve a surface area of 5600 m^2 g^{-1}, showing that, given an appropriate synthetic strategy, these ultra high surface areas could potentially be achieved.

A relatively unexplored area for CMPs is their ability to swell. Woodward *et al.* recently showed that a MOP HCP material can reversibly swell at high pressure to take up exceptional quantities of CO_2 at high pressures.[43] CMPs also have the potential to swell in a similar manner, particularly in solvents and organic molecules such as liquid amines.[91]

Exploration of the electronic structure properties of CMPs is a particularly new area. Recently Xu *et al.* have shown that CMPs can be redox active, meaning that they have great potential for energy storage.[92] The incorporation of monomers that alter the emission spectra and band gaps of CMPs has much potential and may lead to their use in applications such as semiconductors and light harvesting.

The biological applications of CMP materials have been explored, particularly for biosensing,[27b] where their luminescent and fluorescent properties are utilised as extremely sensitive surface sensing for biomarkers. There is also potential for interfacing directly with between soft tissue and metallic devices,[88] and for use as drug delivery devices.[89]

A key limitation in the applications of CMPs is the insolubility and poor processability. Recently, Cheng *et al.* synthesised a conjugated polymer of intrinsic microporosity (C-PIMs) that was soluble and processable.[93] While it cannot be classed as a CMP due to its lack of 3-D network structure, it does open the possibility for similar strategies to be adopted for CMP materials.

7.9 Summary

CMP materials are microporous, amorphous, 3-D nets with extended π-conjugation. This extended conjugation, in combination with inherent microporosity, offers up unique potential for properties that rely upon electronic structure, including conduction, energy storage and optical properties.

CMPs are exceptionally synthetically diverse, with a number of synthetic strategies for chemical functionalisation. Chemical functionalisation will be an essential route to applications in areas including gas storage, gas separation and catalysis. As we explore the vast range of chemical functionalisation available for CMPs there is potential for materials to be discovered with entirely novel properties.

The wide range of monomers available to CMPs also enables the control of the surface areas and pore sizes *via* the use of different chemistries and varying the strut lengths.

References

1. (a) J. M. H. Thomas and A. Trewin, Amorphous PAF-1: Guiding the Rational Design of Ultraporous Materials, *J. Phys. Chem. C*, 2014, **118**(34), 19712–19722; (b) R. Dawson, A. I. Cooper and D. J. Adams, Nanoporous organic polymer networks, *Prog. Polym. Sci.*, 2012, **37**(4), 530–563; (c) L. J. Barbour, Crystal porosity and the burden of proof, *Chem. Commun.*, 2006, **11**, 1163–1168.
2. M. Eddaoudi, D. B. Moler, H. Li, B. Chen, T. M. Reineke, M. O'Keefe and O. M. Yaghi, Modular chemistry: secondary building units as a basis for the design of highly porous and robust metal-organic carboxylate frameworks, *Acc. Chem. Res.*, 2001, **34**, 319–330.
3. (a) A. P. Côté, A. I. Benin, N. W. Ockwig, M. O'Keeffe, A. J. Matzger and O. M. Yaghi, Porous, crystalline, covalent organic frameworks, *Science*, 2005, **310**(5751), 1166–1170; (b) J. R. Hunt, C. J. Doonan, J. D. LeVangie, A. P. Cote and O. M. Yaghi, Reticular synthesis of covalent organic borosilicate frameworks, *J. Am. Chem. Soc.*, 2008, 11872–11873; (c) A. P. Cote, H. M. El-Kaderi, H. Furukawa, J. R. Hunt and O. M. Yaghi, *J. Am. Chem. Soc.*, 2007, **129**, 12914–12915; (d) H. M. El-Kaderi, J. R. Hunt, J. L. Mendoza-Cortes, A. P. Cote, R. E. Taylor, M. O'Keefe and O. M. Yaghi, Designed synthesis of 3D covalent organic frameworks, *Science*, 2007, **316**, 268–272; (e) S. Wan, J. Guo, J. Kim, H. Ihee, D. Jiang and A. Belt-Shaped, Blue Luminescent, and Semiconducting Covalent Organic Framework, *Angew. Chem., Int. Ed.*, 2008, **47**, 1–5.
4. D. V. Soldatov; J. A. Ripmeester, Organic zeolites, *Nanoporous Materials IV*, 2005; vol. 156, pp. 37–54.
5. C. D. Wood, B. Tan, A. Trewin, H. Niu, D. Bradshaw, M. J. Rosseinsky, Y. Z. Khimyak, N. L. Campbell, R. Kirk, E. Stoeckel and A. I. Cooper, Hydrogen Storage in Microporous Hypercrosslinked Organic Polymer Networks, *Chem. Mat.*, 2007, **19**, 2034–2048.

6. N. B. McKeown and P. M. Budd, Polymers of intrinsic microporosity (PIMs): organic materials for membrane separations, heterogenous catalysis and hydrogen storage, *Chem. Soc. Rev.*, 2006, **35**, 675–683.

7. T. Ben, H. Ren, S. Ma, D. Cao, J. Lan, X. Jing, W. Wang, J. Xu, F. Deng, J. Simmons, S. Qiu and G. Zhu, Targeted Synthesis of a Porous Aromatic Framework with High Stability and Exceptionally High Surface Area, *Angew. Chem., Int. Ed.*, 2009, **48**(50), 9457–9460.

8. (a) J. Jiang, F. Su, A. Trewin, C. D. Wood, N. L. Campbell, H. Niu, C. Dickinson, A. Y. Ganin, M. J. Rosseinsky, Y. Z. Khimyak and A. I. Cooper, *Angew. Chem., Int. Ed.*, 2007, **46**, 1–5; (b) J.-X. Jiang, F. Su, A. Trewin, C. D. Wood, H. Niu, J. T. A. Jones, Y. Z. Khimyak and A. I. Cooper, Synthetic Control of the Pore Dimension and Surface Area in Conjugated Microporous Polymer and Copolymer Networks, *J. Am. Chem. Soc.*, 2008, **130**(24), 7710–7720; (c) J.-X. Jiang, A. Trewin, D. J. Adams and A. I. Cooper, Band gap engineering in fluorescent conjugated microporous polymers, *Chem. Sci.*, 2011, **2**(9), 1777–1781.

9. P. Kuhn, M. Antonietti and A. Thomas, *Angew. Chem., Int. Ed.*, 2008, **47**, 3450–3453.

10. S. Ren, M. J. Bojdys, R. Dawson, A. Laybourn, Y. Z. Khimyak, D. J. Adams and A. I. Cooper, Porous, Fluorescent, Covalent Triazine-Based Frameworks Via Room-Temperature and Microwave-Assisted Synthesis, *Adv. Mater.*, 2012, **24**(17), 2357–2361.

11. T. Ben and S. Qiu, Porous aromatic frameworks: Synthesis, structure and functions, *CrystEngComm*, 2013, **15**(1), 17–26.

12. H. Furukawa, N. Ko, Y. B. Go, N. Aratani, S. B. Choi, E. Choi, A. O. Yazaydin, R. Q. Snurr, M. O'Keeffe, J. Kim and O. M. Yaghi, Ultrahigh Porosity in Metal-Organic Frameworks, *Science*, 2010, **329**(5990), 424–428.

13. (a) P. Kuhn, A. Forget, D. Su, A. Thomas and M. Antonietti, From Microporous Regular Frameworks to Mesoporous Materials with Ultrahigh Surface Area: Dynamic Reorganization of Porous Polymer Networks, *J. Am. Chem. Soc.*, 2008, **130**(40), 13333–13337; (b) P. Kuhn, A. Thomas and M. Antonietti, *Macromolecules*, 2009, **42**, 319–326.

14. (a) D. Wang, L. Xue, L. Li, B. Deng, S. Feng, H. Liu and X. Zhao, Rational Design and Synthesis of Hybrid Porous Polymers Derived from Polyhedral Oligomeric Silsesquioxanes via Heck Coupling Reactions, *Macromol. Rapid Commun.*, 2013, **34**(10), 861–866; (b) D. Wang, W. Yang, S. Feng and H. Liu, Constructing hybrid porous polymers from cubic octavinylsilsequioxane and planar halogenated benzene, *Polym. Chem.*, 2014, **5**(11), 3634–3642.

15. J. X. Jiang, F. Su, A. Trewin, C. D. Wood, N. L. Campbell, H. Niu, C. Dickinson, A. Y. Ganin, M. J. Rosseinsky, Y. Z. Khimyak and A. I. Cooper, Conjugated microporous poly (aryleneethynylene) networks, *Angew. Chem., Int. Ed.*, 2007, **46**(45), 8574–8578.

16. (a) M. Rose, N. Klein, W. Bohlmann, B. Bohringer, S. Fichtner and S. Kaskel, New element organic frameworks via Suzuki coupling with

high adsorption capacity for hydrophobic molecules, *Soft Matter*, 2010, **6**(16), 3918–3923; (b) J. Weber and A. Thomas, Toward Stable Interfaces in Conjugated Polymers: Microporous Poly(p-phenylene) and Poly(phenyleneethynylene) Based on a Spirobifluorene Building Block, *J. Am. Chem. Soc.*, 2008, **130**(20), 6334–6335.

17. (a) J. Schmidt, M. Werner and A. Thomas, Conjugated Microporous Polymer Networks via Yamamoto Polymerization, *Macromolecules*, 2009, **42**(13), 4426–4429; (b) T. Ben, H. Ren, S. Q. Ma, D. P. Cao, J. H. Lan, X. F. Jing, W. C. Wang, J. Xu, F. Deng, J. M. Simmons, S. L. Qiu and G. S. Zhu, Targeted Synthesis of a Porous Aromatic Framework with High Stability and Exceptionally High Surface Area, *Angew. Chem., Int. Ed.*, 2009, **48**(50), 9457–9460.

18. W. Lu, D. Yuan, J. Sculley, D. Zhao, R. Krishna and H.-C. Zhou, Sulfonate-Grafted Porous Polymer Networks for Preferential CO2 Adsorption at Low Pressure, *J. Am. Chem. Soc.*, 2011, **133**(45), 18126–18129.

19. S. Yuan, B. Dorney, D. White, S. Kirklin, P. Zapol, L. Yu and D.-J. Liu, Microporous polyphenylenes with tunable pore size for hydrogen storage, *Chem. Commun.*, 2010, **46**(25), 4547–4549.

20. (a) P. Kuhn, M. Antonietti and A. Thomas, Porous, Covalent Triazine-Based Frameworks Prepared by Ionothermal Synthesis, *Angew. Chem., Int. Ed.*, 2008, **47**(18), 3450–3453; (b) S. Ren, M. J. Bojdys, R. Dawson, A. Laybourn, Y. Z. Khimyak, D. J. Adams and A. I. Cooper, Porous, Fluorescent, Covalent Triazine-Based Frameworks Via Room-Temperature and Microwave-Assisted Synthesis, *Adv. Mater.*, 2012, **24**, 2357–2361.

21. (a) M. Rose, N. Klein, I. Senkovska, C. Schrage, P. Wollmann, W. Bohlmann, B. Bohringer, S. Fichtner and S. Kaskel, A new route to porous monolithic organic frameworks via cyclotrimerization, *J. Mater. Chem.*, 2011, **21**(3), 711–716; (b) R. S. Sprick, A. Thomas and U. Scherf, Acid catalyzed synthesis of carbonyl-functionalized microporous ladder polymers with high surface area, *Polym. Chem.*, 2010, **1**(3), 283–285.

22. (a) F. J. Uribe-Romo, J. R. Hunt, H. Furukawa, C. Klock, M. O'Keeffe and O. M. Yaghi, A Crystalline Imine-Linked 3-D Porous Covalent Organic Framework, *J. Am. Chem. Soc.*, 2009, **131**(13), 4570–4571; (b) P. Pandey, A. P. Katsoulidis, I. Eryazici, Y. Wu, M. G. Kanatzidis and S. T. Nguyen, Imine-Linked Microporous Polymer Organic Frameworks, *Chem. Mater.*, 2010, **22**(17), 4974–4979.

23. (a) J. Schmidt, J. Weber, J. D. Epping, M. Antonietti and A. Thomas, Microporous Conjugated Poly(thienylene arylene) Networks, *Adv. Mater.*, 2009, **21**(6), 702–705; (b) J. Schmidt, D. S. Kundu, S. Blechert and A. Thomas, Tuning porosity and activity of microporous polymer network organocatalysts by co-polymerisation, *Chem. Commun.*, 2014, **50**(25), 3347–3349.

24. (a) Q. Chen, M. Luo, P. Hammershøj, D. Zhou, Y. Han, B. W. Laursen, C.-G. Yan and B.-H. Han, Microporous Polycarbazole with High Specific Surface Area for Gas Storage and Separation, *J. Am. Chem. Soc.*, 2012, **134**(14), 6084–6087; (b) Q. Chen, D.-P. Liu, J.-H. Zhu and B.-H. Han,

Mesoporous Conjugated Polycarbazole with High Porosity via Structure Tuning, *Macromolecules*, 2014, **47**(17), 5926–5931; (c) Q. Chen, D.-P. Liu, M. Luo, L.-J. Feng, Y.-C. Zhao and B.-H. Han, Nitrogen-Containing Microporous Conjugated Polymers via Carbazole-Based Oxidative Coupling Polymerization: Preparation, Porosity, and Gas Uptake, *Small*, 2014, **10**(2), 308–315; (d) Y. Zhang, A. Sigen, Y. Zou, X. Luo, Z. Li, H. Xia, X. Liu and Y. Mu, Gas uptake, molecular sensing and organocatalytic performances of a multifunctional carbazole-based conjugated microporous polymer, *J. Mater. Chem. A*, 2014, **2**(33), 13422–13430; (e) L.-J. Feng, Q. Chen, J.-H. Zhu, D.-P. Liu, Y.-C. Zhao and B.-H. Han, Adsorption performance and catalytic activity of porous conjugated polyporphyrins via carbazole-based oxidative coupling polymerization, *Polym. Chem.*, 2014, **5**(8), 3081–3088.
25. S. Qiao, Z. Du and R. Yang, Design and synthesis of novel carbazole-spacer-carbazole type conjugated microporous networks for gas storage and separation, *J. Mater. Chem. A*, 2014, **2**(6), 1877–1885.
26. P. Kuhn, A. l. Forget, D. Su, A. Thomas and M. Antonietti, From Microporous Regular Frameworks to Mesoporous Materials with Ultrahigh Surface Area: Dynamic Reorganization of Porous Polymer Networks, *J. Am. Chem. Soc.*, 2008, **130**(40), 13333–13337.
27. (a) C. Gu, Y. Chen, Z. Zhang, S. Xue, S. Sun, K. Zhang, C. Zhong, H. Zhang, Y. Pan, Y. Lv, Y. Yang, F. Li, S. Zhang, F. Huang and Y. Ma, Electrochemical Route to Fabricate Film-Like Conjugated Microporous Polymers and Application for Organic Electronics, *Adv. Mater.*, 2013, **25**(25), 3443–3448; (b) C. Gu, N. Huang, J. Gao, F. Xu, Y. Xu and D. Jiang, Controlled Synthesis of Conjugated Microporous Polymer Films: Versatile Platforms for Highly Sensitive and Label-Free Chemo- and Biosensing, *Angew. Chem. Int. Ed.*, 2014, **53**(19), 4850–4855.
28. C. Gu, Y. Chen, Z. Zhang, S. Xue, S. Sun, C. Zhong, H. Zhang, Y. Lv, F. Li, F. Huang and Y. Ma, Achieving High Efficiency of PTB7-Based Polymer Solar Cells via Integrated Optimization of Both Anode and Cathode Interlayers, *Adv. Energy Mater.*, 2014, **4**(8), 1301771.
29. A. U. Palma-Cando and U. Scherf, Electrogenerated Thin Films of Microporous Polymer Networks with Remarkably Increased Electrochemical Response to Nitroaromatic Analytes, *ACS Appl. Mater. Interfaces*, 2015, **7**(21), 11127–11133.
30. (a) L. Zhang, T. Lin, X. Pan, W. Wang and T.-X. Liu, Morphology-controlled synthesis of porous polymer nanospheres for gas absorption and bioimaging applications, *J. Mater. Chem.*, 2012, **22**(19), 9861–9869; (b) A. Patra, J.-M. Koenen and U. Scherf, Fluorescent nanoparticles based on a microporous organic polymer network: fabrication and efficient energy transfer to surface-bound dyes, *Chem. Commun.*, 2011, **47**(34), 9612–9614.
31. (a) N. Kang, J. H. Park, M. Jin, N. Park, S. M. Lee, H. J. Kim, J. M. Kim and S. U. Son, Microporous Organic Network Hollow Spheres: Useful Templates for Nanoparticulate Co3O4 Hollow Oxidation Catalysts, *J. Am.*

Chem. Soc., 2013, **135**(51), 19115–19118; (b) J. Jin, B. Kim, N. Park, S. Kang, J. H. Park, S. M. Lee, H. J. Kim and S. U. Son, Porphyrin entrapment and release behavior of microporous organic hollow spheres: fluorescent alerting systems for existence of organic solvents in water, *Chem. Commun.*, 2014, **50**(94), 14885–14888.

32. S. Shi, C. Chen, M. Wang, J. Ma, H. Ma and J. Xu, Designing a yolk-shell type porous organic network using a phenyl modified template, *Chem. Commun.*, 2014, **50**(65), 9079–9082.

33. S. Chakraborty, Y. J. Colon, R. Q. Snurr and S. T. Nguyen, Hierarchically porous organic polymers: highly enhanced gas uptake and transport through templated synthesis, *Chem. Sci.*, 2015, **6**(1), 384–389.

34. A. Laybourn, R. Dawson, R. Clowes, J. A. Iggo, A. I. Cooper, Y. Z. Khimyak and D. J. Adams, Branching out with aminals: microporous organic polymers from difunctional monomers, *Polym. Chem.*, 2012, **3**(2), 533–537.

35. A. Laybourn, R. Dawson, R. Clowes, T. Hasell, A. I. Cooper, Y. Z. Khimyak and D. J. Adams, Network formation mechanisms in conjugated microporous polymers, *Polym. Chem.*, 2014, **5**(21), 6325–6333.

36. X. Du, Y. L. Sun, B. E. Tan, Q. F. Teng, X. J. Yao, C. Y. Su and W. Wang, Troger's base-functionalised organic nanoporous polymer for heterogeneous catalysis, *Chem. Commun.*, 2010, **46**(6), 970–972.

37. M. G. Rabbani, A. K. Sekizkardes, Z. Kahveci, T. E. Reich, R. Ding and H. M. El-Kaderi, A 2D Mesoporous Imine-Linked Covalent Organic Framework for High Pressure Gas Storage Applications, *Chem. – Eur. J.*, 2013, **19**(10), 3324–3328.

38. H. Ren, T. Ben, F. Sun, M. Guo, X. Jing, H. Ma, K. Cai, S. Qiu and G. Zhu, Synthesis of a porous aromatic framework for adsorbing organic pollutants application, *J. Mater. Chem.*, 2011, **21**(28), 10348–10353.

39. Y. Xu, L. Chen, Z. Guo, A. Nagai and D. Jiang, Light-Emitting Conjugated Polymers with Microporous Network Architecture: Interweaving Scaffold Promotes Electronic Conjugation, Facilitates Exciton Migration, and Improves Luminescence, *J. Am. Chem. Soc.*, 2011, **133**(44), 17622–17625.

40. S. Bates, G. Zografi, D. Engers, K. Morris, K. Crowley and A. Newman, *Pharm. Res.*, 2006, **23**(10), 2333–2349.

41. J. R. Holst, E. Stöckel, D. J. Adams and A. I. Cooper, High Surface Area Networks from Tetrahedral Monomers: Metal-Catalyzed Coupling, Thermal Polymerization, and "Click" Chemistry, *Macromolecules*, 2010, **43**, 8531–8538.

42. P. Pandey, O. K. Farha, A. M. Spokoyny, C. A. Mirkin, M. G. Kanatzidis, J. T. Hupp and S. T. Nguyen, A "click-based" porous organic polymer from tetrahedral building blocks, *J. Mater. Chem.*, 2011, **21**(6), 1700–1703.

43. R. T. Woodward, L. A. Stevens, R. Dawson, M. Vijayaraghavan, T. Hasell, I. P. Silverwood, A. V. Ewing, T. Ratvijitvech, J. D. Exley, S. Y. Chong, F. Blanc, D. J. Adams, S. G. Kazarian, C. E. Snape, T. C. Drage and A. I. Cooper, Swellable, Water- and Acid-Tolerant Polymer Sponges for

Chemoselective Carbon Dioxide Capture, *J. Am. Chem. Soc.*, 2014, **136**(25), 9028–9035.

44. Y. Xu and D. Jiang, Structural insights into the functional origin of conjugated microporous polymers: geometry-management of porosity and electronic properties, *Chem. Commun.*, 2014, **50**(21), 2781–2783.

45. R. Dawson, A. Laybourn, Y. Z. Khimyak, D. J. Adams and A. I. Cooper, High Surface Area Conjugated Microporous Polymers: The Importance of Reaction Solvent Choice, *Macromolecules*, 2010, **43**, 8524–8530.

46. R. Dawson, A. Laybourn, R. Clowes, Y. Z. Khimyak, D. J. Adams and A. I. Cooper, Functionalized Conjugated Microporous Polymers, *Macromolecules*, 2009, **42**(22), 8809–8816.

47. S. Hug, M. B. Mesch, H. Oh, N. Popp, M. Hirscher, J. Senker and B. V. Lotsch, A fluorene based covalent triazine framework with high CO2 and H2 capture and storage capacities, *J. Mater. Chem. A*, 2014, **2**(16), 5928–5936.

48. S. Hug, M. E. Tauchert, S. Li, U. E. Pachmayr and B. V. Lotsch, A functional triazine framework based on N-heterocyclic building blocks, *J. Mater. Chem.*, 2012, **22**(28), 13956–13964.

49. A. Bhunia, V. Vasylyeva and C. Janiak, From a supramolecular tetranitrile to a porous covalent triazine-based framework with high gas uptake capacities, *Chem. Commun.*, 2013, **49**(38), 3961–3963.

50. C. Bleschke, J. Schmidt, D. S. Kundu, S. Blechert, A. Thomas and A. Chiral Microporous, Polymer Network as Asymmetric Heterogeneous Organocatalyst, *Adv. Synth. Catal.*, 2011, **353**(17), 3101–3106.

51. R. Dawson, D. J. Adams and A. I. Cooper, Chemical tuning of CO2 sorption in robust nanoporous organic polymers, *Chem. Sci.*, 2011, **2**, 1173–1177.

52. (a) P. Katekomol, J. Roeser, J. Weber and A. Thomas, Covalent Triazine Frameworks prepared from 1,3,5-Tricyanobenzene, *Chem. Mater.*, 2013, **25**(9), 1542–1548; (b) J. Roeser, K. Kailasam and A. Thomas, Covalent Triazine Frameworks as Heterogeneous Catalysts for the Synthesis of Cyclic and Linear Carbonates from Carbon Dioxide and Epoxides, *ChemSusChem*, 2012, **5**(9), 1793–1799.

53. (a) Q. Lin, J. Lu, Z. Yang, X. C. Zeng and J. Zhang, Porphyrinic porous organic frameworks: preparation and post-synthetic modification via demetallation-remetallation, *J. Mater. Chem. A*, 2014, **2**(36), 14876–14882; (b) Z. Wang, S. Yuan, A. Mason, B. Reprogle, D.-J. Liu and L. Yu, Nanoporous Porphyrin Polymers for Gas Storage and Separation, *Macromolecules*, 2012, **45**(18), 7413–7419.

54. (a) L. Chen, Y. Yang and D. Jiang, CMPs as Scaffolds for Constructing Porous Catalytic Frameworks: A Built-in Heterogeneous Catalyst with High Activity and Selectivity Based on Nanoporous Metalloporphyrin Polymers, *J. Am. Chem. Soc.*, 2010, **132**, 9138–9143; (b) L. Chen, Y. Yang, Z. Guo and D. Jiang, Highly Efficient Activation of Molecular Oxygen with Nanoporous Metalloporphyrin Frameworks in Heterogeneous Systems, *Adv. Mater.*, 2011, **23**(28), 3149–3154.

55. A. Sigen, Y. Zhang, Z. Li, H. Xia, M. Xue, X. Liu and Y. Mu, Highly efficient and reversible iodine capture using a metalloporphyrin-based conjugated microporous polymer, *Chem. Commun.*, 2014, **50**(62), 8495–8498.

56. (a) S. Yuan, J.-L. Shui, L. Grabstanowicz, C. Chen, S. Commet, B. Reprogle, T. Xu, L. Yu and D.-J. Liu, A Highly Active and Support-Free Oxygen Reduction Catalyst Prepared from Ultrahigh-Surface-Area Porous Polyporphyrin, *Angew. Chem. Int. Ed.*, 2013, **51**(48), 9805–9808; (b) J. Xia, S. Yuan, Z. Wang, S. Kirklin, B. Dorney, D.-J. Liu and L. Yu, Nanoporous Polyporphyrin as Adsorbent for Hydrogen Storage, *Macromolecules*, 2010, **43**(7), 3325–3330.

57. H. Urakami, K. Zhang and F. Vilela, Modification of conjugated microporous poly-benzothiadiazole for photosensitized singlet oxygen generation in water, *Chem. Commun.*, 2013, **49**(23), 2353–2355.

58. T. Ratvijitvech, R. Dawson, A. Laybourn, Y. Z. Khimyak, D. J. Adams and A. I. Cooper, Post-synthetic modification of conjugated microporous polymers, *Polymer*, 2014, **55**(1), 321–325.

59. V. Guillerm, L. J. Weselinski, M. Alkordi, M. I. H. Mohideen, Y. Belmabkhout, A. J. Cairns and M. Eddaoudi, Porous organic polymers with anchored aldehydes: a new platform for post-synthetic amine functionalization en route for enhanced CO_2 adsorption properties, *Chem. Commun.*, 2014, **50**(16), 1937–1940.

60. (a) T. Hasell, C. D. Wood, R. Clowes, J. T. A. Jones, Y. Z. Khimyak, D. J. Adams and A. I. Cooper, Palladium Nanoparticle Incorporation in Conjugated Microporous Polymers by Supercritical Fluid Processing, *Chem. Mater.*, 2010, **22**(2), 557–564; (b) T. Ishida, Y. Onuma, K. Kinjo, A. Hamasaki, H. Ohashi, T. Honma, T. Akita, T. Yokoyama, M. Tokunaga and M. Haruta, Preparation of microporous polymer-encapsulated Pd nanoparticles and their catalytic performance for hydrogenation and oxidation, *Tetrahedron*, 2014, **70**(36), 6150–6155; (c) L. Li, H. Zhao, J. Wang and R. Wang, Facile Fabrication of Ultrafine Palladium Nanoparticles with Size- and Location-Control in Click-Based Porous Organic Polymers, *ACS Nano*, 2014, **8**(5), 5352–5364.

61. H. Lee, H. Kim, T. Choi, H. W. Park and J. Y. Chang, Preparation of a microporous organic polymer by the thiol-yne addition reaction and formation of Au nanoparticles inside the polymer, *Chem. Commun.*, 2015, **51**(48), 9805–9808.

62. L. Ma, M. M. Wanderley and W. Lin, Highly Porous Cross-Linked Polymers for Catalytic Asymmetric Diethylzinc Addition to Aldehydes, *ACS Catal.*, 2011, **1**(7), 691–697.

63. J.-X. Jiang, C. Wang, A. Laybourn, T. Hasell, R. Clowes, Y. Z. Khimyak, J. Xiao, S. J. Higgins, D. J. Adams and A. I. Cooper, Metal–Organic Conjugated Microporous Polymers, *Angew. Chem., Int. Ed.*, 2011, **50**(5), 1072–1075.

64. Z. Xie, C. Wang, K. E. deKrafft and W. Lin, Highly Stable and Porous Cross-Linked Polymers for Efficient Photocatalysis, *J. Am. Chem. Soc.*, 2011, **133**(7), 2056–2059.

65. A. Patra and U. Scherf, Fluorescent Microporous Organic Polymers: Potential Testbed for Optical Applications, *Chem. – Eur. J.*, 2012, **18**(33), 10074–10080.

66. (a) Q. Chen, J.-X. Wang, F. Yang, D. Zhou, N. Bian, X.-J. Zhang, C.-G. Yan and B.-H. Han, Tetraphenylethylene-based fluorescent porous organic polymers: preparation, gas sorption properties and photoluminescence properties, *J. Mater. Chem.*, 2011, **21**(35), 13554–13560; (b) J.-X. Jiang, A. Trewin, D. J. Adams and A. I. Cooper, Band gap engineering in fluorescent conjugated microporous polymers, *Chemical Science*, 2011, **2**(9), 1777–1781.

67. L. Chen, Y. Honsho, S. Seki and D. L. Jiang, Light-Harvesting Conjugated Microporous Polymers: Rapid and Highly Efficient Flow of Light Energy with a Porous Polyphenylene Framework as Antenna, *J. Am. Chem. Soc.*, 2010, **132**(19), 6742–6748.

68. S. Ren, R. Dawson, D. J. Adams and A. I. Cooper, Low band-gap benzothiadiazole conjugated microporous polymers, *Polym. Chem.*, 2013, **4**(22), 5585–5590.

69. K. V. Rao, S. Mohapatra, T. K. Maji and S. J. George, Guest-Responsive Reversible Swelling and Enhanced Fluorescence in a Super-Absorbent, Dynamic Microporous Polymer, *Chem. – Eur. J.*, 2012, **18**(15), 4505–4509.

70. C. M. Thompson, F. Li and R. A. Smaldone, Synthesis and sorption properties of hexa-(peri)-hexabenzocoronene-based porous organic polymers, *Chem. Commun.*, 2014, **50**(46), 6171–6173.

71. F. Vilela, K. Zhang and M. Antonietti, Conjugated porous polymers for energy applications, *Energy Environ. Sci.*, 2012, **5**(7), 7819–7832.

72. Y. Xu, S. Jin, H. Xu, A. Nagai and D. Jiang, Conjugated microporous polymers: design, synthesis and application, *Chem. Soc. Rev.*, 2013, **42**(20), 8012–8031.

73. R. Dawson, A. I. Cooper and D. J. Adams, Chemical functionalization strategies for carbon dioxide capture in microporous organic polymers, *Polym. Int.*, 2013, **62**(3), 345–352.

74. S. Ren, R. Dawson, A. Laybourn, J.-x. Jiang, Y. Khimyak, D. J. Adams and A. I. Cooper, Functional conjugated microporous polymers: from 1,3,5-benzene to 1,3,5-triazine, *Polym. Chem.*, 2012, **3**(4), 928–934.

75. R. Dawson, D. J. Adams and A. I. Cooper, Chemical tuning of CO_2 sorption in robust nanoporous organic polymers, *Chem. Sci.*, 2011, **2**(6), 1173–1177.

76. Z. Xiang, D. Cao, W. Wang, W. Yang, B. Han and J. Lu, Postsynthetic Lithium Modification of Covalent-Organic Polymers for Enhancing Hydrogen and Carbon Dioxide Storage, *J. Phys. Chem. C*, 2012, **116**(9), 5974–5980.

77. K. Zhang, D. Kopetzki, P. H. Seeberger, M. Antonietti and F. Vilela, Surface Area Control and Photocatalytic Activity of Conjugated Microporous Poly(benzothiadiazole) Networks, *Angew. Chem.*, 2013, **125**(5), 1472–1476.

78. L. Chen, Y. Yang and D. Jiang, CMPs as Scaffolds for Constructing Porous Catalytic Frameworks: A Built-in Heterogeneous Catalyst with High Activity and Selectivity Based on Nanoporous Metalloporphyrin Polymers, *J. Am. Chem. Soc.*, 2010, **132**(26), 9138–9143.

79. T. Hasell, C. D. Wood, R. Clowes, J. T. A. Jones, Y. Z. Khimyak, D. J. Adams and A. I. Cooper, Palladium Nanoparticle Incorporation in Conjugated Microporous Polymers by Supercritical Fluid Processing, *Chem. Mater.*, 2009, **22**(2), 557–564.

80. X. Liu, Y. Xu and D. Jiang, Conjugated Microporous Polymers as Molecular Sensing Devices: Microporous Architecture Enables Rapid Response and Enhances Sensitivity in Fluorescence-On and Fluorescence-Off Sensing, *J. Am. Chem. Soc.*, 2012, **134**(21), 8738–8741.

81. L. Chen, Y. Honsho, S. Seki and D. Jiang, Light-Harvesting Conjugated Microporous Polymers: Rapid and Highly Efficient Flow of Light Energy with a Porous Polyphenylene Framework as Antenna, *J. Am. Chem. Soc.*, 2010, **132**(19), 6742–6748.

82. Y. Kou, Y. Xu, Z. Guo and D. Jiang, Supercapacitive Energy Storage and Electric Power Supply Using an Aza-Fused π-Conjugated Microporous Framework, *Angew. Chem., Int. Ed.*, 2011, **50**(37), 8753–8757.

83. (a) J. Luo, X. Zhang and J. Zhang, Carbazolic Porous Organic Framework as an Efficient, Metal-Free Visible-Light Photocatalyst for Organic Synthesis, *ACS Catalysis*, 2015, **5**(4), 2250–2254; (b) Z. J. Wang, S. Ghasimi, K. Landfester and K. A. I. Zhang, A conjugated porous poly-benzobisthiadiazole network for a visible light-driven photoredox reaction, *J. Mater. Chem. A*, 2014, **2**(44), 18720–18724.

84. Z. J. Wang, S. Ghasimi, K. Landfester and K. A. I. Zhang, Highly porous conjugated polymers for selective oxidation of organic sulfides under visible light, *Chem. Commun.*, 2014, **50**(60), 8177–8180.

85. F. Niu, L. Tao, Y. Deng, H. Gao, J. Liu and W. Song, A covalent triazine framework as an efficient catalyst for photodegradation of methylene blue under visible light illumination, *New J. Chem.*, 2014, **38**(12), 5695–5699.

86. J.-X. Jiang, Y. Li, X. Wu, J. Xiao, D. J. Adams and A. I. Cooper, Conjugated Microporous Polymers with Rose Bengal Dye for Highly Efficient Heterogeneous Organo-Photocatalysis, *Macromolecules*, 2013, **46**(22), 8779–8783.

87. (a) J. H. Park, K. C. Ko, N. Park, H.-W. Shin, E. Kim, N. Kang, J. Hong Ko, S. M. Lee, H. J. Kim, T. K. Ahn, J. Y. Lee and S. U. Son, Microporous organic nanorods with electronic push-pull skeletons for visible light-induced hydrogen evolution from water, *J. Mater. Chem. A*, 2014, **2**(21), 7656–7661; (b) M. G. Schwab, M. Hamburger, X. Feng, J. Shu, H. W. Spiess, X. Wang, M. Antonietti and K. Müllen, Photocatalytic hydrogen evolution through fully conjugated poly(azomethine) networks, *Chem. Commun.*, 2010, **46**(47), 8932–8934.

88. D. C. Martin, Molecular design, synthesis, and characterization of conjugated polymers for interfacing electronic biomedical devices with living tissue, *MRS Commun.*, 2015, **5**(2), 131–153.

89. S. Bhunia, N. Chatterjee, S. Das, K. Das Saha and A. Bhaumik, Porous Polyurea Network Showing Aggregation Induced White Light Emission, Applications as Biosensor and Scaffold for Drug Delivery, *ACS Appl. Mater. Interfaces*, 2014, **6**(24), 22569–22576.

90. (a) G. Cheng, T. Hasell, A. Trewin, D. J. Adams and A. I. Cooper, Soluble Conjugated Microporous Polymers, *Angew. Chem. Int. Ed.*, 2012, **51**(51), 12727–12731; (b) N. Ritter, I. Senkovska, S. Kaskel and J. Weber, Towards Chiral Microporous Soluble Polymers—Binaphthalene-Based Polyimides, *Macromol. Rapid Commun.*, 2011, **32**(5), 438–443; (c) X. Wu, H. Li, B. Xu, H. Tong and L. Wang, Solution-dispersed porous hyperbranched conjugated polymer nanoparticles for fluorescent sensing of TNT with enhanced sensitivity, *Polym. Chem.*, 2014, **5**(15), 4521–4525.

91. X. Liu, Y. Xu, Z. Guo, A. Nagai and D. Jiang, Super absorbent conjugated microporous polymers: a synergistic structural effect on the exceptional uptake of amines, *Chem. Commun.*, 2013, **49**(31), 3233–3235.

92. F. Xu, X. Chen, Z. Tang, D. Wu, R. Fu and D. Jiang, Redox-active conjugated microporous polymers: a new organic platform for highly efficient energy storage, *Chem. Commun.*, 2014, **50**(37), 4788–4790.

93. G. Cheng, B. Bonillo, R. S. Sprick, D. J. Adams, T. Hasell and A. I. Cooper, Conjugated Polymers of Intrinsic Microporosity (C-PIMs), *Adv. Funct. Mater.*, 2014, **24**(33), 5219–5224.

Porous Aromatic Frameworks

8.1 Introduction

In past decades, a great deal of effort has been devoted toward synthesizing materials with high surface area,[1-8] and impressive progress has been made, yielding a wide diversity of porous organic materials such as metal organic frameworks (MOFs),[9,10] covalent organic frameworks (COFs),[11-13] conjugated microporous polymers (CMPs),[14-17] polymers of intrinsic microporosity (PIMs),[18,19] hypercrosslinked polymers (HCPs)[20,21] and crystalline triazine-based frameworks (CTFs).[22-24] Numerous efforts have been employed to develop materials with an ultra-high surface area and physicochemical stability but these attempts failed until the emergence of porous aromatic framework-1 (PAF-1),[25] a dia topology porous framework with the largest surface area of 5600 $m^2\ g^{-1}$ at that time and exceptional physicochemical stability. Since its discovery, PAFs have attracted significant research attention and tremendous progress has been achieved in the development of this kind of material. Many reaction routes were developed to synthesis PAFs, including Yamamoto-type Ullmann cross-coupling reactions,[26-31] Suzuki cross-coupling reactions,[32] the ionothermal reaction[33,34] and so on. Moreover, PAFs show great potential in the field of gas sorption and separation, the adsorption of organic pollutants, and electrochemical and battery applications. In this chapter, we will discuss the latest progress in the design, synthesis and application of PAFs.

8.2 Synthesis of Porous Aromatic Frameworks

8.2.1 Yamamoto-type Ullmann Cross-coupling Reaction

PAF-1, the first porous aromatic framework with dia topology, was proposed based on the structure of diamond.[25] Diamond is a crystal form of carbon

Monographs in Supramolecular Chemistry No. 17
Porous Polymers: Design, Synthesis and Applications
By Shilun Qiu and Teng Ben
© The Royal Society of Chemistry 2016
Published by the Royal Society of Chemistry, www.rsc.org

atoms with very high stability. However, the non-porous property of diamond makes it unsuitable for use in gas storage and separation. Replacement of the C–C bond of diamond with a rigid phenyl ring would provide sufficient exposure of the edges and faces of the phenyl rings and significantly increase the internal surface area, as well as maintain the structural stability. Based on this idea, the structures of PAFs (Figure 8.1) have been constructed by replacing the C–C bond with phenyl rings.[25] Multiscale theoretical methods have been used to predict the porosity. Results showed that the model of P1 exhibited a Langmuir surface area of 2350 $m^2\ g^{-1}$ and model P2 (also known as PAF-1) can increase this value by almost three times to 7000 $m^2\ g^{-1}$. If the C–C bond was replaced by three phenyl rings (model P3), the pore sizes of the structure enlarged to the mesoscale range. Based on the simulation, PAF-1 shows extremely high surface area and the uniform micropore size distribution is expected to be very beneficial for hydrogen storage and CO_2 capture.

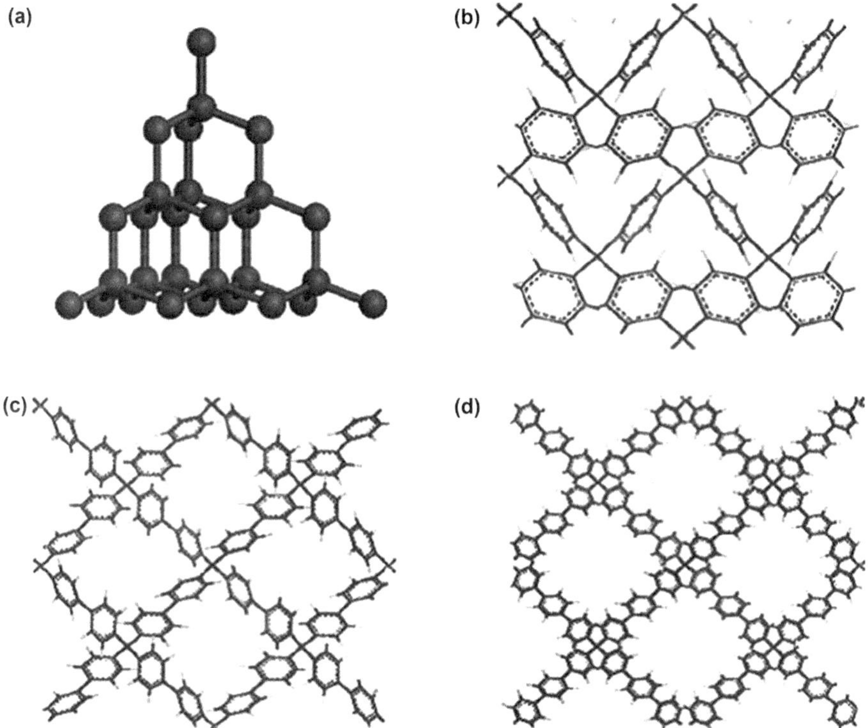

Figure 8.1 (a) structure of diamond; (b) structural model of P1; (c) structural model of P2; (d) structural model of P3.[25]
Reprinted with permission from: T. Ben, H. Ren, S. Ma, D. Cao, J. Lan, X. Jing, W. Wang, J. Xu, F. Deng, J. M. Simmons, S. Qiu and G. Zhu, *Angew. Chem., Int. Ed.*, 2009, **48**, 9457. Copyright 2009 WILEY-VCH Verlag GmbH & Co. KGaA, Weinheim.

The nickel(0)-catalyzed Yamamoto-type Ullmann cross-coupling reaction[35,36] was chosen to synthesize PAF-1 with tetrakis(4-bromophenyl)methane as the monomer. The reaction mechanism is shown in detail elsewhere.[35] Firstly, oxidative addition takes place between $Ni(0)L_m$ (where L is a neutral ligand) and the halogen-functionalized monomers. After the disproportionation of the two (I) and (II) complexes of nickel, reductive elimination of complex (III) leads to regeneration of $Ni(0)L_m$ and an addition product. Nitrogen sorption shows the BET specific surface area of PAF-1 is 5600 m^2 g^{-1}, which is very consistent with the simulated results. Closer examination of the successful synthesis procedure reveals that the great halogen elimination ability of Yamamoto coupling is the key factor of the successful synthesis.[37,38] The prevailing rigid framework of PAF-1 provides very high thermal and chemical stability, making it undisturbed by the vigorous post-treatment required to thoroughly empty the pores of the framework.

Another representative example using the Ullmann reaction is the preparation of JUC-Z1,[27] which is likely to have Linde type A (LTA) topology (Scheme 8.1). It is known that double four-rings (D4R) are the structure building units for the ACO or LTA-type zeolite.[39,40] Connecting D4R along the octant direction has the equal opportunity of generating the LTA or ACO net. By carefully choosing *p*-iodio-octaphenylsilsesquioxane (I8OPS) as the

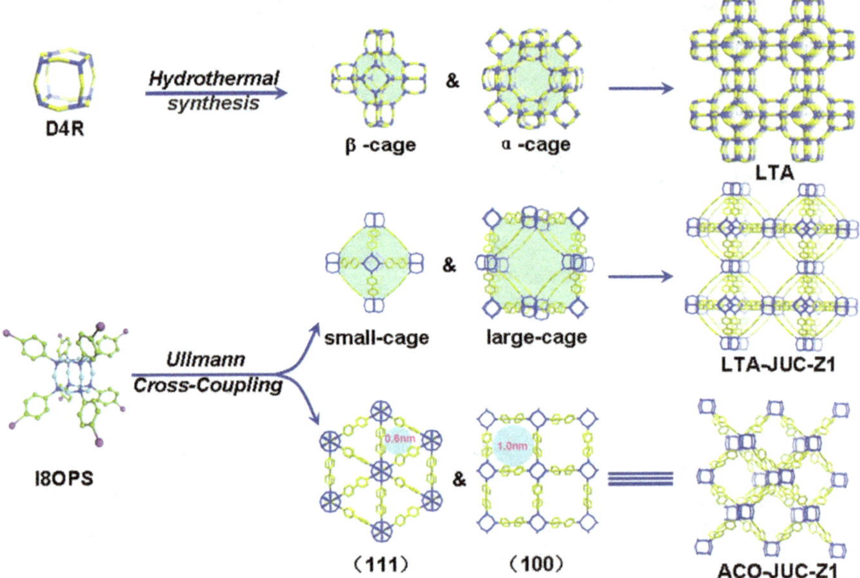

Scheme 8.1   The synthesis strategy of a covalently linked microporous organic–inorganic hybrid framework JUC-Z1 from I8OPS *via* an Ullmann cross-coupling reaction with an LTA or ACO topology (Si is illustrated in blue and O in yellow for D4R, and Si is illustrated in blue, O in cyan, C in green and I in purple for I8OPS).
Reproduced from ref. 27 with permission from The Royal Society of Chemistry.

monomer and the Ullmann reaction as the reaction route, the T–O–T linkage between the D4R in ACO or LTA can be replaced by biphenyl groups. Non-linear density functional theory (NLDFT) pore size distribution analysis shows that two obvious, narrowly distributed peaks of 1.16 and 2.0 nm are present in JUC-Z1, which arise from the beta-cage and alpha-cage in the LTA topology (1.49 and 2.18 nm). Thus, it is believed that JUC-Z1 likely shares the LTA structure. Unlike the hydrophilic LTA zeolite, the aromatic framework makes JUC-Z1 hydrophobic and it can be utilized for the selective adsorption of benzene. Using another cubic D4R siloxane cage and the Yamamoto-type Ullmann coupling reaction, the ACO topology organic–inorganic hybrid framework was also reported by Okubo's group.[41]

8.2.2 Suzuki Coupling Reaction

The Suzuki coupling reaction[42,43] was also used to synthesis PAFs. Using a diboronic acid as a linker and tetrakis(4-bromophenyl)methane (TBPM) as a tetrahedral unit, PAF-11 was successfully synthesized using a Suzuki coupling reaction.[32] The pore size distribution of PAF-11 falls in the mesoporous range, which is also in agreement with the simulated result. Compared with the Yamamoto-type Ullmann cross-coupling reaction, the Suzuki reaction has the advantage of utilizing functional groups, being largely unaffected by the presence of water. However, the relatively low surface area of PAF-11 also indicates that Suzuki coupling is not as efficient as Yamamoto-type Ullmann cross-coupling for halogen elimination and aryl–aryl coupling, although the possibility of interpenetration, which is present in PAF-11, will also decrease its special surface area.

8.2.3 Ionothermal Reaction

The ionothermal reaction reported by Kuhn and Thomas[22] further extended the synthesis routes of PAFs.[33] Using anhydrous $ZnCl_2$ and tetrakis(4-cyanophenyl)methane, PAF-2 was obtained (Figure 8.2). The choice of $ZnCl_2$ is the key to this successful synthesis. At temperatures above 400 °C, $ZnCl_2$ would melt and nitriles readily dissolve in this salt melt, which is beneficial for the reaction. Besides, $ZnCl_2$ is also a good catalyst for the trimerization reaction, by which a C_3N_3 ring can be formed through three nitrile groups. In the case of PAF-2, tetraphenylmethane and the C_3N_3 triazine ring are connected through covalent linkage, forming a three-dimensional organic open framework. The synthesized PAF-2 exhibits a high surface of 1109 m^2 g^{-1} and has high thermal and chemical stability. Due to the dual function of $ZnCl_2$ mentioned above, the reaction temperature and amount (mole equivalent) of $ZnCl_2$ would unavoidably affect the porosity of the products. In another case where tetrakis(4-cyanophenyl)silica (TCPSi) has been used as the monomer, the reaction temperature and mole equivalent of $ZnCl_2$ were investigated systematically.[34] It has been found that the porosity of PAF-16 could be readily tuned by changing the reaction temperature and

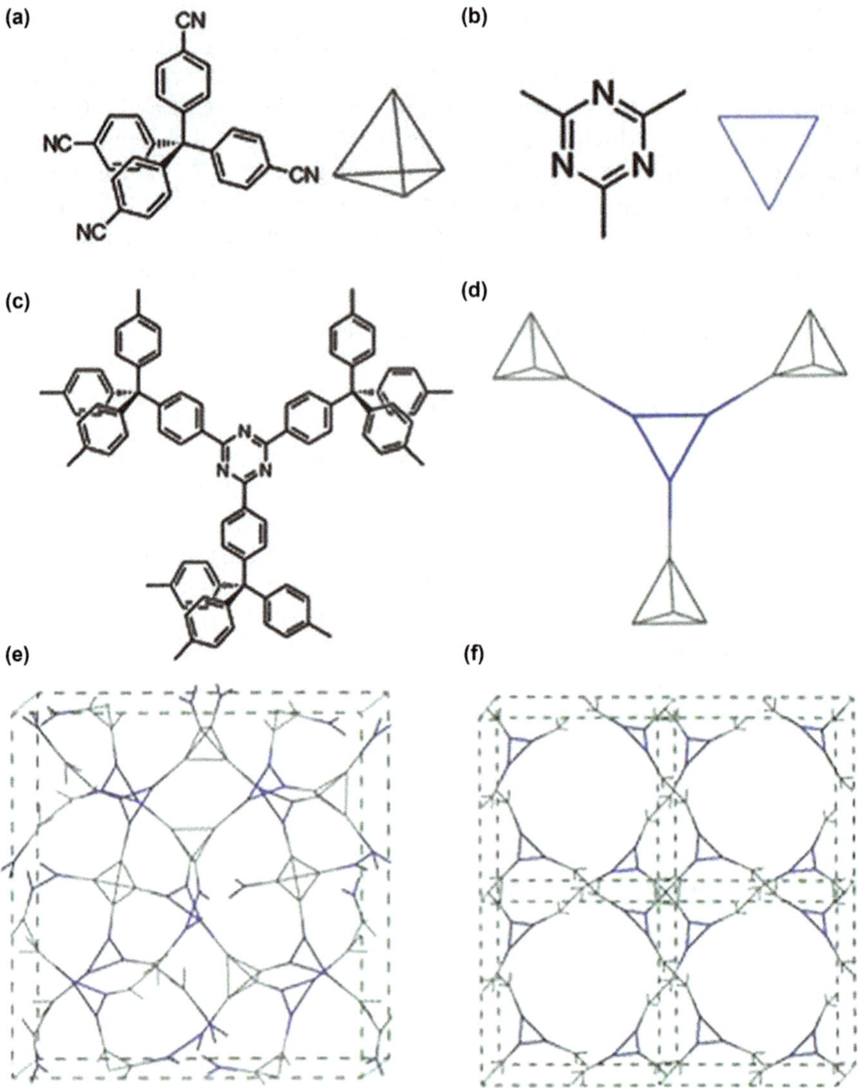

Figure 8.2 The designed process to PAF-2. Tetrakis(4-cyanophenyl)methane was chosen as the tetrahedral building unit (a) and a planar triangular C_3N_3 ring (b) produced by a trimerization reaction is also shown (polyhedron in light grey and triangle in blue). A secondary building unit was formed by linking three tetrahedral units with the planar triangular ring, (c) and (d). These building units connected together can produce the expanded ctn (e) and bor (f) nets.
Reproduced from ref. 33 with permission from The Royal Society of Chemistry.

ratio of monomer and catalyst. The surface area can be enhanced from 190 to 1166 m^2 g^{-1} and the pore size can be adjusted from micro-pores to meso-pores.

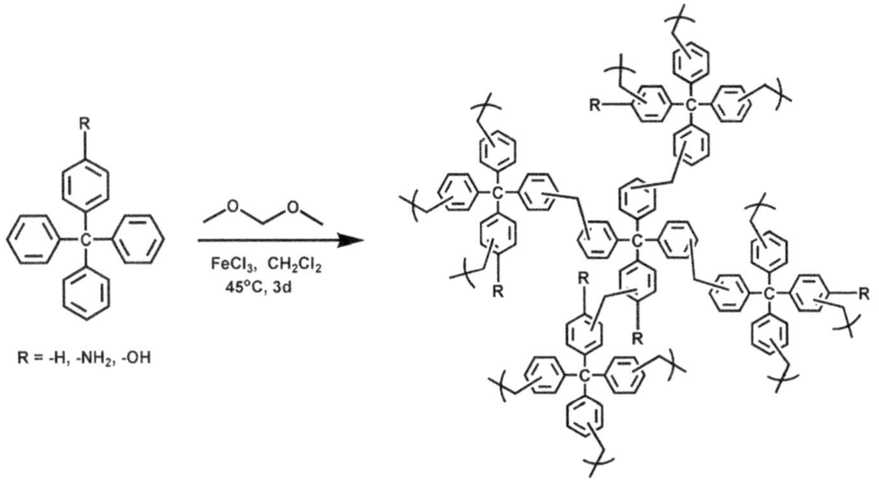

Figure 8.3 Synthesis route of PAF-32 and its functionalized analogues. Reproduced from ref. 47 with permission from The Royal Society of Chemistry.

8.2.4 Friedel–Crafts Alkylation Reaction

The Friedel–Crafts alkylation reaction is a well-developed method to synthesize hypercrosslinked organic polymers.[44,45] Using benzene and aniline as monomers, Cooper has used this method to synthesize copolymers with increased gas selectivity.[46] Recently, the Friedel–Crafts alkylation reaction has also been used to synthesize PAF by Jing *et al.*[47] By choosing tetraphenylmethane and its amino and hydroxyl derivatives as building blocks, and formaldehyde dimethyl acetal (FDA) as an external cross-linker, PAF-32 was obtained in the presence of $FeCl_3$ as a catalyst (Figure 8.3).[47] Unlike PAF-1, due to the present of external cross-linkers, PAF-32 shows a bimodal pore distribution, and due to the use of amino- and hydroxyl-functionalized building blocks, PAF-32-NH_2 and PAF-32-OH show increased CO_2 adsorption.

8.2.5 Summary

Besides the above mentioned methods, many other reaction routes have been developed and used in the preparation of PAFs. By using $AlCl_3$-triggered coupling polymerization, Li *et al.* obtained a series of PAFs, which were structurally similar to PAF-1, PAF-3, PAF-4 and JUC-Z2.[48] Although the surface area is much lower than that obtained by the Ullmann coupling reaction, the low cost and facile procedure are key factors behind the importance of this reaction. Using a condensation reaction, a pyridinium-type PAF, named Cl-PAF-50, has been synthesized by Yuan *et al.*, and AgCl-PAF-50 can be readily obtained due to the size of Cl^- ions (5 Å), which are large enough to accommodate Ag^+ ions (2.52 Å).[49] PAF-50, especially

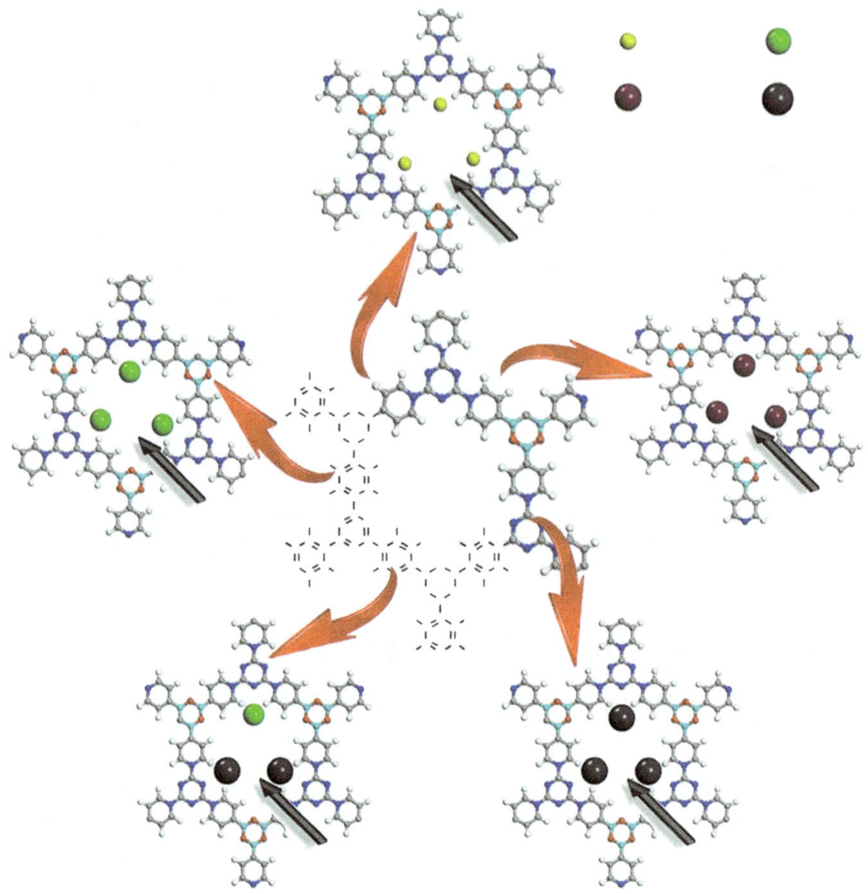

Scheme 8.2   Preparation scheme of X-PAF-50.[50]
Reprinted with permission from Y. Yuan, F. Sun, L. Li, P. Cui and G. Zhu, *Nat. Commun.*, 2014, **5**, 4260. Copyright 2014, Rights Managed by Nature Publishing Group.

AgCl-PAF-50, exhibits very good antibacterial performance. In their further work, through a facile ion exchange method a series of positively charged quaternary pyridinium-type PAFs (X-PAF-50; X = F, Cl, Br, I) was synthesized and their pore size could be tuned from 3.4 to 7 Å (Scheme 8.2).[50] These PAFs exhibit excellent molecular sieving effects for separating multicomponent gas mixtures.

In conclusion, owing to the ever increased molecular design of constituents and the development of advanced organic chemistry, a diverse range of PAFs with defined structures and physical and chemical functions can be easily synthesized, having potential usages in various fields. We believe further progress will be made in the development of PAFs owing to their charming properties and potential applications in molecular recognition, gas storage and separation, and as catalysts.

8.3 Properties and Applications of Porous Aromatic Frameworks

8.3.1 Gas Sorption and Separation

CO_2 emissions arising from fossil fuel combustion may cause detrimental changes to people's life. The worldwide climate and environment problems, such as the greenhouse effect, rising sea levels and the ever-growing acidity of the oceans, caused by excessive greenhouse gas emissions have prompted widespread public concern in recent years. Carbon capture and storage (CCS) is a promising method to reduce the concentration of CO_2.[51,52] Chemical absorption of CO_2 by ethanolamine solutions is a well-established and widely used method to capture CO_2 in industry due to its high CO_2 absorption capacity. However, this method suffers from several drawbacks such as a high energy cost for the regeneration of solutions and equipment corrosion. Thus, alternative, more economical processes need to be developed. Based on this, porous solid adsorbents have been proposed.[53–55] Another alternative route to reduce CO_2 emissions is new clean energy sources. Hydrogen has been considered an ideal clean energy due to the fact that it is carbon-free and has a vast resource in water. However, before the coming of the hydrogen energy era, there are several technical barriers that need to be settled. Among these issues, H_2 storage delivers a great challenge. Usually, H_2 can be stored in the form of liquid H_2 in compressed gas cylinders; however, there are safety issues surrounding the use of compressed gas at high pressures and the high energy cost for liquefaction has limited its application. Reversible physical adsorption and storage hydrogen in porous solids are alternative methods for the safe storage of hydrogen and has been widely investigated.[56,57]

Lots of porous solids, such as porous carbons, porous silica, zeolites and porous organic frameworks, have been reported and investigated for gas sorption. Compared with these porous materials, PAFs show unique advantages of combining high surface area and excellent physicochemical stability. This is especially true for PAF-1, which has a high surface area of $5640\ m^2\ g^{-1}$ and could survive extreme conditions such as boiling in solvent or addition to cold basic or acidic solutions.[25]

The high pressure H_2 storage of PAF-1 shows 7.0 wt% excess hydrogen uptake at 77 K, 48 bar, which equates to 10.7 wt% absolute uptake (Figure 8.4), which is among the highest value of the porous materials.[25] High-pressure CO_2 adsorption also indicates a high uptake of CO_2, with $1300\ mg\ g^{-1}$ at 40 bar and room temperature, suggesting its potential application for carbon dioxide capture. Using an improved Yamamoto-type Ullmann cross-coupling method, Zhou *et al.* reported another porous polymer network, PPN-4 (also known as PAF-3), with a Brunauer–Emmett–Teller (BET) surface area (S_{BET}) $6461\ m^2\ g^{-1}$ and total CO_2, CH_4 and H_2 storage capacities of $2121\ mg\ g^{-1}$ (295 K, 50 bar), $389\ mg\ g^{-1}$ (295 K, 55 bar) and $158\ mg\ g^{-1}$ (77 K, 90 bar), respectively.[58] These values indicated that porous polymers like PPN-4 are very promising in gas storage due to their ultra-high surface area. Although

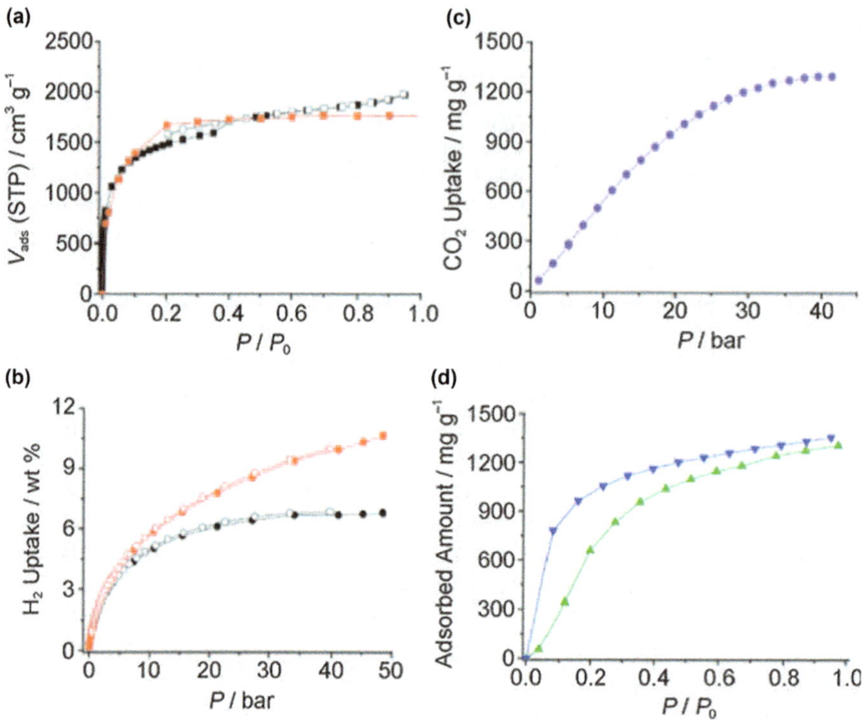

Figure 8.4 (a) Experimental adsorption (black square) and desorption (hollow square), and simulated N_2 adsorption (red square) isotherms of PAF-1 at 77 K. (b) High-pressure hydrogen sorption isotherms at 77 K. Filled black circle: excess adsorption; hollow black circle: excess desorption; filled red circle: total adsorption; hollow red circle: total desorption. (c) Excess high-pressure carbon dioxide adsorption isotherm at 298 K. (d) Benzene (green triangle) and toluene (inverted blue triangle) vapor adsorption isotherms at 298 K.[25]
Reprinted with permission from T. Ben, H. Ren, S. Ma, D. Cao, J. Lan, X. Jing, W. Wang, J. Xu, F. Deng, J. M. Simmons, S. Qiu and G. Zhu, *Angew. Chem., Int. Ed.*, 2009, **48**, 9457. Copyright 2009 WILEY-VCH Verlag GmbH & Co. KGaA, Weinheim.

MOFs with even higher surface area and increased gas adsorption have been synthesized (as shown in Table 8.1), the weak coordination in MOFs makes them rather unstable under harsh conditions and makes them incomparable with PAFs for long time storage and repeated usage.

PAFs with the same topology as PAF-1 but different heteroatoms (Si for PAF-3 and Ge for PAF-4) were also reported to investigate the gas storage of PAFs.[26] Because of the decreased special surface area, the high-pressure H_2 sorptions of PAF-3 (5.5 wt%, 60 bar) and PAF-4 (4.2 wt%, 60 bar) at 77 K were lower than PAF-1 (7.0 wt%, 48 bar). However, PAF-3 and PAF-4 show increased gas uptakes at low pressure. The H_2 uptakes were 186 cm^3 g^{-1}, 232 cm^3 g^{-1} and 169 cm^3 g^{-1} for PAF-1, PAF-3 and PAF-4, respectively, at 77 K.

Table 8.1 Summary of porosity and gas sorption data of high surface area materials.

Compound	Interaction	S_{BET} $(m^2\ g^{-1})$	Pore volume $(cm^3\ g^{-1})$	Excess H_2 uptake $(mg\ g^{-1})$	Excess CO_2 uptake $(mg\ g^{-1})$	Excess CH_4 uptake $(mg\ g^{-1})$
NU-110E[25]	Coordination bond	7140	4.40	–	–	–
NU-109E[25]	Coordination bond	7010	3.75	–	–	–
PPN-4[61]	Covalent bond	6461	3.04	91 (55 bar)	2121 (50 bar)	269 (55 bar)
MOF-210[26]	Coordination bond	6240	3.60	86 (55 bar)	2396 (50 bar)	264 (80 bar)
NU-100[27]	Coordination bond	6143	2.82	99.5 (70 bar)	2315 (40 bar)	–
PAF-1[28]	Covalent bond	5640	3.05	75.3 (48 bar)	1300 (40 bar)	–

Through examination of the heat of adsorption (Q_{st}) it has been found that Q_{st} relates to the H_2 adsorbent interactions $(Q_{stH2}$ for short), which are 5.4 kJ mol^{-1}, 6.6 kJ mol^{-1} and 6.3 kJ mol^{-1} for PAF-1, PAF-3 and PAF-4, respectively. In other words, the interaction of H_2 with PAF-3 or PAF-4 is much higher than PAF-1. Thus, we conclude that at high pressure, the surface area of a material plays a more important role in gas sorption because the total volume determines how much gas molecules can be pushed into the pores; while at low pressure, the interaction between adsorbate and adsorbent (reflected as heat of adsorption) is more important. Especially at ambient conditions, the effect of Q_{st} often surpasses the effect of surface area. So, it must be noted that for the post-combustion capture of CO_2 during CCS, designing materials with higher Q_{st} is more efficient than designing materials with higher surface areas because the CO_2 concentration at this condition is mostly less than 15%.[59]

PAF-3 and PAF-4 did show better selective adsorption of CO_2 than PAF-1. At 1 atm and 273 K, PAF-3 exhibits extraordinarily promising selectivity of 87 : 1 for the adsorption of CO_2 over N_2, and PAF-4 shows a selectivity of 44 : 1, while this value is only 38 : 1 for PAF-1.

Based on the above discussion, it is clear that for clean energy storage, such as H_2 storage at high pressure, high surface areas and large pore volumes are really desirable; meanwhile, for carbon dioxide capture at low pressure or ambient conditions, stronger interactions between the gas and framework are desired. However, it seems that there is an irreconcilable conflict between an ultra-high surface area and the gas–framework inter-action. Open frameworks that are beneficial for gas storage at high pressure usually suffer from low heat of adsorption, making them behave poorly under ambient conditions. Recently, by choosing tris(4-bromophenyl)amine (TBPA) and tetrakis(4-bromophenyl)methane (TBPM) as monomers and a Yamamoto-type Ullmann cross-coupling reaction as the reaction route, Pei and co-workers obtained a series of porous organic frameworks (POFs) with excellent H_2, CH_4 and CO_2 uptakes both at low and high pressures.[60] TBPM is the basic building unit of PAF-1. PAF-1 possesses excellent gas uptake at high pressure but poor adsorption behavior at low pressure. TBPA has been used to synthesize porous aromatic framework JUC-Z2,[28] a layered PAF with relatively lower surface area. However, due to the presence of binding sites on the N atom, its low pressure CO_2 uptake is good.[61] After copolymerization, the obtained copolymerized POFs (C-POFs) inherit the merits of both TBPA and TBPM, and exhibit increased gas uptake at both high and low pressures. Obviously, the increased Q_{st} is responsible for the increased gas uptake at low pressure. On further investigation of the porous character of these C-POFs, it was found that the relative smaller pore size may be another reason for such behavior. It has been reported that materials with pore sizes that match well the kinetic diameter of CO_2 can increase the interactions of the porous framework and carbon dioxide.[62,63]

Introducing basic nitrogen groups into the porous material has been used to increase the interaction with acidic CO_2 gas. Recently, a nitrogen-rich PAF (NPAF) has been synthesized using 1,4-bis(diphenylamino)benzene as the monomer through Yamamoto coupling.[31] The obtained NPAF shows a CO_2 uptake of 3.64 mmol g^{-1} at 273 K and 1 atm, and CO_2/N_2 selectivity of 48 : 1 based on the ideal adsorption solution theory (IAST) using 0.15/0.85 as the mole ratio of CO_2/N_2. Another nitrogen-rich porous aromatic framework, PAF-30, was synthesized by Zhao *et al.*[64] It is very interesting that PAF-30 shows a very small pore size of 3.30–3.54 Å that is beneficial for increasing the heat of adsorption. In the narrow-pore structures, CO_2 molecules can interact with multiple pore surfaces simultaneously. Besides, the high N content in PAF-30 will also favor the strong interaction with acid CO_2. Thus, PAF-30 shows a high CO_2 heat of adsorption with a value of 36.9 kJ mol^{-1}. The value is even higher than that of PPN-6-SO$_3$H and PPN-6-SO$_3$Li (30.4 and 35.7 kJ mol^{-1}, respectively).[65] This provides PAF-30 with high selective adsorption of CO_2 from mixed gases. The IAST prediction shows that the CO_2/CH_4 selectivity for PAF-30 is 24.3–63.2 for 50% CO_2 and 50% CH_4, and in the range of 4–30 for 5% CO_2 and 95% CH_4.

Recent studies have shown that post modification of PAF-1 with functional groups could greatly increase its low pressure CO_2 capture and selectivity. However, post modification usually needed to be done under severe conditions and causes a dramatic decrease of the framework's surface area. By choosing the proper reaction route and starting materials, PAFs with functional groups can be synthesized *in situ via* Sonogashira–Hagihara coupling reactions. For example, functional groups such as –NH$_2$, –OH and –COOH can be incorporated into the framework of PAF materials.[66] Using tetrahedral monomers as building blocks and the Friedel–Crafts alkylation reaction catalysed by FeCl$_3$, hydroxyl and amino functional groups can also been introduced into the networks of PAFs.[47] Compared with PAFs without functional groups, functionalized PAFs show obvious increased carbon capture abilities due to the stronger interaction of CO_2 with the electron-rich functional groups.

The above results reveal the interesting carbon dioxide capture and gas storage properties of PAFs. Among all the reported PAFs, PAF-1 exhibits an extremely high surface area, as well as excellent physicochemical and hydrothermal stabilities. Though some metal organic frameworks reported recently show higher surface areas and high-pressure gas storage than PAF-1, the excellent physicochemical stability of PAF-1 is incomparable and irreplaceable. This will enable it to survive severe post modification to further improve its properties or be target decorated for special applications.

8.3.2 Adsorption of Organic Pollutants

Besides gas sorption and carbon capture, another important application of porous materials is the adsorption of volatile pollutants. Unlike other porous materials, such as zeolites or metal organic frameworks, PAFs are almost totally composed of hydrophobic aromatic frameworks and this makes them especially appropriate and effective for such a kind of application. As shown in Figure 8.4, ultra-high surface area PAF-1 is able to adsorb 1306 mg g^{-1} (16.74 mmol g^{-1}) benzene and 1357 mg g^{-1} (14.73 mmol g^{-1}) toluene at 298 K; it should be noted that these values surpassed all porous materials at their time of publication. Even for PAF-5, whose surface area is 1503 m^2 g^{-1}, it can still adsorb 262.96 cm^3 g^{-1} (11.74 mmol g^{-1}) benzene and 258.22 cm^3 g^{-1} (11.53 mmol g^{-1}) toluene at saturated vapor pressures at 298 K. Besides, PAF-5 shows an extremely high methanol uptake of 653.09 cm^3 g^{-1} (29.16 mmol g^{-1}).

For applications like the adsorption of organic pollutions, selectivity is also an important feature that an absorbent should possess. PAF-2 can adsorb 138 mg g^{-1} benzene vapor at room temperature but little cyclohexane (7 mg g^{-1}; Figure 8.5).[33] This excellent selective sorption of benzene can be directly attributed to the π–π interaction between the aromatic framework of

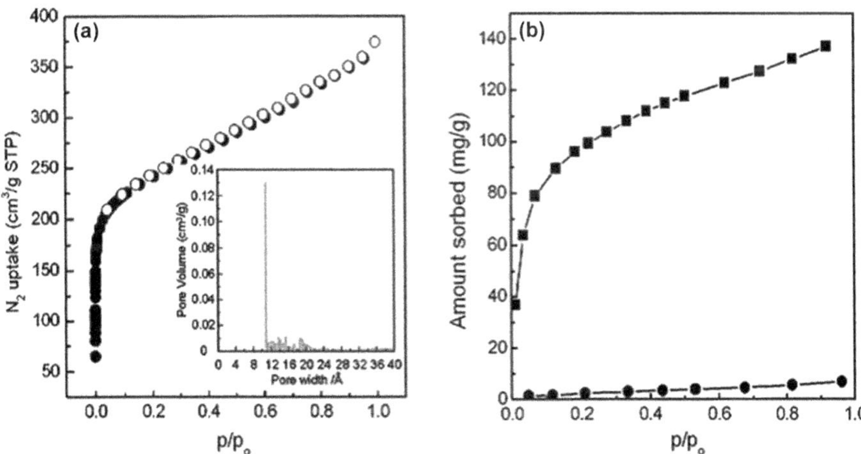

Figure 8.5   (a) Reversible nitrogen gas adsorption–desorption isotherm for PAF-2; inset: the pore size distribution for PAF-2 derived from N$_2$ adsorption calculated by the NLDFT method; (b) benzene (■) and cyclohexane (●) vapor adsorption isotherms at 298 K.
Reproduced from ref. 33 with permission from The Royal Society of Chemistry.

Table 8.2 Summary of organic vapor adsorption in porous aromatic frameworks.

Material	S_{BET} (m^2 g^{-1})	Benzene (mmol g^{-1})	Toluene (mmol g^{-1})	Methanol (mmol g^{-1})
PAF-1	5600	16.74	14.73	–
PAF-5	1503	11.74	11.53	29.16
PAF-11	704	11.2	8.5	20.4

PAF-2 and the benzene molecule. The organic vapor sorption of PAFs are summarized in Table 8.2; it can be clearly seen that PAFs show great potential in the sorption of organic vapor and all are superior to the very famous MOF-5 (10.3 mmol g^{-1} of benzene at 295 K).[67]

8.3.3 Electroactive and Battery Applications

The discovery of conducting polymers has greatly promoted the development of material science. Well-developed conducting polymers, such as polyaniline,[68] polyphenylene[69] and poly(thiophenes),[70] with one-dimensional (1D) linear backbones usually show low surface areas. However, for supercapacitor and battery applications, the porosity plays an important role and much effort has been devoted to developing porous organic-based conducting materials. For example, conducting porous polymers composed of conjugated organic ligands have been investigated and many conjugated microporous polymers, electroactive organic frameworks and lamella covalent organic frameworks have been investigated for photovoltaic and optoelectronic applications.[71–74]

Triphenylamine has been greatly investigated for the construction of conducting polymers as they can then form stable aminium radical cations with relatively high mobility and low ionization potentials.[75,76] In addition, the nitrogen species in triphenylamine is believed to be electroactive. Cooper and co-workers have theoretically predicted that the polytriphenylamine microporous organic polymer is expected to have unique electronic and optoelectronic properties.[77,78] Based on this viewpoint, JUC-Z2, which is composed of triphenylamine as second building units, has been synthesized from the *para*-tribromotribenzylaniline monomer *via* a Yamamoto-type Ullmann coupling reaction.[28] JUC-Z2 exhibits a high surface area of 2081 m^2 g^{-1} and uniform micropore of 1.2 nm. After doping with I$_2$, JUC-Z2 exhibits typical p-type semiconductive properties with a resistivity of 9×10^{-1} Ω cm, a Hall mobility of 0.811 cm^2 V^{-1} s^{-1} and a carrier concentration of 8.26×10^{18} cm^{-3}. Moreover, the cyclic voltammogram of oligomeric JUC-Z2 (oligo-JUC-Z2) further demonstrated its electroactivity, as shown in Figure 8.6.[28] In hydrophilic small anion supporting electrolytes, such as tetrabutylammonium perchlorate (Bu$_4$NClO$_4$), tetrabutylammonium hexafluorophosphate (Bu$_4$NPF$_6$) or tetrabutylammonium tetrafluoroborate (Bu$_4$NBF$_4$), a pair of redox peaks is clearly observed.

What is more interesting for JUC-Z2 is its significant recognition ability towards guest molecules with different shapes and sizes due to its uniform

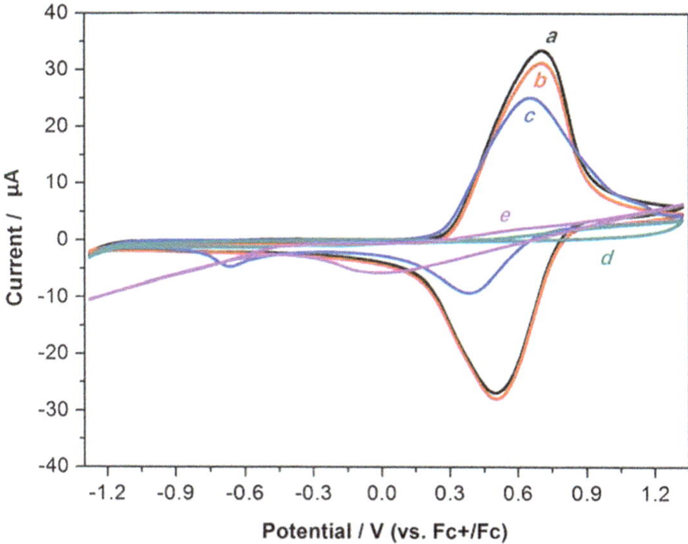

Figure 8.6 Cyclic voltammograms of the oligo-JUC-Z2 powder film on a Pt micro-electrode in a degassed acetonitrile solution containing 0.1 M Bu_4NPF_6 at (a) the first cycle and (b) the fifth cycle, in 0.1 M tetraethylammonium-toluene-4-sulfonate (ET_4TOS) at (c) the first cycle and (d) the fifth cycle, and in (e) 0.1 M camphorsulfonic acid (CSA) at the first cycle. Scan rate: 20 mV s^{-1}.
Reproduced from ref. 28 with permission from The Royal Society of Chemistry.

micropore (Scheme 8.3).[28] The pores of JUC-Z2 enable access and free mobility of anions with a smaller size such as PF_6^- (5.12 Å), ClO_4^- (5.0 Å) and BF_4^- (4.54 Å). However, for ions too large, such as ET_4N^+ or the CSA^- anion (9 Å), the micropore is easily blocked and its reversible electrochemical behavior is lost. This process is shown is Scheme 8.3. It should be noted that this behavior is absolutely different to early conducting polymers and makes the potential usage of JUC-Z2 in sensors possible.

Dai *et al.* further demonstrated the potential application of PAFs in the electrochemical field.[79] By the melting diffusion of sulfur into the framework of JUC-Z2, PAF-S was obtained with sulfur highly dispersed inside JUC-Z2. As a cathode material, the PAF-S composite shows 1083 mA g^{-1} reversible capacity and excellent stability in 1.0 M $LiPF_6$–MiPS. In the ion liquid electrolyte *N*-methyl-*N*-propyl pyrrolidinium bis-(trifluoromethanesulfonyl)imide (MPPY·TFSI), an 830 mA h g^{-1} reversible capacity is achieved and 83% is retained after 50 cycles (Figure 8.7).[79] The authors pointed out that once sulfur is loaded into the aromatic framework, the poor cycle stability of the sulfur electrode can be partially mitigated because of the confinement effect of the porous framework. Though the electronic conductivity of PAFs is low, this research does demonstrate the potential application of PAFs in batteries.

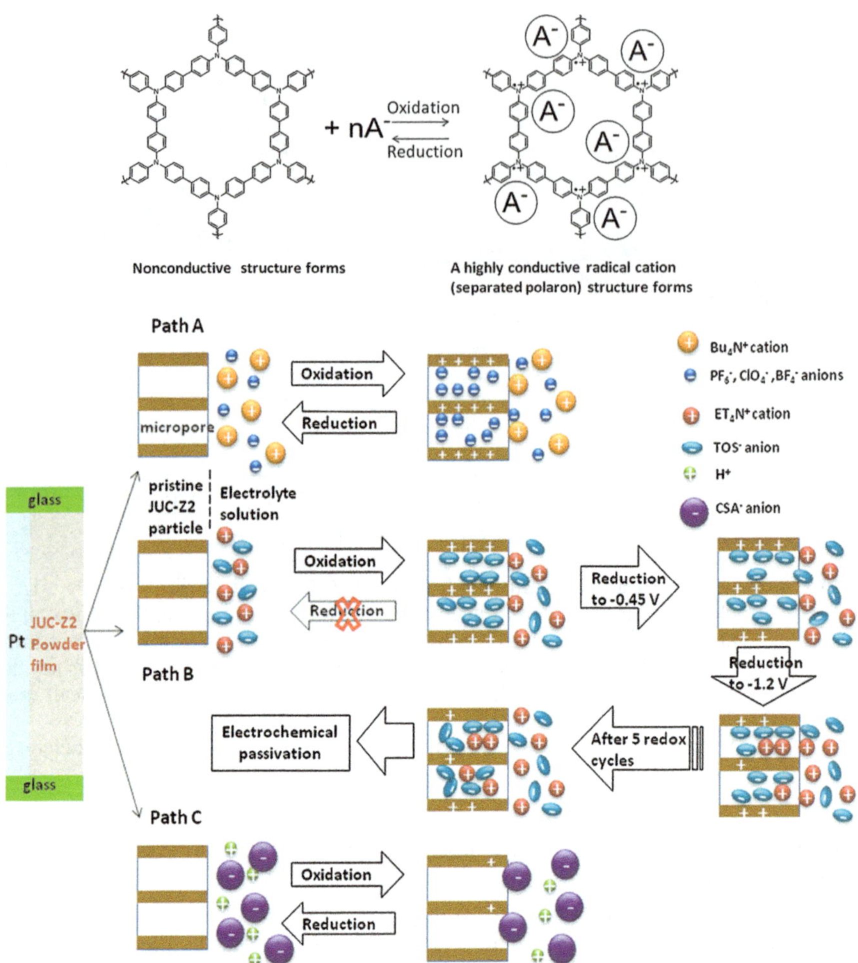

Scheme 8.3 Illustration of the electrochemical redox behavior of pristine JUC-Z2 and its different redox processes occurring in either Bu$_4$NPF$_6$ (Bu$_4$NClO$_4$, Bu$_4$NBF$_4$), ET$_4$TOS or CSA supporting electrolyte solutions. Reproduced from ref. 28 with permission from The Royal Society of Chemistry.

8.3.4　Summary

Beside the above mentioned applications, such as gas storage and carbon capture, the adsorption of organic pollutions, the electrochemical activity and batteries, PAFs are still being investigated for many other applications. Recently, we have reported that PAFs show great potential in the capture of iodine.[80] Two types of PAFs have been investigated in our work. Both of them show extremely high iodine vapor uptake at 298 K and 40 Pa (1.86 g g^{-1} for PAF-1 and 1.44 g g^{-1} for JUC-Z2). Though PAF-1 exhibits much higher iodine vapor uptake due to its high surface and large volume, the N atom in JUC-Z2

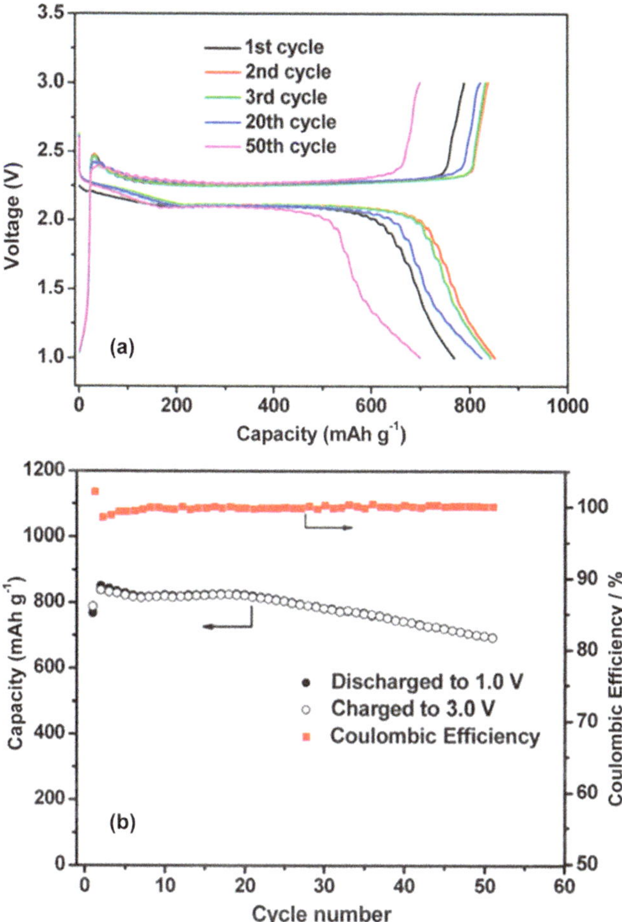

Figure 8.7 Galvanostatic discharge–charge curves (a) and cycling performance (b) of the Li∥PAF-S cell at a rate of 0.05 C in 0.5 M LiTFSI–MPPY · TFSI at 50 °C. Reproduced from ref. 79 with permission from The Royal Society of Chemistry.

provides a much more effective binding site. The heat of adsorption for I_2 is -51.1 kJ mol^{-1}, while it is only -14.9 kJ mol^{-1} for PAF-1. Raman spectroscopy shows that iodine exists in the framework in the form of I_5^-. In addition, gravimetric iodine uptake under dry conditions and humid conditions shows PAF-1 and JUC-Z2 could selectively adsorb iodine over water. As radionuclides ^{129}I and ^{131}I generated from the fission of uranium atoms during the operation of nuclear reactors usually exist in waste streams, this selectivity is really important.

In conclusion, owing to the high surface area and excellent stability, PAFs show great potential in gas storage and separation, the adsorption of organic pollutants, molecular reorganization, as electrode materials for batteries and also in many other fields. Though the investigation of PAFs is still in the initial stages and far away from practical applications, the attractive properties of PAFs make them irreplaceable and raise the expectation of increasing unique applications of PAFs that may be developed in the future.

8.4 Theoretical Simulation and Calculation

8.4.1 Simulation of Gas Storage in PAFs

With the development of computer techniques and molecular simulation, more and more porous aromatic framework-related structures have been designed and theoretical calculations have been used to predict the mechanical, electronic, adsorption, molecular recognition and separation properties. These studies can be classified into two categories. First is the investigation of gas storage in present PAFs or functionalized PAFs. Second is the design of new PAF structures and exploration of their potential applications.

For the functionalization of PAFs, PAF-1 seems to be the most promising host material due to its ultra-high surface area and stability. Different methods, such as lithiation,[81] sulphonation,[65] amine grafting,[82] nitration[83] and carbonization,[84] have been used to modify the structure of PAF-1 and improvement of the gas storage and separation have been observed. Details of this can be found in Chapter 9. Besides these experimental functionalization methods, computer simulations and theoretical calculations have been employed to predict the properties of PAFs. The multiscale simulation method was used to estimate the hydrogen storage properties of PAF-30X ($X = 1$–4).[85] Results show that the densities and free volumes are the two main factors that affect the hydrogen uptakes of PAFs. Among the investigated PAFs, PAF-303 and PAF-304 show larger gravimetric hydrogen adsorption due to their larger pore size and tremendous pore volume. PAF-304, with the smallest density (0.0998 g cm^{-3}), exhibits the highest hydrogen uptake of about 6.53 wt% at temperature $T = 298$ K and pressure $p = 100$ bar. However, for experimental synthesis, it is difficult to get the ideal PAF-304 structure due to the possibilities of interpenetration of the aromatic rings. For example, PAF-11, which shares the same structure as PAF-304, synthesized *via* Suzuki cross-coupling only shows a surface area of 704 m^2 g^{-1}, far lower than the simulated result. Thus, more effective, optical synthesis procedures should be explored to gain more accurate structures.

Cossi and co-authors have predicated theoretically high-pressure methane adsorption in PAF-30X ($X = 1$–4) using the Grand Canonical Monte Carlo (GCMC) technique based on ideally crystalline materials (Figure 8.8).[86] All the investigated materials show a high affinity toward methane. Both PAF-301 and PAF-302 exceed the target set by the U.S. Department of Energy [DOE; 180 cm^3 (STP)/cm^3 at 35 bar and 298 K (STP, standard temperature and pressure); Figure 8.9].[86] PAF-301 is very efficient at pressures below 10 bar due to its high density of aromatic residues, while PAF-302 (PAF-1)

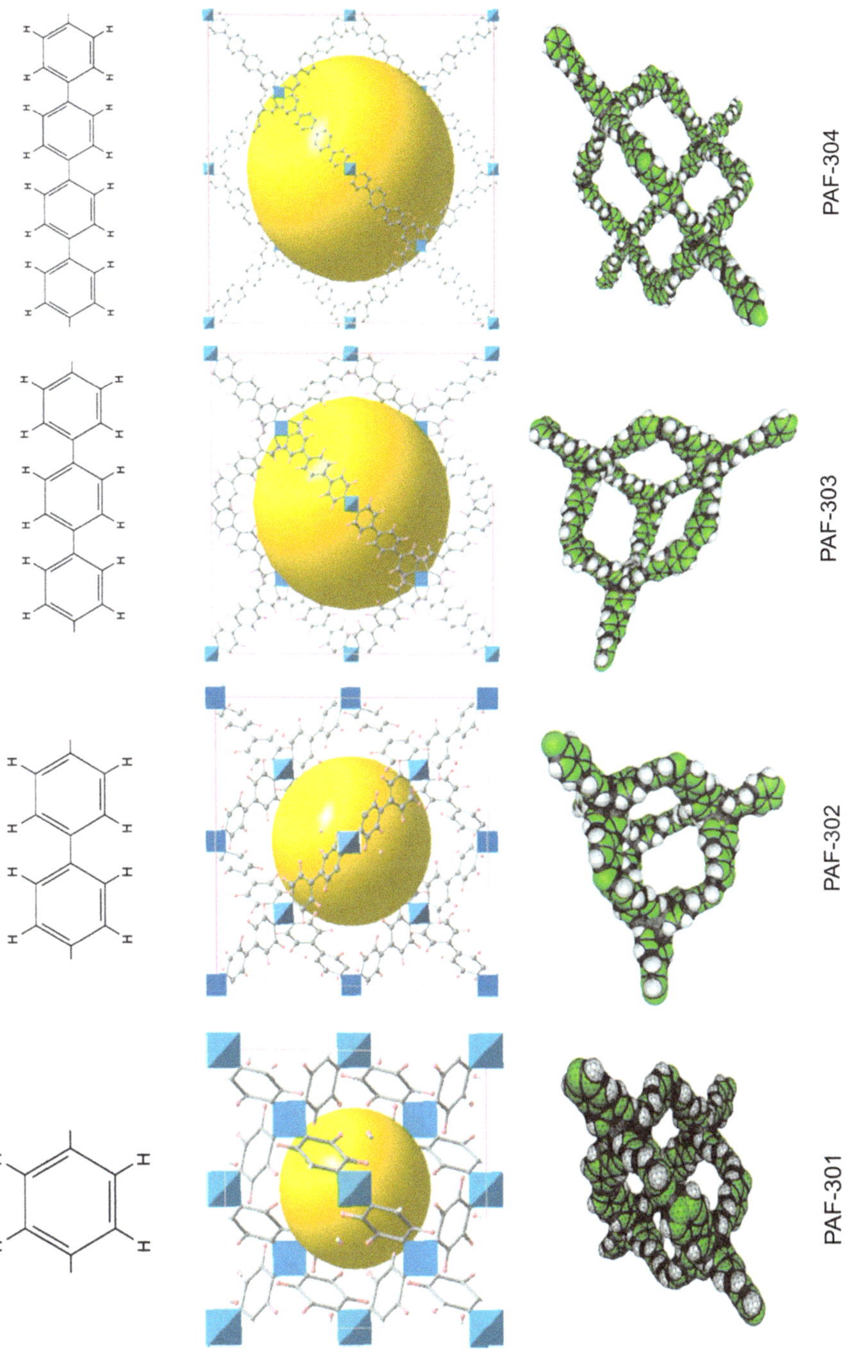

Figure 8.8 Top to bottom: aromatic building blocks, three-dimensional structures of the unit cells and the skeletal volumes defined as a collection of atomic spheres by the GEPOL procedure for PAF-30X ($X = 1 - 4$).[86] Reprinted with permission from M. Cossi, G. Gatti, L. Canti, L. Tei, M. Errahali and L. Marchese, *Langmuir*, 2012, **28**, 14405. Copyright 2012, American Chemical Society.

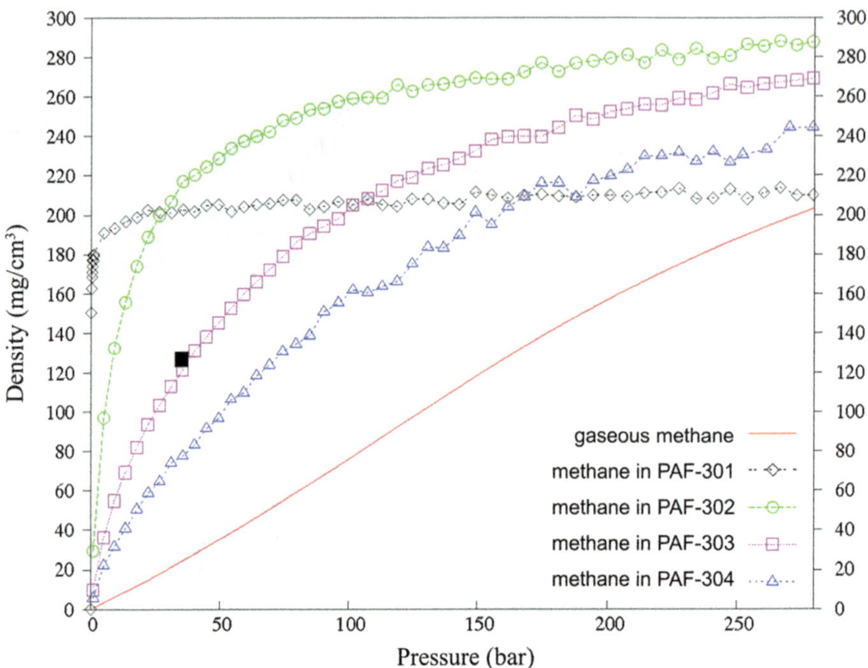

Figure 8.9 GCMC adsorption isotherms of methane in PAF-30X at 298 K. The density of free gaseous methane (from EOS) is shown for comparison. The black square indicates the storage target proposed by the DOE at 35 bar, 298 K.[86]
Reprinted with permission from K. M. Cossi, G. Gatti, L. Canti, L. Tei, M. Errahali and L. Marchese, *Langmuir*, 2012, **28**, 14405. Copyright 2012, American Chemical Society.

combines the advantage of a high surface area and narrow micropore, showing the best efficiency at pressures above 10 bar. The pore size of PAF-303 and PAF-304 is so large that the adsorbed molecules cannot interact with the phenyl rings efficiently. As these results come from ideal crystalline materials and the synthesized materials are basically amorphous, it doubtfully demonstrated the potential application of PAFs in methane storage. In further work, Fourier transform infrared (FTIR) and Raman spectroscopy have been used to investigate the system after methane adsorption.[87] Results show that adsorbed CH_4 interacts with two aromatic rings.

Carbon dioxide adsorption in PAFs was also stimulated by the GCMC technique.[88] Results obtained from PAF-30X ($X = 1$–4) are similar to that of methane, and PAF-302 shows the best performance. In addition, a series of PAF-302 frameworks modified with different functional groups and with different degrees of substitution were investigated. As already seen is some studies, all the functionalized PAFs show obvious enhanced CO_2 capture at low pressure and a higher substitution degree will result in higher CO_2

uptake. However, their high-pressure uptake decreased accordingly due to the decreased surface area. So, it is important to search for a balance between the two factors to design materials with high gas uptake at both high and low pressure, or rationally design proper materials for special applications under certain conditions.

Theoretical calculation of CO_2 storage and separation of PAF-30X ($X = 1$–4) was also conducted by Cao *et al.*[89] PAF-301 shows the highest CO_2 uptake of 275 mg g^{-1} at room temperature and 1 atm, as well as the highest selectivity for CO_2/N_2, CO_2/H_2, CO_2/CH_4 and CH_4/H_2 among the four PAFs due to its relatively narrow pore size (5.2 Å) and high isosteric heat of adsorption. These results once again highlight the potential applications of PAFs for carbon capture and storage.

8.4.2 Predication of Gas Storage in New Designed PAFs

Designing new PAFs is another important field in the search for new energy storage materials. By incorporating two Li($^+$)-CHN$_4$($^-$) groups to the middle aromatic ring of the building units, a new PAF was constructed based on the structure of PAF-1 (Figure 8.10).[90] The new PAF is very promising for H_2 storage due to the increased binding sites originating from lithium tetrazolide. GCMC simulations predict that the hydrogen adsorption at 233 K and 10 MPa will exceed the 2010 DOE target (4.9 wt% relative to 4.5 wt%). In another study, a series of new PAFs were computationally designed by introducing polar organic groups to the biphenyl unit of PAF-1 to investigate the CO_2 capture ability of these PAFs (Figure 8.11).[91] Results show functionalized PAF-1 shows a high gas uptake and selectivities for CO_2/CH_4, CO_2/N_2, and CO_2/H_2 mixtures, even higher than the amine functionality. All these examples show that incorporation of functional groups into the building blocks of PAFs or the post-synthetic modification may lead to enhanced gas storage and separation abilities of PAFs.

In another recent report, a series of nitrogen-containing PAFs were designed and their CO_2 uptakes were investigated.[92] The topological structures of the materials used are shown in Figure 8.12. The selectivity for CO_2 over CH_4, H_2 and N_2 is similar at low pressure in the order of PAF-NCC > PAF-N1 > PAF-N2 > PAF-N4 = PAF-NC > PAF-N0, which is in accordance with the calculated interaction energy. The results show groups with a single nitrogen atom, such as pyridine, are able to provide a stronger interaction with CO_2, while more nitrogen atoms in heterocycles reduce the interaction and this is much more obvious at low pressure. Additionally, the –COOH groups play more important roles for capture of CO_2 than the –NH$_2$ group. Whether the structures can be effectively obtained or not, this report obviously provides more information and clues for the functionalization of PAFs for CO_2 capture and on the development of more attractive PAFs.

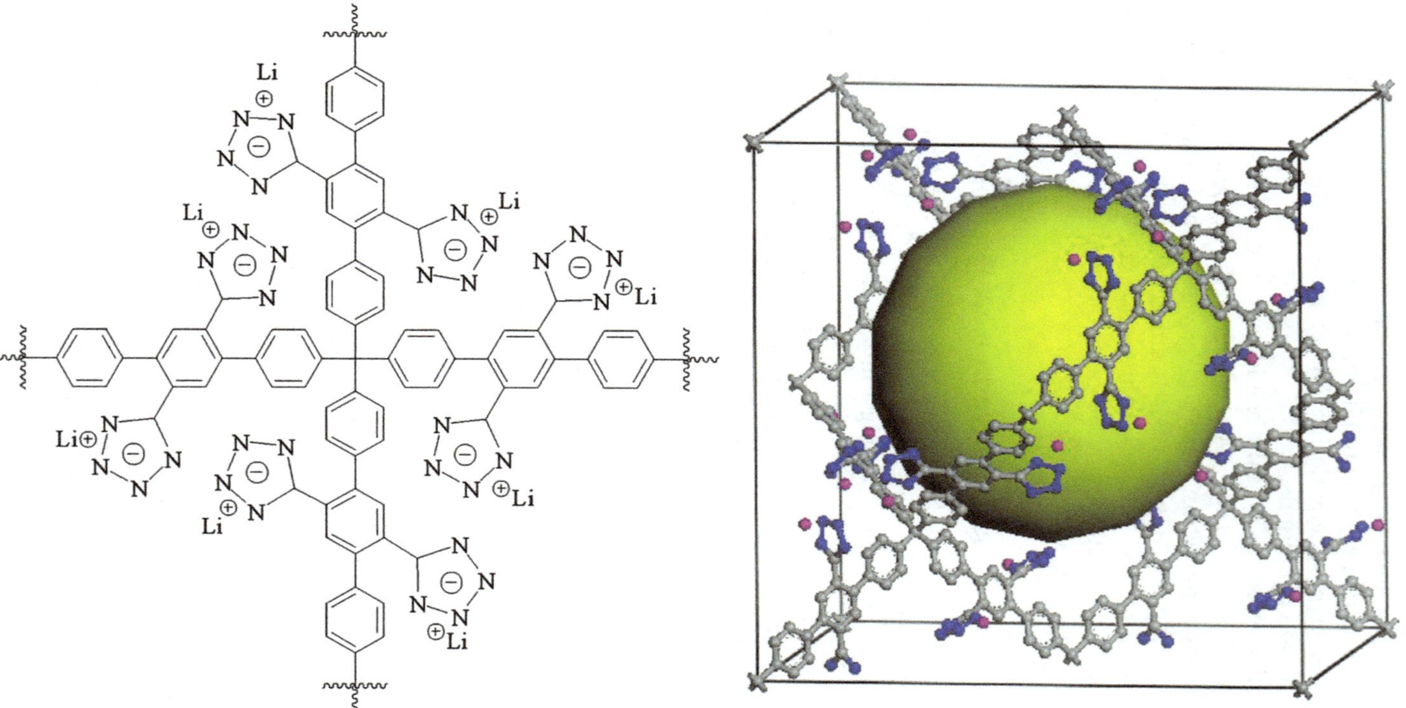

Figure 8.10 The building block (left) and unit cell (right) of the proposed porous aromatic framework (PAF-4) containing $Li(^+)$-$CHN_4(^-)$ moieties. The yellow ball denotes the free volume; grey spheres indicate carbon, blue spheres indicate nitrogen and purple spheres indicate lithium.[90]
Reprinted with permission from Y. Sun, T. Ben, L. Wang, S. Qiu and H. Sun, *J. Phys. Chem. Lett.*, 2010, **1**, 2753. Copyright 2010, American Chemical Society.

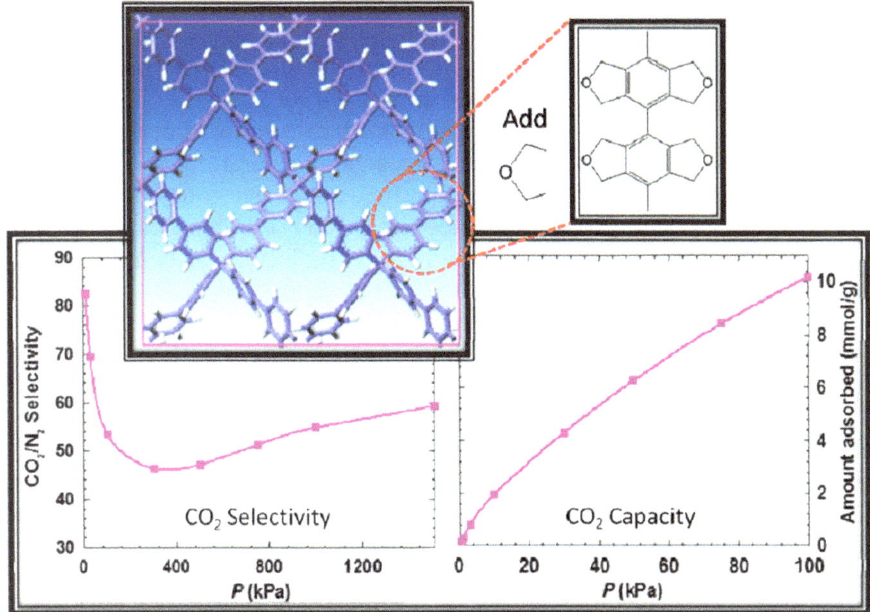

Figure 8.11 The unit cell of the proposed porous aromatic framework containing tetrahydrofuran-like moieties and the simulation pattern of CO_2/N_2 selectivity and CO_2 capacity.[91]
Reprinted with permission from R. Babarao, S. Dai and D. Jiang, *Langmuir*, 2011, **27**, 3451. Copyright 2011, American Chemical Society.

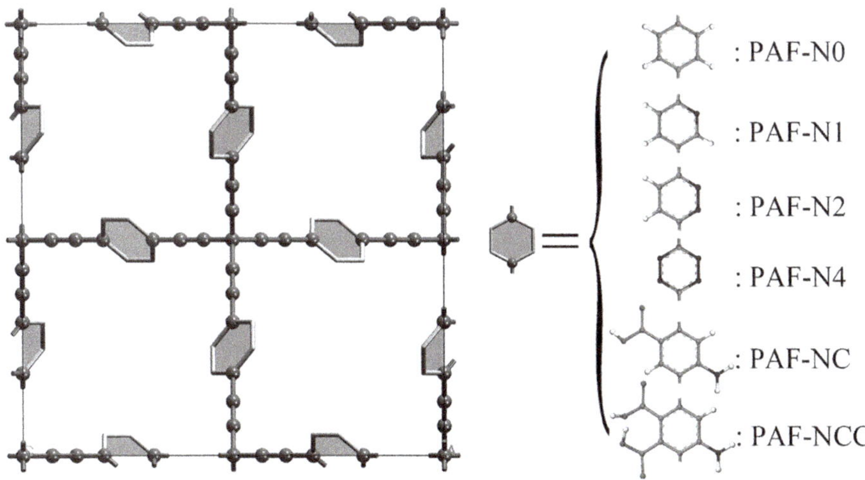

Figure 8.12 Structure topology of the porous diamond-like frameworks with alkynyl and selected rings.[92]
Reprinted with permission from W. Li, H. Shi and J. Zhang, *ChemPhysChem*, 2014, **15**, 1772. Copyright 2014 WILEY-VCH Verlag GmbH & Co. KGaA, Weinheim.

8.5   Conclusion and Perspectives

In conclusion, as a new generation of porous materials, PAFs are a class of
material with intrinsic micropores composed of covalent bonds. Both
experiments and theoretical simulation and calculation have shown the
potential applications of PAFs in gas storage and separation, the adsorption
of organic pollutions and many other fields. Besides, due to their high
physicochemical stability and surface area, PAFs are good candidates as
host materials that can be loaded with guest molecules or be modified by
incorporation of functional groups for specific applications.

Though great progress has been made in the synthesis and applications of
PAFs, there are still some aspects that require further investigation before
the applications of PAFs become a reality. First, for the synthesis of PAFs, the
Yamamoto-type Ullmann cross-coupling reaction seems to be the most
effective method. However, the high cost and harsh synthetic conditions of
this reaction are limitations with regards to the scale up to make it
industrially valuable. Though lots of PAF structures have been designed
theoretically showing large surface areas, their proper reaction routes need
to be elucidated to avoid the possibility of interpenetration during synthesis.
Secondly, investigations on gas storage in PAFs are usually conducted at low
temperature and do not meet practical applications. Incorporation of novel
and useful functions seems to be an effective method to increase the
interaction of gases with the framework, which is also very challenging.
Besides, most of the PAFs synthesized nowadays show very low conductivity
and this limits their use in the fields of supercapacitors and batteries, and
constructing conductive PAFs is another exciting opportunity in this field.

References

1. R. R. Xu, W. Q. Pang, J. H. Yu, Q. S. Huo and J. S. Chen, *Chemistry of Zeolites and Related Porous Materials*, Wiley-Inter Science, 2007.
2. C. T. Kresge, M. E. Leonowicz, W. J. Roth, J. C. Vartuli and J. S. Beck, *Nature*, 1992, **359**, 710.
3. J. S. Beck, J. C. Vartuli, W. J. Roth, M. E. Leonowicz, C. T. Kresge, K. D. Schmitt, C. T.-W. Chu, D. H. Olson, E. W. Sheppard, S. B. McCullen, J. B. Higgins and J. L. Schlenker, *J. Am. Chem. Soc.*, 1992, **114**, 10834.
4. D. Zhao, J. Feng, Q. Huo, N. Melosh, G. H. Fredrickson, B. F. Chmelka and G. D. Stucky, *Science*, 1998, **279**, 548.
5. O. M. Yaghi, M. O'Keeffe, N. W. Ockwig, H. K. Chae, M. Eddaudi and J. Kim, *Nature*, **423**, 705.
6. S. Kitagawa, R. Kitaura and S. Noro, *Angew. Chem., Int. Ed.*, 2004, **43**, 2334.
7. A. P. Cote, A. I. Benin, N. W. Ockwig, M. O'Keeffe, A. J. Matzger and O. M. Yaghi, *Science*, 2005, **310**, 1166.
8. H. M. El-Kaderi, J. R. Hunt, J. L. Medoza-Cortes, A. P. Cote, R. E. Taylor, M. O'Keeffe and O. M. Yaghi, *Science*, 2007, **316**, 268.

9. M. O'Keeffe and O. M. Yaghi, *Chem. Rev.*, 2012, **112**, 675.
10. H. C. Zhou and S. Kitagawa, *Chem. Soc. Rev.*, 2014, **43**, 5415.
11. X. Feng, X. Ding and D. Jiang, *Chem. Soc. Rev.*, 2012, **41**, 6010.
12. S. Y. Ding and W. Wang, *Chem. Soc. Rev.*, 2013, **42**, 548.
13. C. J. Doonan, D. J. Tranchemontagne, T. G. Glover, J. R. Hunt and O. M. Yaghi, *Nat. Chem.*, 2010, **2**, 235.
14. J. Jiang, F. Su, A. Trewin, C. D. Wood, N. L. Campbell, H. Niu, C. Dickinson, A. Y. Ganin, M. J. Rosseinsky, Y. Z. Khimyak and A. I. Cooper, *Angew. Chem., Int. Ed.*, 2007, **46**, 8574.
15. Y. Xu, S. Jin, H. Xu, A. Nagai and D. Jiang, *Chem. Soc. Rev.*, 2013, **42**, 8012.
16. S. Wan, J. Guo, J. Kim, H. Ihee and D. Jiang, *Angew. Chem., Int. Ed.*, 2008, **47**, 8826.
17. J. X. Jiang, F. Su, A. Trewin, C. D. Wood, H. Niu, T. A. Jones, Y. Z. Khimyak and A. I. Cooper, *J. Am. Chem. Soc.*, 2008, **130**, 7710.
18. N. B. McKeown and P. M. Budd, *Chem. Soc. Rev.*, 2006, **35**, 675.
19. N. B. McKeown, P. M. Budd, K. J. Msayib, B. S. Ghanem, H. J. Kingston, C. E. Tattershall, S. Makhseed, K. J. Reynolds and D. Fritsch, *Chem. – Eur. J.*, 2005, **11**, 2610.
20. M. P. Tsyurupa and V. A. Davankov, *React. Funct. Polym.*, 2002, **53**, 193.
21. J. Y. Lee, C. D. Wood, D. Bradshaw, M. J. Rosseinsky and A. I. Cooper, *Chem. Commun.*, 2006, **25**, 2670.
22. P. Kuhn, M. Antonietti and A. Thomas, *Angew. Chem. Int. Ed.*, 2008, **47**, 3450.
23. R. Palkovits, M. Antonietti, P. Kuhn, A. Thomas and F. Schuth, *Angew. Chem., Int. Ed.*, 2009, **48**, 6909.
24. C. E. Chan-Thaw, A. Villa, L. Prati and A. Thomas, *Chem. – Eur. J.*, 2011, **17**, 1052.
25. T. Ben, H. Ren, S. Ma, D. Cao, J. Lan, X. Jing, W. Wang, J. Xu, F. Deng, J. M. Simmons, S. Qiu and G. Zhu, *Angew. Chem., Int. Ed.*, 2009, **48**, 9457.
26. T. Ben, C. Pei, D. Zhang, J. Xu, F. Deng, X. Jing and S. Qiu, *Energy Environ. Sci.*, 2011, **4**, 3991.
27. Y. Peng, T. Ben, J. Xu, M. Xue, X. Jing, F. Deng, S. Qiu and G. Zhu, *Dalton Trans.*, 2011, **40**, 2720.
28. T. Ben, K. Shi, Y. Cui, C. Pei, Y. Zuo, H. Guo, D. Zhang, J. Xu, F. Deng, Z. Tian and S. Qiu, *J. Mater. Chem.*, 2011, **21**, 18208.
29. H. Ren, T. Ben, F. Sun, M. Guo, X. Jing, H. Ma, K. Cai, S. Qiu and G. Zhu, *J. Mater. Chem.*, 2011, **21**, 10348.
30. C. Pei, T. Ben, H. Guo, J. Xu, F. Deng, Z. Xiang, D. Cao and S. Qiu, *Philos. Trans. R. Soc., A.*, 2013, **371**, 20120312.
31. D. E. Demirocak, M. K. Ram, S. S. Srinivasan, D. Y. Goswamia and E. K. Stefanakosa, *J. Mater. Chem. A.*, 2013, **1**, 13800.
32. Y. Yuan, F. Sun, H. Ren, X. Jing, W. i. Wang, H. Ma, H. Zhaob and G. Zhu, *J. Mater. Chem.*, 2011, **21**, 13498.
33. H. Ren, T. Ben, E. Wang, X. Jing, M. Xue, B. Liu, Y. Cui, S. Qiu and G. Zhu, *Chem. Commun.*, 2010, **46**, 291.

34. W. Wang, H. Ren, F. Sun, K. Cai, H. Ma, J. Du, H. Zhao and G. Zhu, *Dalton Trans.*, 2012, **41**, 3933.

35. T. Yamamoto, *Bull. Chem. Soc. Jpn.*, 1999, **72**, 621.

36. G. Zhou, M. Baumgarten and K. Mllen, *J. Am. Chem. Soc.*, 2007, **129**, 12211.

37. W. Lu, D. Yuan, D. Zhao, C. I. Schilling, O. Plietzsch, T. Muller, S. Brase, J. Guenther, J. Blumel, R. Krishna, Z. Li and H. Zhou, *Chem. Mater.*, 2010, **22**, 5964.

38. A. Trewin and A. I. Cooper, *Angew. Chem., Int. Ed.*, 2010, **49**, 1533.

39. S. Sugiyama, S. Yamamoto, O. Matsuoka, H. Nozoye, J. Yu, G. Zhu, S. Qiu and O. Terasaki, *Microporous Mesoporous Mater.*, 1999, **28**, 1.

40. M. O'Keeffe and O. M. Yaghi, *Chem. – Eur. J.*, 1999, **5**, 2796.

41. W. Chaikittisilp, A. Sugawara, A. Shimojima and T. Okubo, *Chem. Mater.*, 2010, **22**, 4841.

42. N. Miyaura, K. Yamada and A. Suzuki, *Tetrahedron Lett.*, 1979, **36**, 343.

43. N. Miyaura, T. Yanagi and A. Suzuki, *Synth. Commun.*, 1981, **11**, 513.

44. Y. Luo, B. Li, W. Wang, K. Wu and B. Tan, *Adv. Mater.*, 2012, **24**, 5703.

45. R. Dawson, L. A. Stevens, T. C. Drage, C. E. Snape, M. W. Smith, D. J. Adams and A. I. Cooper, *J. Am. Chem. Soc.*, 2012, **134**, 10741.

46. R. Dawson, T. Ratvijitvech, M. Corker, A. Laybourn, Y. Z. Khimyak, A. I. Cooper and D. J. Adams, *Poly. Chem.*, 2012, **3**, 2034.

47. X. Jing, D. Zou, P. Cui, H. Ren and G. Zhu, *J. Mater. Chem. A*, 2013, **1**, 13926.

48. L. Li, H. Ren, Y. Yuan, G. Yu and G. Zhu, *J. Mater. Chem. A*, 2014, **2**, 11091.

49. Y. Yuan, F. Sun, F. Zhang, H. Ren, M. Guo, K. Cai, X. Jing, X. Gao and G. Zhu, *Adv. Mater.*, 2013, **25**, 6619.

50. Y. Yuan, F. Sun, L. Li, P. Cui and G. Zhu, *Nat. Commun.*, 2014, **5**, 4260.

51. D. M. D'Alessandro, B. Smit and J. R. Long, *Angew. Chem., Int. Ed.*, 2010, **49**, 6058.

52. IPCC (Intergovernmental Panel on Climate Change), *CO$_2$ Capture and Storage, a Special Report of the Intergovernmental Panel on Climate Change*, Cambridge University Press, Cambridge, 2005.

53. Q. Wang, J. Luo, Z. Zhong and A. Borgan, *Energy Environ. Sci.*, 2011, **4**, 42.

54. K. Sumida, D. L. Rogow, J. A. Mason, T. M. McDonald, E. D. Bloch, Z. R. Herm, T.-H. Bae and J. R. Long, *Chem. Rev.*, 2012, **112**, 724.

55. An-H. Lu and S. Dai, *Porous Materials for Carbon Dioxide Capture*, Springer, London, 2014.

56. M. P. Suh, H. J. Park, T. K. Prasad and D.-W. Lim, *Chem. Rev.*, 2012, **112**, 782.

57. L. J. Murray, M. Dinca and J. R. Long, *Chem. Soc. Rev.*, 2009, **38**, 1294.

58. D. Yuan, W. Lu, D. Zhao and H. C. Zhou, *Adv. Mater.*, 2011, **23**, 3723.

59. D. D'Alessandro, B. Smit and J. Long, *Angew. Chem., Int. Ed.*, 2010, **49**, 6058.

60. C. Pei, T. Ben, Y. Li and S. Qiu, *Chem. Commun.*, 2014, **50**, 6134.

61. C. Pei, T. Ben, Y. Cui and S. Qiu, *Adsorption*, 2012, **18**, 375.
62. J. McDonough, S. Yeon and Y. Gogotsi, *Energy Environ. Sci.*, 2011, **4**, 3059.
63. T. Ben, Y. Li, L. Zhu, D. Zhang, D. Cao, Z. Xiang, X. Yao and S. Qiu, *Energy Environ. Sci.*, 2012, **5**, 8370.
64. H. Zhao, Z. Jin, H. Su, J. Zhang, X. Yao, H. Zhao and G. Zhu, *Chem. Commun.*, 2013, **49**, 2780.
65. W. Lu, D. Yuan, J. Sculley, D. Zhao, R. Krishna and H. Zhou, *J. Am. Chem. Soc.*, 2011, **133**, 18126.
66. R. Yuan, H. Ren, Z. Yan, A. Wang and G. Zhu, *Polym. Chem.*, 2014, **5**, 2266.
67. H. Li, M. Eddaoudi, M. O'Keeffe and O. M. Yaghi, *Nature*, 1999, **402**, 276.
68. W. S. Huang, B. D. Humphrey and A. G. MacDiarmid, *J. Chem. Soc., Faraday Trans.*, 1986, **82**, 2385.
69. P. Kovacic and A. Kyriak, *J. Am. Chem. Soc.*, 1963, **85**, 454.
70. G. B. Street and T. C. Clarke, *IBM J. Res. Dev.*, 1981, **25**, 51.
71. J. W. Colson and W. R. Dichte, *Nat. Chem.*, 2013, **5**, 453.
72. J. W. Colson, A. R. Woll, A. M. Mukherjee, P. Levendorf, E. L. Spitler, V. B. Shields, M. G. Spencer, J. Park and W. R. Dichtel, *Science*, 2011, **332**, 228.
73. S. Wan, F. Gandara, A. Asano, H. Furukawa, A. Saeki, S. K. Dey, L. Liao, M. W. Ambrogio, Y. Y. Botros, X. F. Duan, S. Seki, J. F. Stoddart and O. M. Yaghi, *Chem. Mater.*, 2011, **23**, 4094.
74. G. H. V. Bertrand, V. K. Michaelis, T.-C. Ong, R. G. Griffin and M. Dinca, *Proc. Natl. Acad. Sci.*, 2013, **110**, 4923.
75. Y. Shirota, *J. Mater. Chem.*, 2000, **10**, 1.
76. Y. Shirota, *J. Mater. Chem.*, 2005, **15**, 75.
77. J. X. Jiang, A. Trewin, F. Su, C. D. Wood, H. Niu, J. T. A. Jones, Y. Z. Khimyak and A. I. Cooper, *Macromolecules*, 2009, **42**, 2658.
78. J. X. Jiang, A. Laybourn, R. Clowes, Y. Z. Khimyak, J. Bacsa, S. J. Higgins, D. J. Adams and A. I. Cooper, *Macromolecules*, 2010, **43**, 7577.
79. B. Guo, T. Ben, Z. Bi, G. M. Veith, X.-G. Sun, S. Qiu and S. Dai, *Chem. Commun.*, 2013, **49**, 4905.
80. C. Pei, T. Ben, S. Xu and S. Qiu, *J. Mater. Chem. A.*, 2014, **2**, 7179.
81. K. Konstas, J. s. W. Taylor, A. W. Thornton, C. M. Doherty, W. Lim, T. J. Bastow, D. F. Kennedy, C. D. Wood, B. J. Cox, J. M. Hill, A. J. Hill and M. R. Hill, *Angew. Chem., Int. Ed.*, 2012, **51**, 6639.
82. W. Lu, J. P. Sculley, D. Yuan, R. Krishna, Z. Wei and H.-C. Zhou, *Angew. Chem., Int. Ed.*, 2012, **51**, 7480.
83. E. Kang, S. B. Choi, N. Ko and J. K. Yang, *Bull. Korean Chem. Soc.*, 2014, **35**, 283.
84. Y. Li, T. Ben, B. Zhang, Y. Fu and S. Qiu, *Sci. Rep.*, 2013, **3**, 2420.
85. J. Lan, D. Cao, W. Wang, T. Ben and G. Zhu, *J. Phys. Chem. Lett.*, 2010, **1**, 978.
86. M. Cossi, G. Gatti, L. Canti, L. Tei, M. Errahali and L. Marchese, *Langmuir*, 2012, **28**, 14405.

87. M. Errahali, G. Gatti, L. Tei, L. Canti, A. Fraccarollo, M. Cossi and L. Marchese, *J. Phys. Chem. C.*, 2014, **118**, 10053.
88. A. Fraccarollo, L. Canti, L. Marchese and M. Cossi, *Langmuir*, 2014, **30**, 4147.
89. Z. Yang, X. Peng and D. Cao, *J. Phys. Chem. C.*, 2013, **117**, 8353.
90. Y. Sun, T. Ben, L. Wang, S. Qiu and H. Sun, *J. Phys. Chem. Lett.*, 2010, **1**, 2753.
91. R. Babarao, S. Dai and D. Jiang, *Langmuir*, 2011, **27**, 3451.
92. W. Li, H. Shi and J. Zhang, *ChemPhysChem.*, 2014, **15**, 1772.

Functionalization of Porous Polymers

9.1 Introduction

Controlled functionalization and pre-desirable tuning of pore size, as well as the shape of porous materials, can be achieved by introduction or modification of the building blocks of porous materials. Different functional groups or doping agents can be introduced onto these polymers *via* the pre- or post-synthetic modification strategy to either enhance the already existing properties or introduce a new property. Characteristics like pore size, surface area and micropore volume of the porous framework are principally dependent on the dimensions and structures of the monomers or building blocks. Thus, tuning of the physical properties of such polymers can be achieved by using monomers of various sizes, shapes and also by linking functional moieties onto them. Pore dimension and surface area can be governed by controlling the strut length of the monomers. Increased strut length results in a lower surface area, while a smaller strut gives a smaller pore size and higher micropore volume. This is due to the higher interpenetration of longer struts into the framework than the shorter ones. Organic moieties often provide designed networks with higher surface areas. Manipulation of monomer strut length and controlled functionalization empowers one to control the dimensions of pores and the surfaces of such porous frameworks. This approach holds ground for the proposed structure–property relationship and enlightens researchers with regards to porous polymer frameworks and the possible exploitation of their properties. This strategy helps the understanding of the trends and connecting links among different classes of polymers with respect to the effect of monomer strut length, the side chains on the pore surface area, pre- or post-synthetic functionalization to enhance the surface area and the selection of aromatic

Monographs in Supramolecular Chemistry No. 17
Porous Polymers: Design, Synthesis and Applications
By Shilun Qiu and Teng Ben
© The Royal Society of Chemistry 2016
Published by the Royal Society of Chemistry, www.rsc.org

monomers for tailoring stable pores owing to their rigidity. Now we shall discuss various porous frameworks in the light of functionalization in detail.

9.2 Pre-modification of Porous Organic Frameworks

9.2.1 Pre-modification of Hyper-crosslinked Polymers

Hyper-crosslinked polymers (HCPs) were the first reported porous polymeric materials[1] and carbenol HCPs were the first non-styrenic HCPs that were synthesized by linking biphenyls together with C–OH groups.[2,3] Polystyrene-based HCPs, also known as 'Davankov resins',[1] are the other type, which are obtained *via* Friedel–Crafts alkylation reactions to get even higher surface areas.[4,5] Both types of HCPs have gone through several post-synthetic functionalization to obtain better or new properties. Fontanals *et al.* has reported that micro-sized hyper-crosslinked polystyrene networks were synthesized employing a vinylbenzyl chloride (VBC)–divinylbenzene (DVB) co-polymer resin, which can be used to obtain spherical ultra-high specific surface area, monodisperse polymer particles for significant absorption of hydrocarbon solvents and water.[6]

Pore size tuning is another great aspect in HCPs with pre-designed strategies. Control of the pore size of a hyper-crosslinked polymer, *i.e.*, poly(divinylbenzene-co-vinylbenzyl chloride) (HCP-DVB–VBC) has been demonstrated by adjusting the DVB content in poly(divinylbenzene-co-vinylbenzyl chloride) (DVB–VBC) precursors and then hyper-crosslinking these DVB–VBC precursors (Figure 9.1).[7] The pore structure of HCP-DVB–VBC can be tuned by varying the DVB content. When the DVB concentration is less than 7%, a sheer microporous structure results. When the DVB content is higher than 7%, HCP-DVB–VBC changes to a pure microporous organic polymer.

It was also reported that HCP networks can be used as stabilizers to produce metal nanoparticles. A hyper-crosslinked polystyrene network (HPS),[1] obtained from a styrene–divinylbenzene copolymer, acts as a matrix to synthesize Co nanoparticles owing to its nano-sized rigid cavities of comparable size. These cavities can consume several Co complexes and can control the growth, particle size and size distribution of Co nanoparticles.[8]

Various Brunauer–Emmett–Teller (BET) surface areas have been obtained for hyper-crosslinked networks of polypyrrole using diiodomethane, triiodomethane and triiodoborane by crosslinking with CH_2, CH and B crosslinkers, respectively.[9] It was also observed that HCPs reduced with metal particles like Li possess increased selectivity towards CO_2 over CH_4, although there was a decrease in surface area as metal particles fill the pores[10] (not post-synthesis).

9.2.2 Polymers of Intrinsic Microporosity

Polymers of intrinsic microporosity (PIMs) consist of phthalocyanines and porphyrins containing metal ions or H^+ in the cavities and they exhibit variation in surface area due to the varying metal ions or H^+.[11,12] On tuning

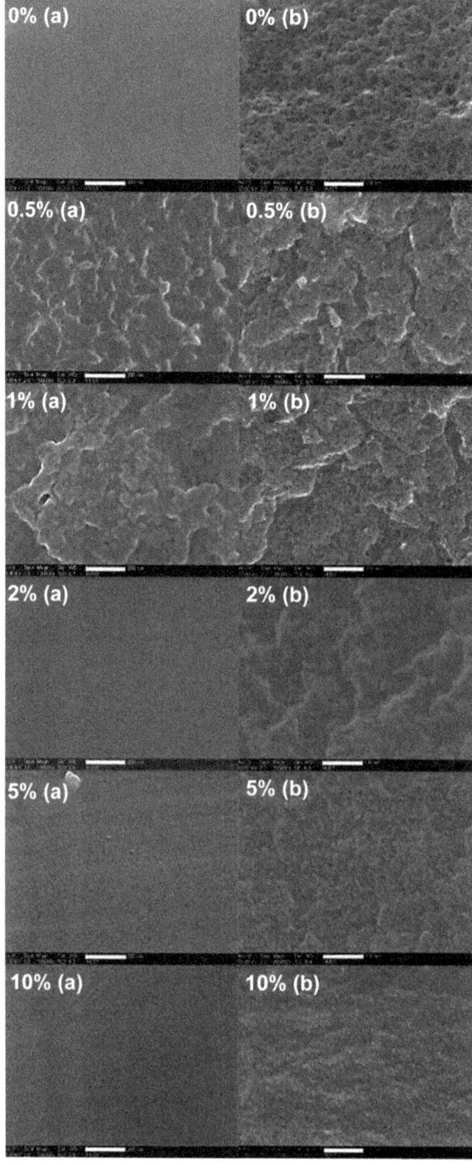

Figure 9.1 Field emission scanning electron microscopy (FE-SEM) image of the fracture section of HCP-DVB–VBC with an increase in DVB content, with 0%, 0.5%, 1%, 2%, 5% and 10% DVB before (a, scale 200 nm) and after (b, scale 100 nm) the hyper-crosslinking reaction.
Reproduced from ref. 7 with permission from The Royal Society of Chemistry.

the alkyl chain length of the triptycene monomer of hexahydroxytriptycene-based PIMs, the surface area of the whole network can be successfully controlled. The shortest alkyl chain, methyl, gives rise to the highest surface

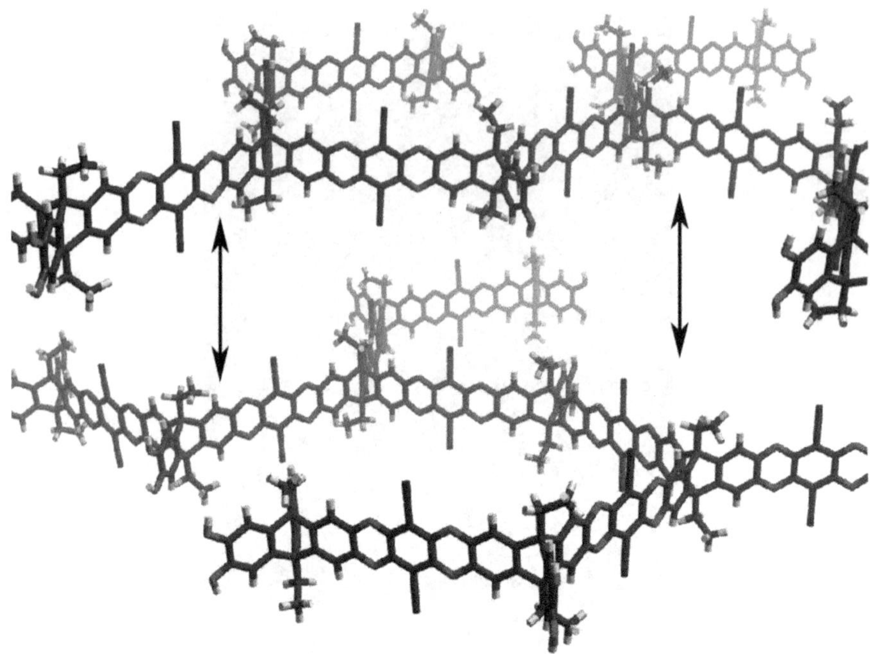

Figure 9.2   Representation of two ideal fragments of the network of Trip(Et)-PIM, showing how the shape of each macromolecule, as dictated by the architecture of the triptycene (Trip) units, prevents close intermolecular interactions between the planar 'struts'. The loose network that arises from the ideal layered structure may account for the tendency of the materials to swell in organic solvents or during nitrogen adsorption, as indicated by the arrows, especially when bridgehead alkyl chains block interpenetration of the nitrile groups.[14]
Reprinted with permission from B. S. Ghanem, M. Hashem, K. D. M. Harris, K. J. Msayib, M. Xu, P. M. Budd, N. Chaukura, D. Book, S. Tedds, A. Walton and N. B. McKeown, *Macromolecules*, 2010, **43**, 5287. Copyright 2010, American Chemical Society.

area, whereas the longest possible chain, octyl, results in the lowest surface area (Figure 9.2).[13,14] Rearrangement of chain conformation and topology can create microporosity, which can be used in the selective separation of gas molecules. Such post-modification of PIM-1 has been reported *via* a [2 + 3] cycloaddition reaction from a tetrazole derivative of PIM-1, providing excellent gas separation properties.[15] Tuning the length and branching of alkyl chains of triptycene-based PIMs empowers researchers to vary the apparent surface area of the polymeric network, and thus the gas adsorption properties. Higher microporosity has been obtained from shorter and branched alkyl chains, whereas linear and longer alkyl chains produce small pores as they cause blockage of the networks. The exceptional hydrogen adsorption capacity of PIMs with methyl (shorter chain) and isopropyl (branched chain) groups verifies this fact.[14] PIMs consisting of phthalocyanine, porphyrin or hexaazatrinaphthylene show high catalytic activity on insertion of certain transition metal ions into their network.[16] Such a study has been demonstrated in the case of a metal-containing porphyrin or phthalocyanine network, which shows excellent catalytic properties towards H_2O_2 degradation.[16] Pd^{2+}-loaded network PIMs of hexaazatrinaphthylene provide tremendous potential as heterogeneous catalysts and chemoselective adsorbents.[17] Such applications have deliberately been shown exploiting a palladium-containing microporous material in the efficient heterogeneous catalysis of aryl–aryl coupling Suzuki reactions.[16] Porous polymeric networks with very high specific surface area can be obtained from PIM-7 as its phenazine subunits coordinate strongly with metal ions.

Exploiting the design principle of PIMs, polyimide derivatives of PIMs (PIM-PIs) can be obtained with gas separation properties better than conventional polyimides with permeation properties. Permeation of PIM-PIs depends on the specific structure of the diamine units, variation of which empowers prediction of permeation parameters for other PIM-PI structures.[18]

9.2.3 Covalent Organic Frameworks

Impressive progress in the functionalization of covalent organic frameworks (COFs) is found using three-dimensional (3D) monomers like tetra(4-dihydroxyborylphenyl)methane and the silicon-centered analogue,[19] replacing 2D monomers like hexahydroxytriphenylene (COF-102, 103) or using them together (COF-105, 108).[20,21] Resultant materials are crystalline and exhibit higher surface areas than 2D-COFs.

Pore engineering of COFs is an important aspect to explore their applicability towards a wide field of uses. A covalently linked porphyrin-COF catalyst has been synthesized using imine groups[22] by attaching organocatalytic sites onto the pore walls of the synthesized COF (Figure 9.3). Such a kind of surface activation provides enhanced activity in the stereoselective asymmetric Michael addition reaction. Recyclability and continuous catalytic ability are the key features of this COF catalyst.

The pore size of COFs can be tuned by varying the dimension of the alkyl chain of the monomer used. The highest pore sizes have been found in the absence of any alkyl group; small length chains, like methyl, cause reduction of the pore size, whereas using alkyl groups of higher length (propyl) reduce the pore size further.[23] Such a type of pore size tuning is useful for selective gas separations as small pores exhibit higher permeation for small gas molecules, like hydrogen, than bigger ones.

The applicability of porous materials lies principally on the porosity and crystallinity of the materials. Tuning of the π-electronic interlayer interaction among different layers empowers one to hold a control over the porosity and crystallinity of 2D COFs.[24] Arenes and fluoro-substituted arenes have been incorporated into the edges of imine-linked copper-porphyrin COFs. A tetrafluoro-substituted arene linked copper-porphyrin COF shows the strongest self-complementary electronic interactions. For such a π-stacked

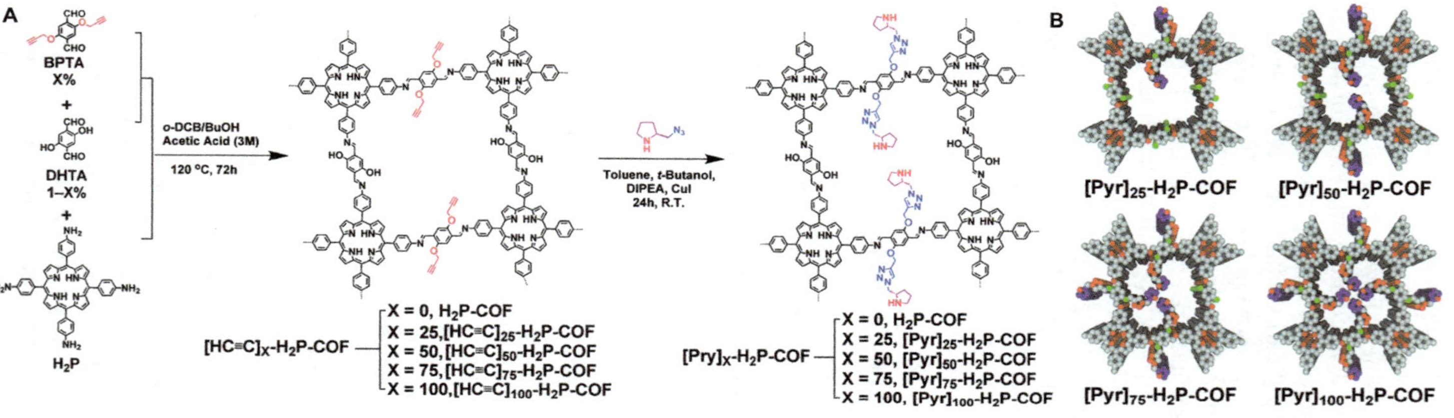

Figure 9.3 Strategy of pore surface engineering (A) of imine-linked COFs (B) *via* a condensation reaction and click chemistry (BPTA, 2,5-bis(2-propynyloxy)terephthaladehyde; DHTA, 2,5-dihydroxyterephthaladehyde; o-DCB, *ortho*-dichlorobenzene; DIPEA, *N,N*-diisopropylethylamine).
Reproduced from ref. 22 with permission from The Royal Society of Chemistry.

COFs, the crystalline nature of the COFs is dependent on the improved interlayer π-interactions.[25]

High crystallinity and large surface area are also obtained by incorporating triphenylene and diimide as donor and acceptor building blocks, respectively, in COFs. COFs with large pore dimensions, resulting from variation of the donor–acceptor structure, play an important role in charge transfer and separation in lattices.[26]

Introduction of dyes, like squaraine, in the 2D-COF framework to construct a tetragonal mesoporous skeleton (squaraine-based COF) opens up a huge possibility of applications towards imaging, nonlinear optics, photovoltaics, photodynamic therapy, ion sensing *etc*.[27] These COFs have a zig-zagged conformation, making them highly stable and providing an extended π-conjugation over the 2D network. Properties, like low band gap energy and high absorbance capacity, make them promising in new functional scopes.

9.2.4 Conjugated Microporous Polymers

The Sonogashira–Hagihara crosscoupling reaction has been employed for the synthesis of several conjugated microporous poly(aryleneethynylene) (PAE) networks by the Cooper group.[28,29] These conjugated microporous polymer (CMP) networks can be tuned to vary the surface area, pore volume and pore size by starting with 1,4- or 1,3,5-substituted monomers of different strut lengths. Free rotation of the groups along the alkyne bond creates the 3D network. CMP-1 is basically hydrophobic due to its aromatic network. Incorporation of fluorinated groups makes CMP-1 even more hydrophobic. Hydrophilicity can be introduced into the CMP-1 network by incorporating alcohol groups.[30,31] The CMP networks can be further modified by the introduction of nitrogen atoms into the networks, which can accommodate various functional molecules, like metal atoms, owing to their ability to bind *via* the lone pairs of electrons. Such functionalization is used especially in photovoltaic devices.[32]

Polymer semiconductors generally exhibit lower charge carrier mobility, which is the principle drawback regarding their applicability. Porous conjugated macromolecules act in a tremendous role here when crosslinked with π-conjugated molecules for the design of semiconducting polymers with improved charge transport properties.[33] Another such approach is to introduce metal ions into conjugated polymers for potential applications through the manipulation of the electronic properties of these materials.[34–39] PAE-derived CMPs can be functionalized with various moieties, such as organometallic complexes,[40] for different applications, including heterogeneous catalysis. π-Conjugated metallopolymers are mainly designed by two techniques: incorporation of metal centers into the polymer to co-ordinate with the conjugated network backbone, or by attaching metals as side groups *via* conjugated or non-conjugated spacer units.[41] One such functionalization *via* the formation of organometallic conjugated networks by ligand-exchange was reported by Wright, where he proposed the crosslinking reaction of poly(aryleneethynylene)s with a $Cr(CO)_3$–benzene moiety, with the formation of phenylene–$Cr(CO)_2$–ethynyl moieties.[42] Many other such kinds of metallo-organic crosslinking reactions have also been reported, such as by Hirao *et al.*, who coordinated Pd^{2+} or Cu^{2+} with poly(*o*-toluidine).[43] Iron(III) porphyrin-based CMP[44] and an Ir-loaded CMP[45] have been incorporated for the oxidation of sulfides to sulfoxides and a reductive amination reaction, respectively.

Microporous conjugated spirobifluorene polymers have been synthesized by Thomas *et al.* by coupling 2,2′,7,7′-tetrabromo-9-9′-spirobifluorene with 1,4-diethynylbenzene, providing a BET surface area of 510 $m^2\ g^{-1}$ by utilizing the Sonogashira–Hagihara reaction.[46] A much decreased surface area is obtained employing the Suzuki reaction for the same monomer reacted with benzene-1,4-diboronic acid and 4,4′-biphenyldiboronic acid, yielding BET surface areas of 450 and 210 $m^2\ g^{-1}$, respectively. Adsorption and emission wavelengths of networks derived from 2,5-thiophene diboronic acid with a spiro monomer can be controlled by changing the thiophene and benzene diboronic acid ratio.[47]

The high physical surface area in CMPs can be utilized and filled with other conjugated polymers. Such a property of CMPs leads to the synthesis of an interpenetrating network of more than one polymer, which otherwise would be incompatible for blending and would separate out from the phase immediately.

CMPs can be tuned to explore their various photocatalytic and absorptive properties *via* the incorporation of a dye molecule into the network. CMPs with high porosity (surface area >830 $m^2\ g^{-1}$) and high activity for heterogeneous photocatalytic aza-Henry reactions at room temperature are obtained by introducing the Rose Bengal dye into the CMP-1 and CMP-2 network.[48] Such CMP photocatalysts are highly reusable in catalytic cycles and are eco-friendly as they are free from noble metals. There are also a lot of reports exploiting the photocatalytic activity of such porous networks by post-synthetic modifications. The aza-Henry reaction can also be catalyzed by Ir- and Ru-incorporated CMPs.[49] Photocatalytic coupling of primary amines can be effectively catalyzed by benzodifuran-functionalized CMPs.[50] Kiskan *et al.* have synthesized the phenolphthalein-functionalized CMP ion used as a photosensitizer for the heterogeneous polymerization of methyl methacrylate.[51]

Using selective monomers, CMPs can be modified to hold control over the physical properties, which in turn can be channelled in various applications. Controlling the dye sorption behavior by tuning the hydrophobicity is such an example that has been illustrated by the Cooper group,[31] where they have prepared porous PAE networks with a range of functionalities and with high surface areas of the functionalized dibromobenzenes with 1,3,5-triethynylbenzene. Variation of monomer selection produces differences in the hydrophobicity of the resulting polymers, thus their dye sorption behavior can be controlled. Such an approach can be utilized especially in separation and catalysis.

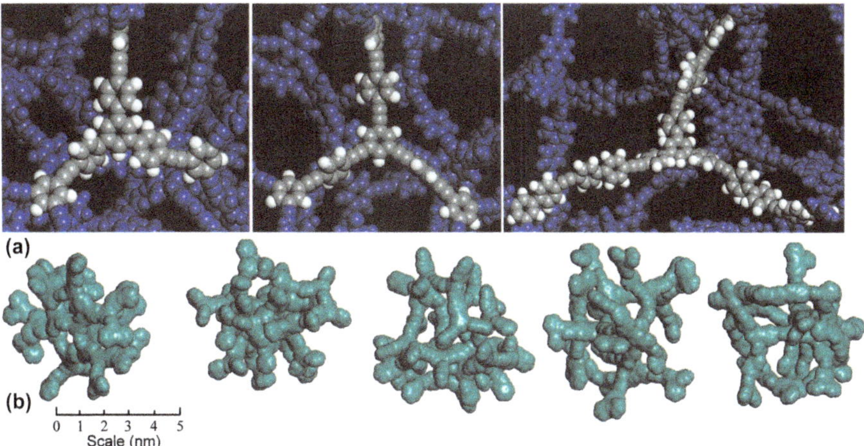

Figure 9.4 Atomistic simulations for PAE networks with different strut lengths. (a) The node–strut topology for simulated network fragments is shown in deep blue for CMP-0, CMP-1 and CMP-5 (left to right) with a 1,3,5-connected benzene node connecting three other nodes via rigid struts highlighted in grey/white. (b) From left to right, atomistic simulations of network fragments showing the solvent-accessible surface for CMP-0, CMP-1, CMP-2, CMP-3 and CMP-5.[30]
Reprinted with permission from R. Dawson, A. Laybourn, R. Clowes, Y. Z. Khimyak, D. J. Adams and A. I. Cooper, *Macromolecules*, 2009, **42**, 8809. Copyright 2009, American Chemical Society.

Incorporating statistical copolymerization of monomers with different strut lengths, the micropore properties of CMPs, such as the pore dimension, surface area and hydrogen uptake ability, can be tuned in a controlled fashion, as reported by Cooper *et al.* (Figure 9.4).[29] They employed the Sonogashira–Hagihara coupling reaction to synthesize a series of rigid microporous PAE networks with surface areas of over 1000 $m^2\ g^{-1}$. Resulting polymeric networks have shown good stability with retention of microporosity under various physical and chemical conditions. This shows that extended long-range order is not a necessary condition to control the micropore properties.

CMPs can be made soluble in organic solvents by introducing soluble alkyl groups by hyperbranching.[52] Resulting polymers can be processed from solution to form films. It should be noted that the porosity is a function of the processing conditions and originates from rigidity combined with non-interpenetrating cavities.

9.3 Post-synthetic Modification

Porous organic frameworks (POFs), such as metal–organic frameworks (MOFs),[53–56] covalent organic frameworks (COFs),[57,58] conjugated microporous polymers (CMPs),[59,60] polymers of intrinsic microporosity (PIMs),[61] crystalline triazine-based organic frameworks (CTFs),[62] hyper-crosslinked polymers (HCPs)[63,64] and porous aromatic frameworks (PAFs),[65–68] are characterized by tailorable organic building blocks, tuneable porosity and modifiable frameworks. They exhibit diverse applications in gas storage,[69] molecule separation,[70] catalysis,[71–73] sensing,[74] light harvesting,[75–79] ionic conductivity,[80–83] nonlinear optical behavior[84] *etc.* Access to specialized and sophisticated applications requires the integration of a wide range of chemical functional groups into POFs. Despite that many diverse functional POFs,[85] such as MIL-101,[86,87] IRMOF-3[88] and JUC-Z3,[89] have been successfully prepared *via* a direct condensation reaction, the scope of functional groups introduced on POFs has been largely limited by the synthesis method. For instance, if the functional groups on the monomer are incompatible with or unstable under the assembly conditions, they sometimes interfere with the formation of the polymers. Furthermore, finding an appropriate method to condense the monomer with functionalized groups is a time-consuming and non-trivial process, which limits the use of the pre-modification strategy to functionalize POFs.

9.3.1 Strategies for the Post-synthetic Modification of Porous Polymers

Faced with the requirement of introducing various functional groups into POFs, the potential advantage of post-synthetic modification (PSM) has been highlighted. PSM is defined as the chemical synthesis and modification of POFs in a heterogeneous manner after the formation of POFs.[90–92] It means that chemical bonds form and break during the functionalization process. PSM was first mentioned by Hoskins and Robson in 1990 by 'relatively unimpeded migration of species throughout the lattice may allow chemical functionalization of the rods subsequent to construction of the framework'.[93] After that, it has been extensively reviewed by Cohen *et al.*[91,92] and by Burrows *et al.*[94] It has been proven that PSM is a general, practical approach for the functionalization of POFs since it can control the type and the number of functional groups in the polymer frameworks. There are several strategies for PSM on POFs. Herein, these are divided into five categories: covalent PSM, dative PSM, post-synthetic deprotection (PSD), tandem PSM and carbonization. The first four are to modify frameworks without any effect on the overall stability of the network, while the latter is to partly break the bonds and reform a novel porous framework.

As shown in Figure 9.5, covalent PSM is described as the modification of organic building blocks in POFs with covalent bonds. At present, this strategy is the most extensively investigated one among all the PSM methods. It's superior as it is a powerful and versatile method that introduces a broad range of functional groups into POFs. Modification of organic linkers, including direct aromatic substitution, and functionalization of existed functional groups in POFs has also been investigated. In particular, sulfonation[95,96] and amine-grafting[97] are two effective methods to directly functionalize aromatic units, while amino[98–100] and hydroxyl[101,102] functional groups in POFs are two popular and reactive candidates for further diverse decoration. As a result, the change of polarity and electron density influences the host–guest interactions and makes POFs exhibit specific properties.

Dative PSM is defined as the modification of POFs through dative bonds (previously referred to as coordinate bonds, which generally appear between metal ions and ligands) without altering the framework topology. It can be divided into two cases, as shown in Figure 9.5. In dative PSM I, the organic building block supports the metal coordinate sites, such as catechol,[103–106] and bipyridyl unit.[49] After the formation of POFs, these hooks can catch any metal ions matching them in scale and charge to metallize the network. On the other hand, dative PSM II refers to POFs with unsaturated metal sites, in which the metal nodes coordinate with ligands, such as alkylamine and pyridines, to functionalize the frameworks.

Protection and deprotection of functional groups is an important strategy in organic synthesis. Particularly, trimethylsilyl ether (TMS), methoxymethyl acetal (MOM) and *tert*-butoxy carbamate (Boc) are three frequently used effective reagents in the protection of alcohols, phenols, amines, carboxylic acids, aldehydes and ketones.[107–110] Hence, there are many optional reactions and schemes in this PSM strategy. Utilized protection and deprotection of functional groups to modify POFs is named post-synthetic deprotection (PSD). In the PSD strategy, protection of sensitive units is generally carried out before the condensation of monomers, while the target product is obtained by cleavage of the chemical bonds to deprotect the functional groups after the formation of the polymer. Apart from sensitive functional group protection, the protecting groups could be utilized as templates during the assembly process to prevent interpenetration and collapse of the framework. Thus, after the deprotection, removing the large armour, an open porous framework with low density may possibly be obtained, especially for the irreversible, condensed product and highly stable materials.

It should be noted that these PSM strategies are not used in isolation to obtain a specific property, but are usually used in combination with other strategies, and this approach is named tandem PSM. Figure 9.5 exhibits just one case; however, tandem PSM extends POFs further into the realm of functional diversity since the functionalization is not limited to unsaturated metal sites or the unreactive functional groups.[111] The advantage is that you can combine any optimal factors you want into one POF material. At this point, we wonder how many functional groups can be incorporated into one POF and how the spatial localization can be controlled. For instance, up to eight distinct functional groups may be introduced into multivariate MOFs (MTV-MOFs), while four sequential additions of anhydrides and isocyanates containing a mixture of alkyl, unsaturated hydrocarbons and aromatic substituents could be added to IRMOF-3. This proves that a well-designed functionalization scheme and carefully controlled modification conditions could allow the construction of multi-functional Schweizer Messer-like POFs, which adapt to the requirements of complex applications.

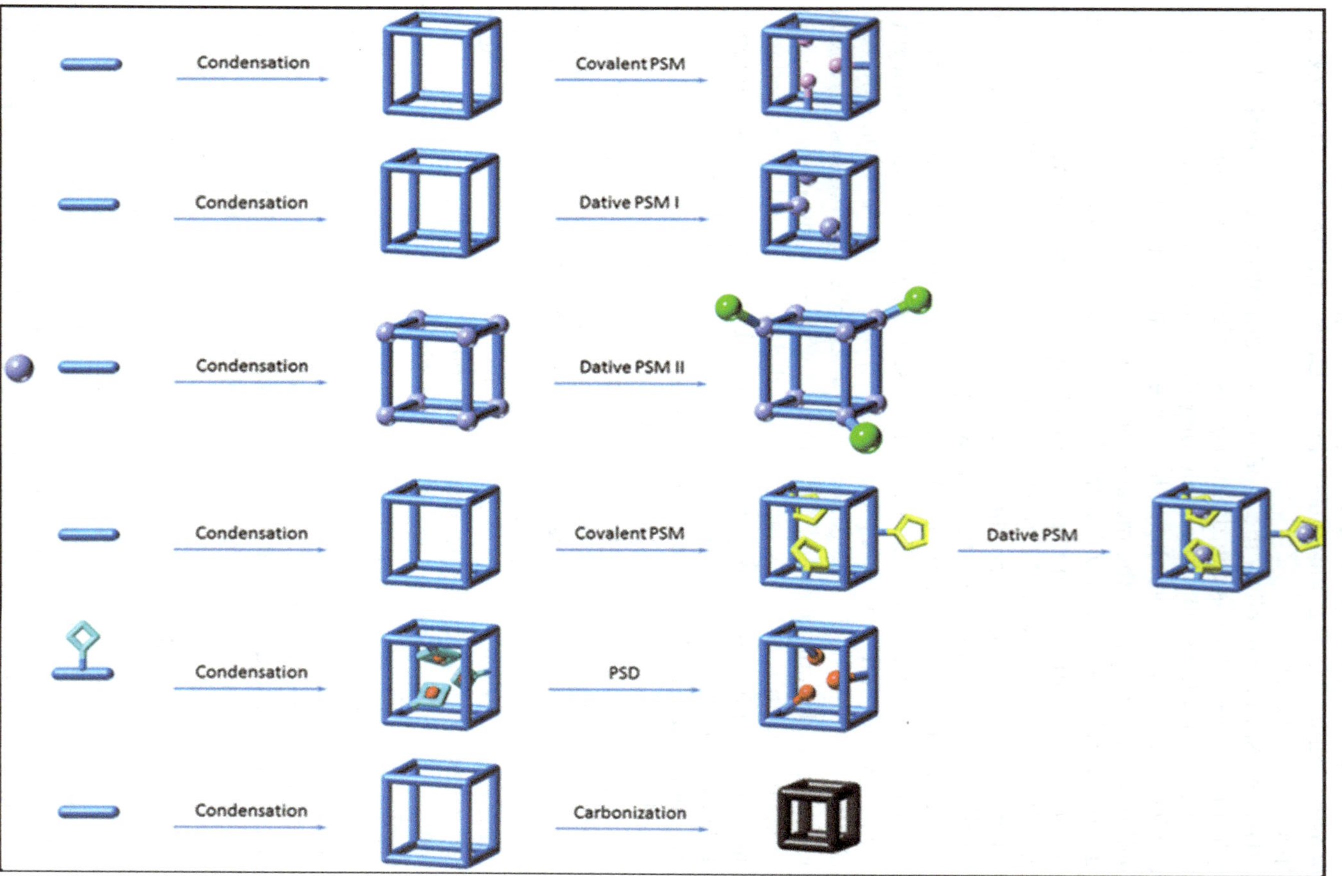
Condensation
Covalent PSM
Condensation
Dative PSM I
Condensation
Dative PSM II
Condensation
Covalent PSM
Dative PSM
Condensation
PSD
Condensation
Carbonization

Carbonization is another PSM strategy. This approach fabricates porous carbon by thermal treatment of POFs with or without a template. In the carbonization process, chemical bond breaking accompanies the formation of a novel porous framework, as well as shrinking of the pore size. Generally, the carbonized porous material exhibits strong host–guest interactions due to the overlapped force field and the enhanced polarity.[112]

9.3.2 Advantages and Disadvantages of PSM of Porous Polymers

9.3.2.1 Tailoring the Pore Size

Tuning the physical properties of POFs can be achieved by using monomers of various sizes, shapes and also by linking functional moieties onto them. The pore dimension and surface area can be governed by controlling the strut length of the monomers. Increased strut length results in lower surface area, while a smaller strut gives smaller pore sizes and higher micropore volumes. This is due to the higher interpenetration of longer struts into the framework than the shorter ones. Organic moieties often provide designed networks with higher surface areas. Manipulation of monomer strut length and controlled functionalization empowers one to get control over the dimensions of the pores and surfaces of such porous frameworks.

Along with experimentally developed functionalization of porous aromatic frameworks, several computational and theoretical methods have been introduced to investigate modified networks to obtain improved properties. Lan *et al.* have shown the hydrogen adsorption capacity in PAF networks by replacing the C–C bond in the diamond structure with several phenyl rings.[113] Another PAF with lithium tetrazolide linkers has been proposed with a good hydrogen storage capacity using GCMC simulations.[114]

9.3.2.2 Introduction of a Hybrid Component

Jiang and his co-workers have designed new PAFs computationally using grand-canonical Monte Carlo (GCMC) simulations by introducing polar

---

Figure 9.5   Generic schemes for covalent post-synthetic modification, dative post-synthetic modification, tandem post-synthetic modification, post-synthetic deprotection and carbonization (blue sticks are organic building blocks, pink balls are functional groups linked with covalent bonds, purple balls are metal ions, green balls are functional groups linked with coordinate bonds, yellow five-membered rings are functional groups with metal coordinate sites, red balls are sensitive functional groups, cyan four-membered rings are units protecting sensitive functional groups and black sticks are carbonized building blocks).[111] Reprinted with permission from K. Oisaki, Q. Li, H. Furukawa, A. U. Czaja and O. M. Yaghi, *J. Am. Chem. Soc.*, 2010, **132**, 9262. Copyright 2010, American Chemical Society.

organic groups into the biphenyl unit and then investigated their gas separating abilities.[22] They have introduced various functional moieties in the PAF, like acetonitrile, acetone, methanol, methyl acetate, dimethylformamide, dimethyl ether, ethylamine, tetrahydrofuran *etc.* Among all these functional molecules, they have found that tetrahydrofuran, like ether-functionalized PAF-1, shows the highest adsorption capacity for CO_2 at ambient conditions. It also exhibits better selectivity than amine functionalities towards CO_2/CH_4, CO_2/N_2 and CO_2/H_2 gas mixtures. Such high selectivity is owing to the electrostatic interactions of the functional PAFs.

Considering the fact that the interaction energy between hydrogen and the adsorbent materials can be enhanced by doping lithium into the adsorbent,[114] recently, a study from our group was dedicated to the incorporation of a lithium tetrazolide moiety into porous materials by introducing a salt group consisting of an aromatic tetrazolide anion (CHN_4^-) and Li^+ cation *via* redox reactions to theoretically simulate the hydrogen storage capacity. The predicted hydrogen uptake using first principles GCMC simulations was found to be 4.9 wt% at 233 K and 10 MPa, which exceeded the 2010 DOE target of 4.5 wt%. High stability and polarizability, along with having 14 binding sites for hydrogen molecules with modest interaction energies, have made the lithium tetrazolide group an ideal candidate for doping into PAFs with improved absorptivity.

With the help of RIMP2/TZVPP calculations, Sun *et al.* have proposed that PAFs with large numbers of binding sites by inclusion of metal ions having relatively low binding energies and are remarkably capable of hydrogen storage (Figure 9.6).[115] PAF-Mg and PAF-Ca have been designed with a 1,2,4,5-benzenetetroxide anion to predict their hydrogen adsorption isotherms and isosteric heats of adsorption. The maximum prediction for hydrogen uptake was 6.8 wt% for PAF-Mg and 6.4 wt% for PAF-Ca at 233 K and 10 MPa, exceeding the 2015 DOE target of 5.5 wt%. Such high adsorption capacities can be explained by very stable and highly polarized ionic bonds between the metal cations and the organic anions in the networks.

Quaternary pyridinium-type PAF composites with AgCl have been introduced as antimicrobials to kill bacteria efficiently.[116] PAF-50 with AgCl loaded in the pore displays excellent antibacterial properties, even better than commonly used antibacterial materials. Formation of a dispersion with various solvents and compatibility with a large number of polymers, such PAFs can lead to large scale industrial production for different medicinal applications.

9.3.2.3 Carbonization

Carbonization is another strategy for the PSM of POFs. Generally, breaking chemical bonds in POFs leads to pore size shrinking and the framework collapsing, which makes the pore size better match the dynamic diameter of guest molecules and results in enhanced sorption enthalpy. On the other hand, template or hybrid components introduced in the carbonization

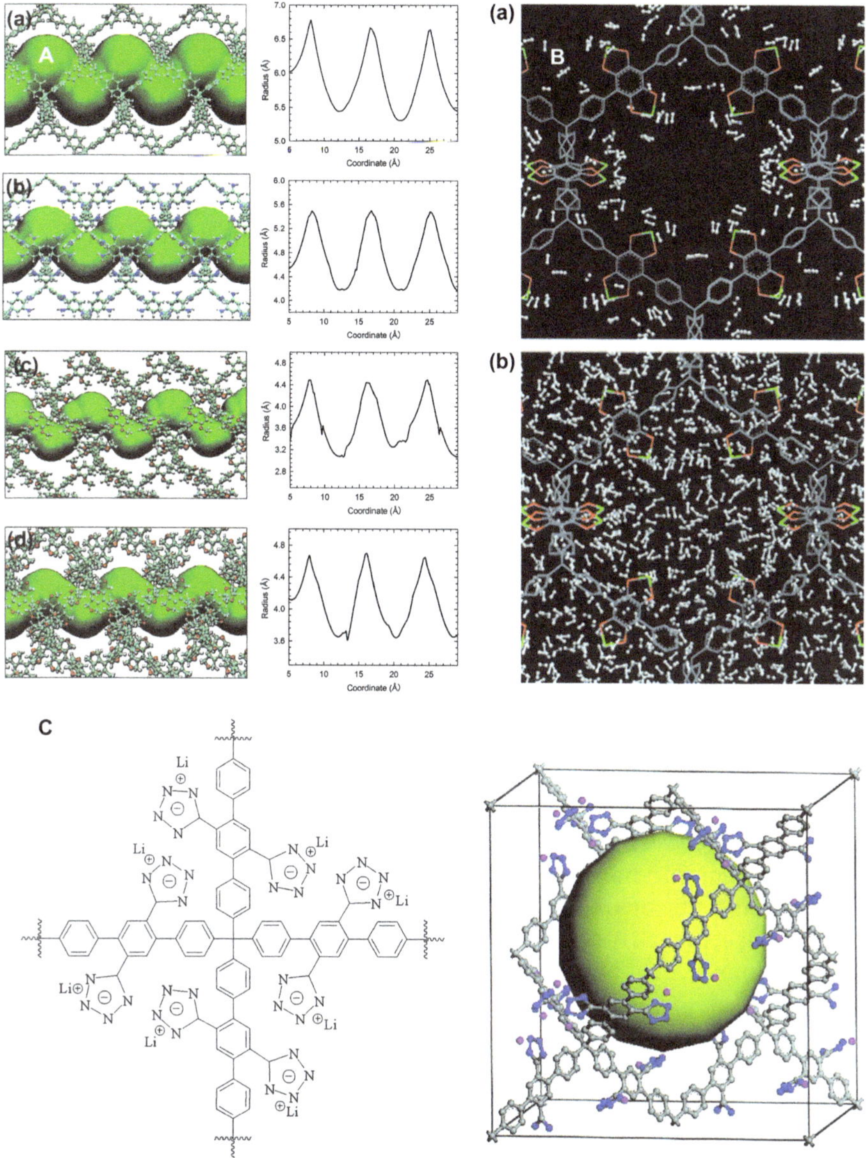

Figure 9.6 (A) Comparison of channel morphologies and radius of the channel along the XY plane in the crystalline model of PAF-1 with polar-group-functionalized PAF-1: (a) PAF-1, (b) NH_2-PAF-1, (c) CH_3O-PAF-1 and (d) dihydrofuran-PAF-1.[81] Reprinted with permission from A. Shigematsu, T. Yamada and H. Kitagawa, *J. Am. Chem. Soc.*, 2011, **133**, 2034. Copyright 2011, American Chemical Society. (B) Snapshots of the GCMC simulation along the small channel in PAF-Mg at 77 K at (a) low pressure and (b) high pressure. Reproduced from ref. 84 with permission from The Royal Society of Chemistry. (C) (Left) building block (PAF-4) containing Li^+-CHN_4^- and (right) its unit cell. The yellow ball denotes the free volume.[80] Reprinted with permission from Y. Sun, T. Ben, L. Wang, S. Qiu and H. Sun, *J. Phys. Chem. Lett.*, 2010, **1**, 2753. Copyright 2010, American Chemical Society.

process vary the porosity and electron density of the whole network. As a result, hierarchical pores and excess binding sites are exhibited in carbonized POFs. In addition, the carbon skeleton enhances the tolerance of carbonized POFs to humid, acidic or other organic solvent conditions. Hence, the low cost, large number of raw materials, high porosity and stability contribute to the wide application area and increased attention on a strategy for the modification of POFs.

The potential advantages of the PSM of POFs may be appreciated by the following considerations:

1) A large number of functional groups and reactions are available to be chosen in this process and freedom from restriction is contributed to by POFs stability. In addition, the functional units could be decorated on both the interior and exterior of the channels.
2) Great control over the type and the number of functional groups makes it possible to incorporate multiple substituents into one framework and flexibly tune the properties.
3) Many POFs with open frameworks could be decorated by functional groups.
4) It is easy to separate modified POFs from the reaction system.

There is one disadvantage that cannot be ignored in both the PSM and pre-modification, which is the difficult diffusion of oversized functional units into the POF channel for modification of the parent frameworks. In this case, mesoporous or macroporous materials are preferred.

In fact, desolvated or activated samples (*e.g.*, the supercritical carbon dioxide method),[117] exchange of guest particles (including molecules, cations and ions) from host materials,[118–120] micropatterning[121] (*i.e.*, introduction of large guests[122,123] or the formation of metal nanoparticles[124,125]), building block replacement (BBR),[126–130] *etc.* have all been performed in a post-synthetic manner for specific POFs.

9.3.3 Significance of the PSM of Porous Polymers

Functionalization of POFs *via* the PSM strategy enhances the binding capacity to small molecules (*e.g.*, hydrogen and methane in clean energy storage,[131] and carbon dioxide in carbon capture and sequestration), improves the catalytic activity and selectivity[132–134] and promotes drug delivery and release.[135]

On the other hand, PSM is used as a valuable tool to study the reaction mechanism. This is because the steric confinement of POFs controls the reaction carried out in the nano-vessel and obtains distinctive products compared with the one prepared under free conditions. For example, an IRMOF-9 analogue with stilbene as an organic link obtains only the meso product after the bromination reaction (Figure 9.7).[136] This proves that the rotation of the C–C bond is prevented by the immobilized stilbene unit.

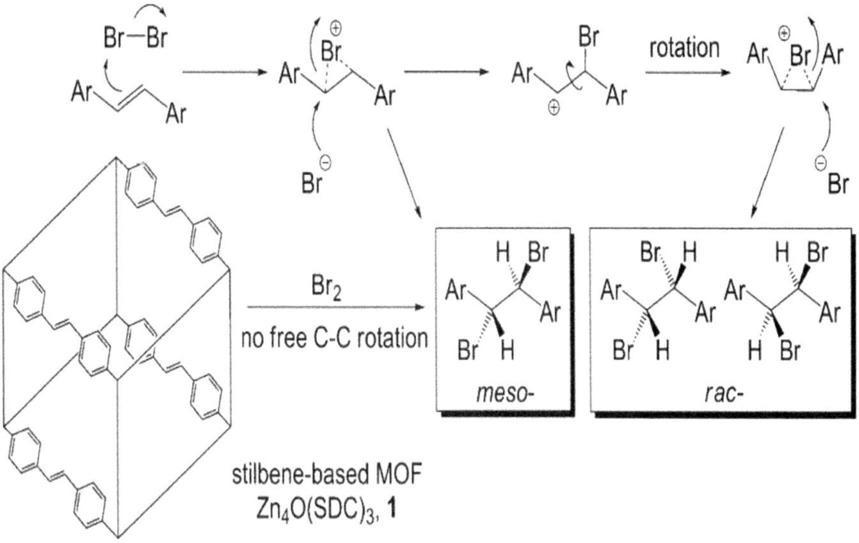

Figure 9.7   Polar mechanism for the electrophilic addition of Br_2 to stilbenes. Reproduced from ref. 130 with permission from The Royal Society of Chemistry.

So, after the formation of the intermediate, the bromide ion attacks it only from one side, which finally results in the inhibition of the formation of enantiomer products.

9.4   Typical PSM of Porous Polymers

9.4.1   Lithiation

Modification of POFs with metal particles has become a mainstream approach for functionalized materials. The strategy to introduce metal atoms or ions into the framework can mainly be divided into two parts: (1) the vaporized fusion or mechanical mixing of the metal atoms with the host network; (2) ion exchange or post-coordination of the metal cation with the binding sites of the host skeleton. The advantage of dative PSM is that it effectively prevents particle aggregation and flexibly controls the doping content. The simulation result demonstrates that alkaline-earth metal atoms exhibit weak affinity to the framework, while transition metals are to the contrary. However, the binding affinity is not strong, which is better because excess interaction energy brings desorption problem to materials. Hence, among the alkali metals, the Li atom stands out due to its light weight, stable binding to the framework and because it easily loses its valence electrons to form the Li^+ cation, which in turn increases the electropositive force. In addition, the formation of dative bonds between electrons of the H_2 σ bond and the empty Li 2s orbital contributes to the strong hydrogen affinity of lithiated POFs.[137] For example, Li-doped POFs exhibit stronger interactions with quadrupole CO_2 molecules, which should result in higher gravimetric CO_2 uptake, especially in the low pressure range. When it comes to hydrogen sorption, electropositive metal cations strengthen the van der Waals forces between hydrogen molecules and the framework. However, the highly reactive and flammable nature of lithium brings big challenges to the lithiation of POFs. Herein, we will highlight Li-doped MOFs, COFs, CMPs and PAFs. Based on these examples, experimental methods and modification effects will be discussed.

In 2007, Han and Goddard reported the hydrogen storage simulation results of lithium-doped MOFs (Figure 9.8).[138] Five lithium-doped MOF structures have been constructed with quantum mechanics (QM) calculations (X3LYP flavor of density functional theory, DFT).[139] The structure of MOF-C6 is the same as IRMOF-1, while MOF-C10 is the same as IRMOF-8.[140] With the first-principles-derived force field and grand canonical ensemble Monte Carlo (GCMC) simulations, the hydrogen uptake at 1 bar, 20 bar and 50 bar coincides well with the experimental results.[141,142] However, the hydrogen sorption capacity of pure MOFs is still far away from the target set by US Department of Energy (DOE). This is because the hydrogen binding energy of metal oxide clusters and the aromatic linkers is just 1.5 and 0.9 kcal mol^{-1}, respectively. Comparatively, Li$^+$ in Li-MOFs, which effectively separates the charge, enhances the stability of hydrogen molecules and leads to high binding energies of 4.0 kcal mol^{-1}. The improvement is obvious, especially for the high-temperature hydrogen uptake. As a result, Li-MOF-C30 is expected to store hydrogen at 5.16 wt% (300 K), 5.57 wt% (273 K) and 5.99 wt% (243 K) at 100 bar, which almost reaches the 2010 DOE target of 6.0 wt%. Even at 1 bar and 300 K, it can adsorb hydrogen at 1.98 wt%. If the

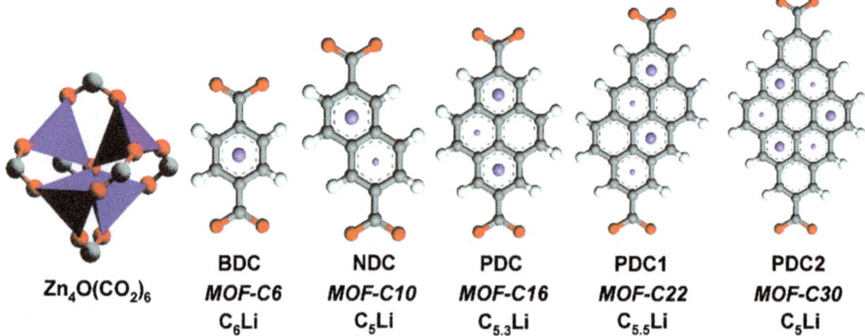

Figure 9.8　Li-doped MOFs. In each case the Zn$_4$O(CO$_2$)$_6$ connector couples to six aromatic linkers through the O–C–O common to each linker. The MOFs are named according to the number of aromatic carbon atoms. The large violet atoms in the linkers represent Li atoms above the linkers, while small violet Li atoms lie below the linkers. The C$_x$Li ratio considers only aromatic carbon atoms.[138]
Reprinted with permission from S. S. Han and W. A. Goddard III, *J. Am. Chem. Soc.*, 2007, **129**, 8422. Copyright 2004, American Chemical Society.

pressure is increased to 100 bar, this value could be as high as 5.2 wt%. These data show that the hydrogen uptake increases with the surface area, while the slope increases with the ratio of Li to C. In addition, the effect of the Li concentration is obvious at room temperature.

Mulfort and Hupp obtained another Li-doped MOF with chemical reduction methods in the same year.[143] The organic building block in this work is similar to MOF-C16 in Figure 9.8. The direct reduction with lithium metal in dimethylformamide (DMF) is better than the other reduction strategies such as employing a redox shuttle and interaction with the solvated electron in liquid ammonia. Powder X-ray diffraction (PXRD), thermogravimetric analysis (TGA) and inductively coupled plasma (ICP) are chosen to monitor the doping ratio. The hydrogen sorption showed that a lithium amount of around 5 mol% is optimal for improving gas uptake. Although the surface area of lithium-doped MOFs is inevitably decreased, the introduction of lithium has apparently rendered the interpenetrated networks mobile in the solid state. As a result, the structure changes, with more space for guest molecules, and exhibits a hysteresis loop in the N_2 sorption isotherm. As expected, the hydrogen uptake at 77 K/1 bar of this Li-MOF rises twice as high as that of the corresponding MOF (H_2 uptake at 77 K/1 bar is 0.93 wt% and 1.63 wt% for MOF and Li-doped MOF, respectively). This equates to 60 hydrogen molecules adsorbed per additional Li^+, but it's still far away from DOE target. Finally, the authors would like to point out that the enhancement of hydrogen sorption capacity is caused by not only one, but multiple factors. (*e.g.*, framework displacement, polarizability and additional binding sites).

The hydrogen sorption capacity of lithium-doped covalent organic frameworks has also been studied in the simulation step. Based on the low density, large surface areas and high stability of COF-102, COF-103, COF-105 and COF-108, the hydrogen sorption capacity of Li-doped items is expect to improve (Figure 9.9).[137] First, we have to introduce the structure of these four polymers. Three monomers, tetra(4-dihydroxyborylphenyl)methane (TBPM), tetra(4-dihydroxyborylphenyl)silane (TBPS) and hexahydroxytriphenylene (HHTP), are used to construct the four polymers. Self-condensation of TBPM and TBPS synthesizes COF-102 and COF-103, respectively, while co-condensation of TBPM or TBPS and HHTP obtains COF-105 and COF-108, respectively. In the simulation, the binding energy of Li on TBPM/TBPS and HHTP is -24.90 kcal mol^{-1} and -14.71 kcal mol^{-1}, respectively, which makes the design of Li-COFs reasonable and acceptable. On the other hand, considering that only the positively charged Li atoms contribute to the enhancement of hydrogen adsorption, the Li-COF model is built with charge transfer from Li atoms to COFs. In addition, the transfer degree could be determined by the ratio and distribution of Li dopants. The simulation of hydrogen adsorption isotherms demonstrates that the hydrogen sorption capacity of the Li-doped COF is more than double that of the corresponding COF. Particularly, the gravimetric hydrogen uptake of Li-doped COF-105 and Li-doped COF-108 reaches 6.84 wt% and 6.73 wt% at 298 K/100 bar,

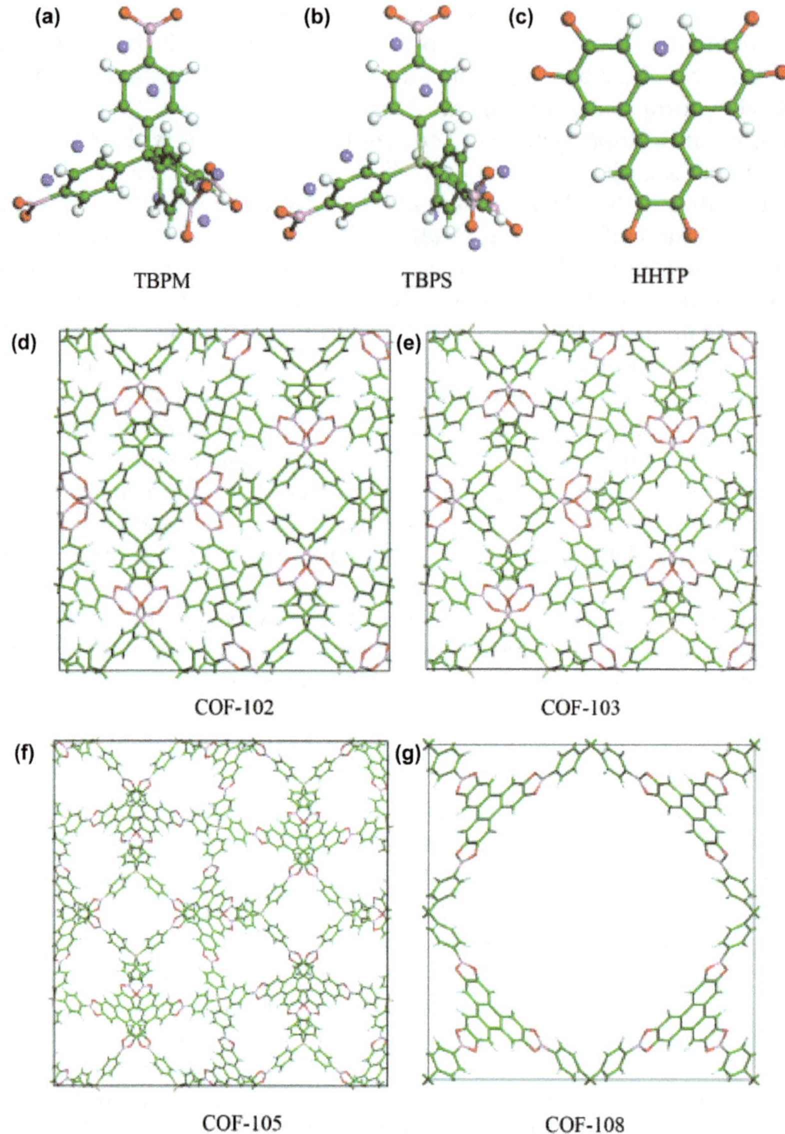

Figure 9.9 (a–c) Li-doped building blocks and (d–g) unit cells of 3D covalent organic frameworks. H atoms linked to oxygen in (a–c) are omitted for clarity. The scheme of Li doping on 3D COFs is determined from calculations. Li, violet; H, white; B, pink; C, green; O, red; Si, gold.[137]
Reprinted with permission from D. Cao, J. Lan, W. Wang and B. Smit, *Angew. Chem., Int. Ed.*, 2009, **48**, 4730. Copyright 2009 WILEY-VCH Verlag GmbH & Co. KGaA, Weinheim.

respectively, while the volumetric adsorption capacity of Li-doped COF-102 and Li-doped COF-103 is 25.98 g L^{-1} and 25.00 g L^{-1}, respectively, which is higher than all the other Li-doped MOFs.

In 2010, an experimental study was reported that showed the hydrogen sorption of a Li-doped CMP.[10] A solution of the naphthalene anion radical salt ($Li^+C_{10}H_8{}^{\bullet-}$) in tetrahydrofuran (THF) was used to introduce Li to the CMP.[144,145] The hydrogen sorption isotherms of the Li-CMP with various Li contents at 77 K/1 bar showed: 1) hydrogen uptake of the Li-CMP with 0.5 wt% Li content was as high as 6.1 wt% at 77 K/1 bar, which is four times that of the CMP; 2) among all the Li-CMPs, 0.5 wt% Li-CMP exhibited the highest hydrogen uptake due to the appropriate surface area and agglomeration degree. Although the hydrogen sorption enthalpy of Li-CMP is up to 8.1 kJ mol^{-1} at zero coverage, it decreases sharply with the increase of hydrogen coverage. Hence, the hydrogen sorption capacity of Li-CMP is still below the DOE target.

With the same method, Li@PAF-1 was prepared in 2012.[146] The hydrogen heat of sorption increased to 9 kJ mol^{-1} after lithiation. As a result, the hydrogen uptake of Li@PAF-1 with 5% Li content was 2.7 wt% at 77 K/1 bar. Better performance can be expected for elevated pressure hydrogen storage applications due to the high surface area of PAF-1. The topographically integrated mathematical thermodynamic adsorption model (TIMTAM) result reveals a total uptake of 11 wt% at 77 K/50 bar, which would be the highest hydrogen uptake among POFs. Besides, the methane and carbon dioxide uptake of Li@PAF-1 increased dramatically due to the lithiation of PAF-1.

In conclusion, the hydrogen uptake of Li-doped PAF-1 at 77 K/50 bar, and COF-105 and COF-108 at 298 K/100 bar exceed the 6 wt% target in simulations. However, only the hydrogen uptake of 0.5 wt% Li-CMP is over the 6 wt% target in experiments. In addition, the experimental conditions of 77 K/1 bar are still too harsh to apply in the real world. One of the big challenges in lithiation is lithium atom agglomeration when excess atoms arrange on the surface of adsorbents. How to solve this problem is a bottleneck of lithium PSM development.

9.4.2 Grafting

Grafting is one PSM strategy that could introduce various specific functional groups into POFs *via* covalent PSM, PSD and tandem PSM routes. Whether *via* an amine, aldehyde or in the absence of any chemical tags in POFs, functional units can be grafted onto them. In 2008, amine-tagged IRMOF-3, which is a derivative of MOF-5, was reacted with acetic anhydride to obtain IRMOF-3-AM1.[147] The degree of conversion confirmed by nuclear magnetic resonance (NMR) is over 90%. Soon after, IRMOF-3-AMR,[148] IRMOF-3-AM3,[141] IRMOF-3-URR (UR = urea),[141] IRMOF-3-AM3Br$_2$,[149] IRMOF-3b,[98] IRMOF-3-3c,[98] IRMOF-3-AMMal,[150,151] IRMOF-3[Mn][152] and IRMOF-3-FITC (FITC = fluorescein isothiocyanate),[153] which were prepared by grafting strategies on amine-tagged IRMOF-3, have been reported (Figure 9.10). It has been proven that the amine group is compatible to construct MOFs. So, besides IRMOF-3, other amine-tagged MOFs, such as MIL-53(Al) (MIL = Material Institute Lavoisier), MIL-101(Fe), MIL-101(Al),[154] CAU-1

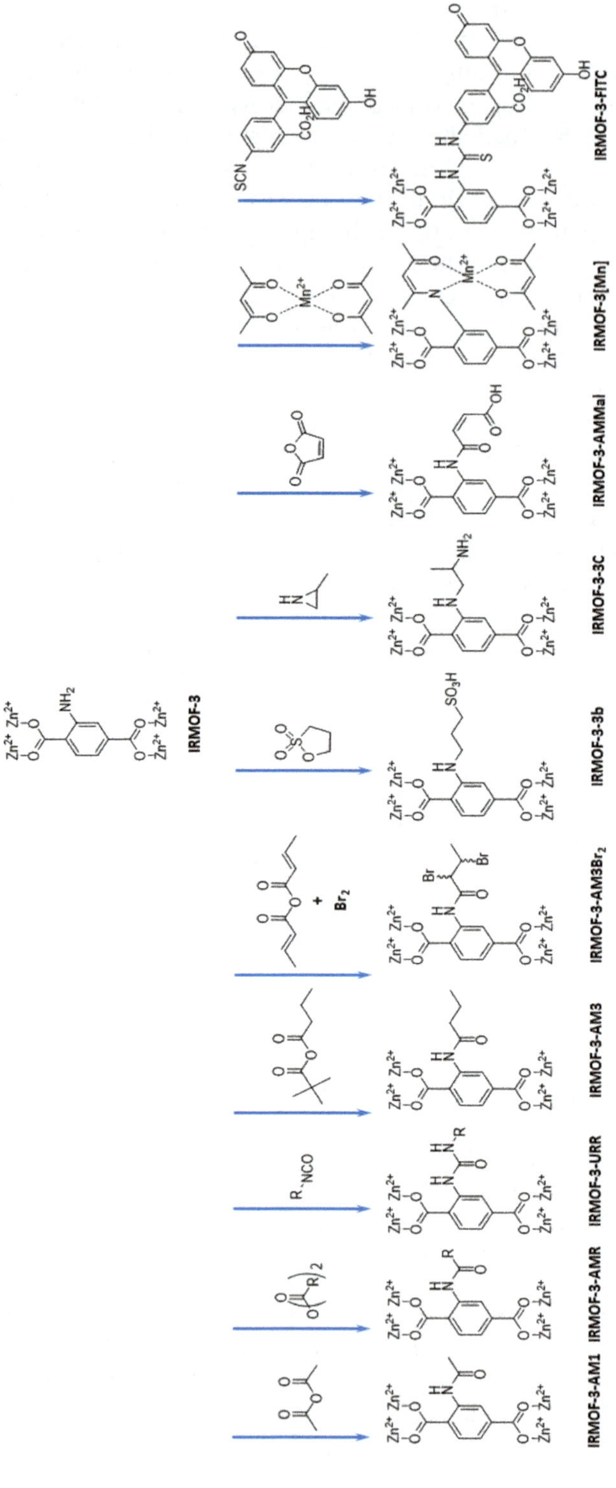

Figure 9.10 PSM on IRMOF-3.

(CAU = Christian-Albrechts University)[155] and UiO-66 (UiO = University of Oslo),[156–158] have opened up the possibility of performing an even wider range of PSM reactions.

The first PSM study on aldehyde-tagged POFs was reported by Burrows and co-workers in 2008.[159] An IRMOF-9 analogue was chosen to produce hydrazine MOFs. It has been proven that the stability of zeolitic imidazolate frameworks (ZIFs) is better than many other MOFs. So, modification of ZIFs has attracted scientists' interest. For example, $NaBH_4$ reduces the aldehyde of ZIF-90 to obtain the corresponding alcohol product ZIF-91, while ethanolamine modifies the aldehyde to generate an imine substituent terminated with an alcohol group, yielding ZIF-92.[160] The ZIF-90 film modified by the same strategy showed enhanced physicochemical stability, as well as selectivity for H_2 over CO_2.[161]

Post-synthetic deprotection (PSD), which masks the functional group of the organic linker during the incorporation process and then removes the protecting group to reveal the desired functionality, is another effective grafting PSM strategy. For example, in 2010, Telfer *et al.* reported the utility of PSD on IRMOF-12-NH_2 to protect the amine group with a bulky *tert*-butylcarbamate (Boc) protecting group. In addition, the bigger grafting Boc units also played a role as a template, which effectively prevented interpenetration.[110]

PSM on POFs with no chemical tags is a big challenge. However, the outstanding properties of novel POFs, such as CMPs, COFs and PAFs, prompts us to explore modification methods for the functionalization of them.

MOFs remain pioneers in this field. In 2010, Mir *et al.* reported $[2+2]$ cycloadditions of the nearest olefin bonds by photochemical reactions.[162] Besides the photochemical cyclization reactions, directly substituting hydrogen on phenyl units with a functional group has been reported recently. For example, MIL-101(Cr) could be nitrated to MIL-101(Cr)-NO_2,[163] ammoniated to MIL-101(Cr)-NH_2 or amidated to MIL-101(Cr)-UR_2.[155]

PAF-1 was first reported in 2009.[17] It resolved the thermal and chemical stability problem, which bothered POFs for a long time, with covalent bonds and the construction of a hybrid *dia* topology framework. Soon after, polyamine-tethered PAF-1 was reported, as shown in Figure 9.11.[164] Four polyamine-modified POFs and their chlorinated precursor were synthesized. N_2 sorption isotherms showed a decrease of the BET surface area and pore size compared with PAF-1. However, the CO_2 affinity dramatically increased in the order of DETA > TAEA > TETA > EDA. In particular, PPN-6-CH_2TETA exhibits the highest zero point carbon dioxide heat of sorption (around 64 kJ mol^{-1}), which is into the chemical adsorption range (*ca.* 50–100 kJ mol^{-1}). In addition, PPN-6-CH_2DETA shows the highest carbon dioxide uptake, as well as the exceptional selectivity of CO_2 over N_2. The ultra-high carbon dioxide capture capacity and its reversibility make amine-grafted PAF-1 effective for the adsorption of carbon dioxide from air (400 ppm CO_2, 78.96% N_2 and 21% O_2).[165]

Amine modification on POFs is an ideal method to enhance the CO_2 sorption capacity. However, primary and secondary amines, triamines, and

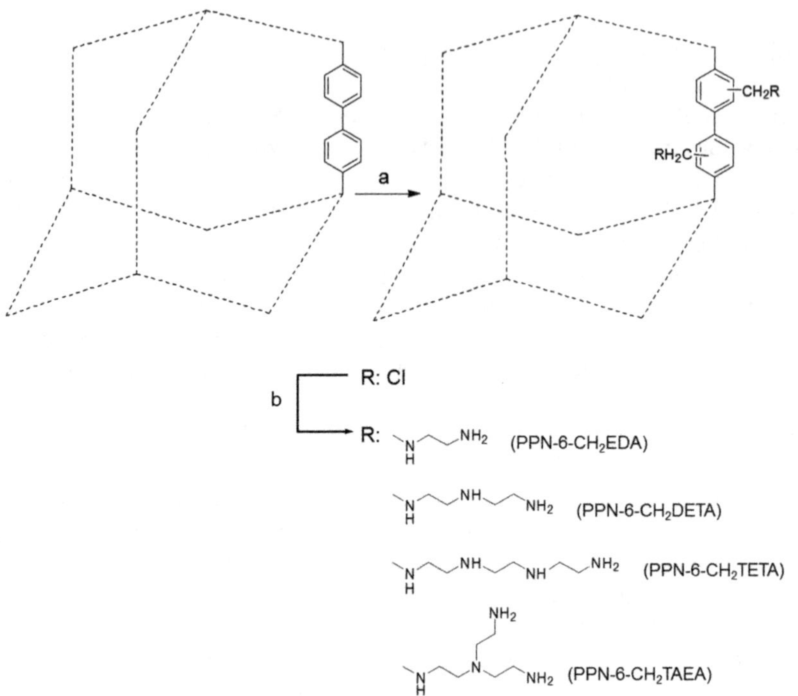

Figure 9.11 Synthetic route to polyamine-tethered porous polymer networks (PPNs). (a) CH₃COOH/HCl/H₃PO₄/HCHO, 90 °C, 3 days; (b) amine, 90 °C, 3 days (DETA, diethylenetriamine; TAEA, tris(aminoethyl)amine; TETA, triethylenetetramine; EDA, ethylenediamine).[164]
Reprinted with permission from W. Lu, J. P. Sculey, D. Yuan, R. Krishna, Z. Wei and H. Zhou, *Angew. Chem., Int. Ed.*, 2012, **51**, 7480. Copyright 2012 WILEY-VCH Verlag GmbH & Co. KGaA, Weinheim.

polyamines are deactivated in the presence of dry CO_2 under mild conditions, with presentation of the decomposition product urea.[166] This is still a big challenge for the functionalization of POFs with amine derivatives.

For carbon dioxide sorption applications, apart from the amine grafting of POFs, other functional groups with different polarities also contribute to the improvement of the carbon dioxide sorption capacity. Combined with the simulations, the isosteric heats of adsorption (Q_{st}) of modified POFs are in the order of –COOH > –(OH)₂ > –NH₂ > –(CH₃)₂ > non-functionalized framework.[85] Selecting the optimal reaction route to graft these specific units on POFs is one of the hot directions of the PSM of POFs.

9.4.3 Sulfonation

Sulfonation of POFs demonstrates two effects: (1) it provides a coordinate site for further functionalization of the frameworks; (2) it provides hydrophilic building groups to tune the acidity of the channels. One typical

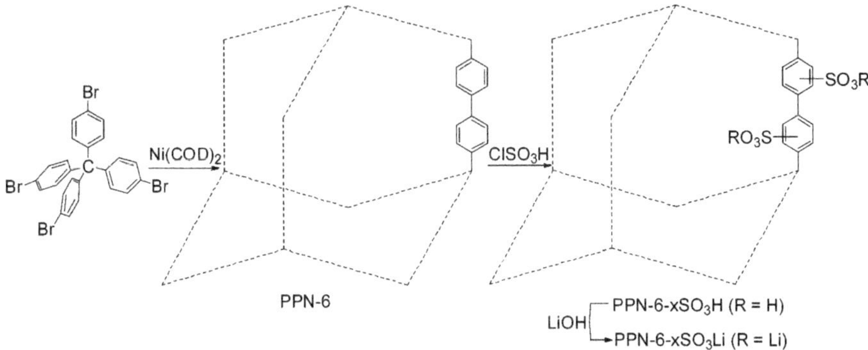

Figure 9.12 Synthesis and grafting of PPN-6.[95]
Reprinted with permission from W. Lu, D. Yuan, J. Sculley, D. Zhao, R. Krishna and H. Zhou, *J. Am. Chem. Soc.*, 2011, **133**, 18126. Copyright 2011, American Chemical Society.

example of the first issue is the sulfonation of PPN-6 (Figure 9.12).[95] Based on the open framework and high physicochemical stability, PPN-6 reacts with chlorosulfonic acid to obtain PPN-6-SO_3H and then neutralizes with LiOH to get PPN-6-SO_3Li (Figure 9.12). It combines covalent PSM with dative PSM, which may be defined as tandem PSM. Sulfonic acid grafting produces PPN-6-SO_3H and its lithium salt PPN-6-SO_3Li exhibits significantly increased carbon dioxide enthalpy and uptake (Q_{stCO2} of PPN-SO_3H and PPN-SO_3Li is 30.4 and 35.7 kJ mol^{-1}, respectively, and the CO_2 uptakes of them at 295 K/1 bar are 13.1 and 13.5 wt%, respectively). In addition, the IAST calculation demonstrates the exceptionally high CO_2/N_2 selectivity of PPN-6-SO_3H (adsorption selectivity, $S_{ads} = 150$) and PPN-6-SO_3Li ($S_{ads} = 414$). These values surpass the selectivity of the NaX zeolite.

As shown in Figure 9.13, sulfonic of PPAF, which is characterized by a 9,9′-spiro-bisfluorene unit, is another example.[167] Further grafting with amine makes PPAF-SO_3H-NH_2, which may be used as a bifunctional catalyst in a model cascade reaction. In particular, this catalyst can be used several times without obvious loss of the activity.

The examples shown above are all sulfonated directly on the phenyl units of POFs. In addition, it can be grafted on functional groups *via* various reactions and with diverse sulfonate reagents. This strategy is superior due to the more controllable nature of the modified sites. For instance, Burrows *et al.* reported an IRMOF-9 analogue with a sulfur functional group oxidized to its sulfonic analogue with a conversion yield of 77%.[168] One year later, sulfonation of IRMOF-3 by the amine functional group with 1,3-propanesultone in $CHCl_3$ at 45 °C was reported.[98] Elemental analysis showed 57% of the content of amine units were converted to sulfonate groups.

In conclusion, sulfonation of POFs tunes the porosity, which influences the interaction between the host and guest, and further expands the application areas. The increase of polarity and electron density improves the CO_2

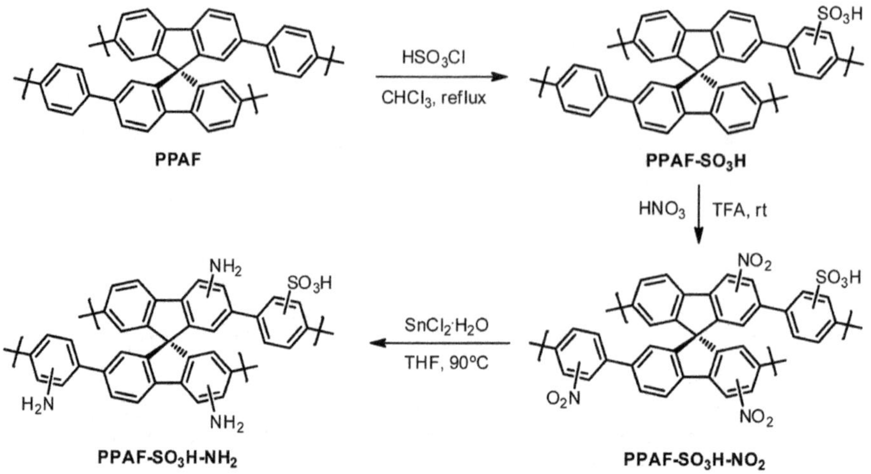

Figure 9.13 Synthesis of PPAF-SO$_3$H-NH$_2$ (TFA, trifluoroacetic acid).[167]
Reprinted with permission from E. Merino, E. Verde-Sesto, E. M. Maya,
M. Iglesias, F. Sanchez and A. Corma, *Chem. Mater.*, 2013, **25**, 981.
Copyright 2013, American Chemical Society.

sorption capacity of POFs. The enhanced acidity makes the material catalytic
and may be applied in the catalytic pyrolysis of hydrocarbon. The sulfonate-
modified POF, which is tuneable with regards to the hydrophilic : hydrophobic ratio, exhibits a potential application in the separation of small
molecules. Hence, sulfonation is anticipated to be one of the effective PSM
strategies for the functionalization of POFs.

9.4.4   Carbonization

Porous carbon is one of the most attractive materials due to its full carbon
skeleton. First, the framework with only the carbon element in it builds the
foundation for a competitive, lightweight sorbent, which contributes to the
high specific surface area of porous carbon. Second, the full carbon skeleton
brings a specific electrical property, which expands the application areas of
porous carbon to the electrochemical field and supercapacitors. Third, the
wild range of sources and the easy preparation of porous carbon make its
high mass productivity possible at relatively low cost. Forth, porous carbon
exhibits good compatibility with common solvents, water, acid and base.
Fifth, the pore size and ordered degree can be flexibly tuned by templates.
Hence, porous carbon can be a nano-vessel that may be doped, modified and
applied in optical, electrical and sorption fields. Pyrolysis of POFs is a convenient method to prepare porous carbon, as they may broaden the library of
porous carbons with novel structures and properties. This strategy can be
mainly divided into two parts: directly carbonized POFs and carbonized
hybrid POFs after the template or specific components have been mixed.

In fact, utilization of POFs as sacrificial templates to anneal the carbon source in channels is another effective method for porous carbon preparation. But it is not classed as PSM of POFs, so it will not be discussed here.

The synthesis of PAF-1-x belongs to the first issue mentioned above.[169] Annealing PAF-1 at different temperatures obtains PAF-1-x (x stands for the temperature from 350 to 450 °C). The shrink pore matches the CO_2 molecule better and enhances the van der Waals forces between them. As a result, the CO_2 enthalpy of PAF-1-x is dramatically increased to over 21 kJ mol^{-1}, particularly for PAF-1-450, which can be up to 27.8 kJ mol^{-1}. In addition, the CO_2 uptakes of PAF-1-x at 273 K/1 bar are all larger than that of PAF-1, especially the value of PAF-1-450, which is twice that of PAF-1. Interestingly, the carbonized counterparts not only increase the carbon dioxide sorption capacity but also increase the selectivity of CO_2/N_2, CO_2/CH_4 and CO_2/H_2, which is very important in carbon capture and sequestration. K-PAF-1-x (x stands for the temperature from 500 to 900 °C) result from carbonized PAF-1 after KOH chemical activation.[112] KOH as a template retains part of the relatively larger pore in PAF-1, and the obtained novel porous carbon has a unique bimodal micropore. It makes K-PAF-1-x exhibit significant gas storage properties in both the low pressure and high pressure ranges. For honeycomb-type COFs, carbonizing them to graphene or hybrid-element-doped graphene results in impressive electrochemical properties.[170,171] Considering the nature of MOFs, which contain hybrid metals in the frameworks, the product of carbonizing MOFs is a porous carbon complex. For functionalization, specific particles, such as sulfur,[172] Fe_3O_4 nanoparticles,[173] N_2[174,175] and NH_3,[176] are introduced into the frameworks before the annealing process. These strategies have greatly enhanced the diversity of the porous carbons.

9.4.5 Doping Ions

Triazine-based polymeric networks with tailorable porosity have been reported by a reversible polymerization reaction of various aromatic nitriles by the group of Kuhn (Figure 9.14).[177] They employed different functional molecules, like pyridine, thiophene and bipyridine, to coordinate with metal ions, like Zn^{2+}, to form a firm porous network expected to be a matrix in metal-organic catalyzed reactions. It has been shown that the pore size of these networks can be tailored by fixing a proper N/C ratio within certain monomers rather than varying the geometry and dimensions of the monomers. Use of 2,6-dicyanonaphthalene with an increased amount of Zn^{2+} (5 equivalents) enhances the surface area of the formed network to 2255 m^2 g^{-1} compared to 90 m^2 g^{-1} using only 1 equivalent of Zn^{2+}.[178] The CTF obtained by the trimerization of 1,3,5-tricyanobenzene in the presence of Zn^{2+} shows variety in its structure and pore dimensions as the Zn^{2+}:monomer concentration ratio is changed.[179] At high reaction temperatures or increasing Zn^{2+}: monomer ratio, the resulting networks exhibit very high surface area[180] with applications in catalysis.

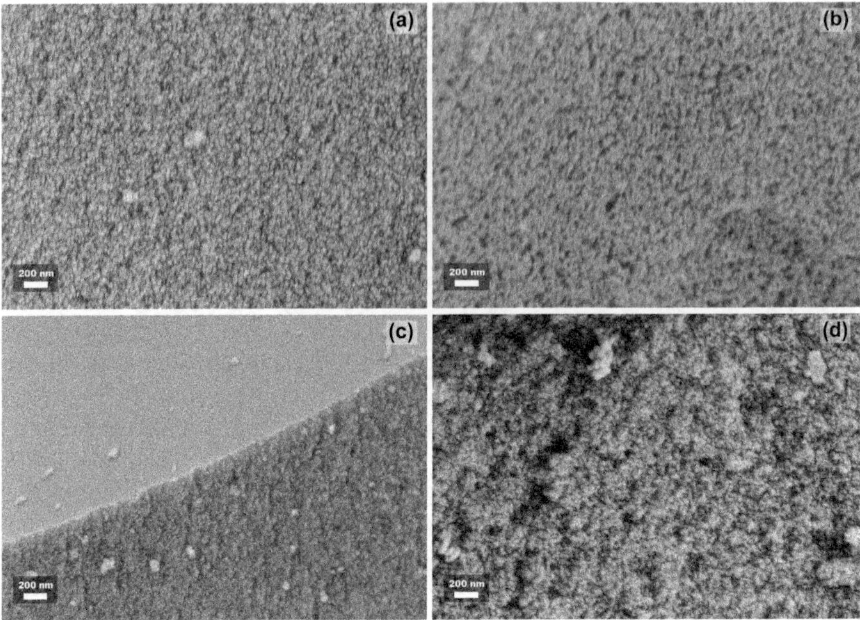

Figure 9.14 Scanning electron micrographs of (a) 4,4′-dicyanobiphenyl (DCBP; 5 equiv. of $ZnCl_2$, 600 °C), (b) DCBP (20 equiv. of $ZnCl_2$, 600 °C), (c) thiocyanuric acid (TCA; 4 equiv. of $ZnCl_2$, 400 °C) and (d) 4,4″-dicyanoterphenyl (DCTP; 10 equiv. of $ZnCl_2$, 400 °C).[58]
Reprinted with permission from H. Furukawa and O. M. Yaghi, *J. Am. Chem. Soc.*, 2009, **131**, 8875. Copyright 2009, American Chemical Society.

Metal-functionalized CTFs have other important aspects. The Pt-loaded CTF shows high catalytic activity with recyclability towards methane oxidation at an elevated temperature in acidic media. A triazine network of 1,4-dicyanobenzene loaded with Pd nanoparticles has been shown to exhibit a better catalytic activity and lifetime than Pd-loaded activated carbon in the glycerol oxidation reaction. N-heterocyclic moieties of CTFs provide the required stability to Pd nanoparticles, while the rigid framework facilitates the selectivity of the catalyst.[181]

9.5 Computational Methods Involving Functionalization

Along with experimentally developed functionalization of porous aromatic frameworks, several computational and theoretical methods have been introduced to investigate modified networks to obtain improved properties. Lan *et al.* have shown the hydrogen adsorption capacity of PAF networks by replacing the C–C bond in the diamond structure with several phenyl rings.[113] Another PAF with lithium tetrazolide linkers has been proposed with good hydrogen storage capacity using GCMC simulations.[114]

Jiang and co-workers have designed new PAFs computationally using GCMC simulations by introducing polar organic groups into the biphenyl unit and have then investigated their gas separating abilities.[182] They have introduced various functional moieties into the PAF framework, like acetonitrile, acetone, methanol, methyl acetate, dimethylformamide, dimethyl ether, ethylamine, tetrahydrofuran *etc*. Among all these functional molecules, they have found that tetrahydrofuran-like ether-functionalized PAF-1 shows the highest adsorption capacity for CO_2 at ambient conditions. It also exhibits better selectivity than amine functionalities towards CO_2/CH_4, CO_2/N_2 and CO_2/H_2 gas mixtures. Such high selectivity is owing to the electrostatic interactions of functional PAFs.

References

1. V. A. Davankov and M. P. Tsyurupa, *React. Polym.*, 1990, **13**, 27.
2. O. W. Webster, F. P. Gentry, R. D. Farlee and B. E. Smart, *Makromol. Chem., Macromol. Symp.*, 1992, **54–55**, 477.
3. C. Urban, E. F. McCord, O. W. Webster, L. Abrams, H. W. Long, H. Gaede, P. Tang and A. Pines, *Chem. Mater.*, 1995, **7**, 1325.
4. M. P. Tsyurupa and V. A. Davankov, *React. Funct. Polym.*, 2002, **53**, 193.
5. J. Y. Lee, C. D. Wood, D. Bradshaw, M. J. Rosseinsky and A. I. Cooper, *Chem. Commun.*, 2006, **25**, 2670.
6. N. Fontanals, P. Manesiotis, D. C. Sherrington and P. A. G. Cormack, *Adv. Mater.*, 2008, **20**, 1298.
7. B. Li, R. Gong, Y. Luo and B. Tan, *Soft Matter*, 2011, **7**, 10910.
8. S. N. Sidorov, L. M. Bronstein, V. A. Davankov, M. P. Tsyurupa, S. P. Solodovnikov and P. M. Valetsky, *Chem. Mater.*, 1999, **11**, 3210.
9. J. Germain, J. M. J. Frechet and F. Svec, *Chem. Commun.*, 2009, **12**, 1526.
10. O. K. Farha, Y. S. Bae, B. G. Hauser, A. M. Spokoyny, R. Q. Snurr, C. A. Mirkin and J. T. Hupp, *Chem. Commun.*, 2010, **46**, 1056.
11. N. B. McKeown, S. Makhseed and P. M. Budd, *Chem. Commun.*, 2002, **23**, 2780.
12. N. B. McKeown, S. Hanif, K. Msayib, C. E. Tattershall and P. M. Budd, *Chem. Commun.*, 2002, **23**, 2782.
13. B. S. Ghanem, K. J. Msayib, N. B. McKeown, K. D. M. Harris, Z. Pan, P. M. Budd, A. Butler, J. Selbie, D. Book and A. Walton, *Chem. Commun.*, 2007, **1**, 67.
14. B. S. Ghanem, M. Hashem, K. D. M. Harris, K. J. Msayib, M. Xu, P. M. Budd, N. Chaukura, D. Book, S. Tedds, A. Walton and N. B. McKeown, *Macromolecules*, 2010, **43**, 5287.
15. N. Du, H. B. Park, G. P. Robertson, M. M. Dal-Cin, T. Visser, L. Scoles and M. D. Guiver, *Nat. Mater.*, 2011, **10**, 372.

16. N. B. McKeown, P. M. Budd, K. J. Msayib, B. S. Ghanem, H. J. Kingston, C. E. Tattershall, S. Makhseed, K. J. Reynolds and D. Fritsch, *Chem. - Eur. J.*, 2005, **11**, 2610.

17. T. Ben, H. Ren, S. Ma, D. Cao, J. Lan, X. Jing, W. Wang, J. Xu, F. Deng, J. M. Simmons, S. Qiu and G. Zhu, *Angew. Chem., Int. Ed.*, 2009, **48**, 9457.

18. B. S. Ghanem, N. B. McKeown, P. M. Budd, N. M. Al-Harbi, D. Fritsch, K. Heinrich, L. Starannikova, A. Tokarev and Y. Yampolskii, *Macromolecules*, 2009, **42**, 7881.

19. H. M. El-Kaderi, J. R. Hunt, J. L. Mendoza-Cortes, A. P. Cote, R. E. Taylor, M. O'Keeffe and O. M. Yaghi, *Science*, 2007, **316**, 268.

20. R. W. Tilford, W. R. Gemmill, H. C. zur Loye and J. J. Lavigne, *Chem. Mater.*, 2006, **18**, 5296.

21. A. P. Cote, H. M. El-Kaderi, H. Furukawa, J. R. Hunt and O. M. Yaghi, *J. Am. Chem. Soc.*, 2007, **129**, 12914.

22. H. Xu, X. Chen, J. Gao, J. Lin, M. Addicoat, S. Irleb and D. Jiang, *Chem. Commun.*, 2014, **50**, 1292.

23. R. W. Tilford, S. J. Mugavero, P. J. Pellechia and J. J. Lavigne, *Adv. Mater.*, 2008, **20**, 2741.

24. X. Chen, M. Addicoat, S. Irle, A. Nagai and D. Jiang, *J. Am. Chem. Soc.*, 2013, **135**, 546.

25. C. Dai, P. Nguyen, T. B. Marder, A. J. Scott, W. Clegg and C. Viney, *Chem. Commun.*, 1999, **24**, 2493.

26. S. Jin, K. Furukawa, M. Addicoat, L. Chen, S. Takahashi, S. Irle, T. Nakamuraa, D. Jiang, *Chem. Sci.*, 2013, **4**, 4505.

27. W. Ziegenbein and H. E. Sprenger, *Angew. Chem. Int. Ed. Engl.*, 1966, **5**, 893.

28. J. X. Jiang, F. Su, A. Trewin, C. D. Wood, N. L. Campbell, H. Niu, C. Dickinson, A. Y. Ganin, M. J. Rosseinsky, Y. Z. Khimyak and A. I. Cooper, *Angew. Chem., Int. Ed.*, 2007, **46**, 8574.

29. J. X. Jiang, F. Su, A. Trewin, C. D. Wood, H. Niu, J. T. A. Jones, Y. Z. Khimyak and A. I. Cooper, *J. Am. Chem. Soc.*, 2008, **130**, 7710.

30. R. Dawson, A. Laybourn, R. Clowes, Y. Z. Khimyak, D. J. Adams and A. I. Cooper, *Macromolecules*, 2009, **42**, 8809.

31. R. Dawson, A. Laybourn, Y. Z. Khimyak, D. J. Adams and A. I. Cooper, *Macromolecules*, 2010, **43**, 8524.

32. M. Ishikawa, M. Kawai and Y. Ohsawa, *Synth. Met.*, 1991, **40**, 231.

33. V. N. Prigodin and K. B. Efetov, *Phys. Rev. Lett.*, 1993, **70**, 2932.

34. I. Manners, *Angew. Chem., Int. Ed. Engl.*, 1996, **35**, 1062.

35. M. Rehahn, *Acta Polym.*, 1998, **49**, 201.

36. P. G. Pickup, *J. Mater. Chem.*, 1999, **9**, 1641.

37. U. S. Schubert and C. Eschbaumer, *Angew. Chem., Int. Ed.*, 2002, **41**, 2892.

38. M. O. Wolf, *Adv. Mater.*, 2001, **13**, 545.

39. R. P. Kingsborough and T. M. Swager, *Prog. Inorg. Chem.*, 1999, **48**, 123.

40. C. Weder, *Chem. Commun.*, 2005, **43**, 5378.

41. B. J. Holliday and T. M. Swager, *Chem. Commun.*, 2005, **1**, 23.
42. M. E. Wright, *Macromolecules*, 1989, **22**, 3256.
43. T. Hirao, S. Yamaguchi and S. Fukuhara, *Tetrahedron Lett.*, 1999, **40**, 3009.
44. L. Chen, Y. Yang and D. L. Jiang, *J. Am. Chem. Soc.*, 2010, **132**, 9138.
45. J. X. Jiang, C. Wang, A. Laybourn, T. Hasell, R. Clowes, Y. Z. Khimyak, J. L. Xiao, S. J. Higgins, D. J. Adams and A. I. Cooper, *Angew. Chem., Int. Ed.*, 2011, **50**, 1072.
46. J. Schmidt, J. Weber, J. D. Epping, M. Antonietti and A. Thomas, *Adv. Mater.*, 2009, **21**, 702.
47. J. Brandt, J. Schmidt, A. Thomas, J. D. Epping and J. Weber, *Polym. Chem.*, 2011, **2**, 1950.
48. J. X. Jiang, Y. Li, X. Wu, J. Xiao, D. J. Adams and A. I. Cooper, *Macromolecules*, 2013, **46**, 8779.
49. Z. Xie, C. Wang, K. E. deKrafft and W. Lin, *J. Am. Chem. Soc.*, 2011, **133**, 2056.
50. N. Kang, J. H. Park, K. C. Ko, J. Chun, E. Kim, H.-W. Shin, S. M. Lee, H. J. Kim, T. K. Ahn, J. Y. Lee and S. U. Son, *Angew. Chem., Int. Ed.*, 2013, **52**, 6228.
51. B. Kiskan, M. Antonietti and J. Weber, *Macromolecules*, 2012, **45**, 1356.
52. G. Cheng, T. Hasell, A. Trewin, D. J. Adams and A. I. Cooper, *Angew. Chem., Int. Ed.*, 2012, **51**, 12727.
53. H. Furukawa, N. Ko, Y. B. Go, N. Aratani, S. B. Choi, E. Choi, A. Ö. Yazaydin, R. Q. Snurr, M. O'Keeffe, J. Kim and O. M. Yaghi, *Science*, 2010, **329**, 424.
54. O. K. Farha, I. Eryazici, N. C. Jeong, B. G. Hauser, C. E. Wilmer, A. A. Sarjeant, R. Q. Snurr, S. T. Nguyen, A. Ö. Yazaydin and J. T. Hupp, *J. Am. Chem. Soc.*, 2012, **134**, 15016.
55. P. L. Llewellyn, S. Bourrelly, C. Serre, A. Vimont, M. Daturi, L. Hamon, G. D. Weireld, J. Chang, D. Hong, Y. K. Hwang, S. H. Jhung and G. Férey, *Langmuir*, 2008, **24**, 7245.
56. D. Saha, Z. Bao, F. Jia and S. Deng, *Environ. Sci. Technol.*, 2010, **44**, 1820.
57. A. P. Côté, A. I. Benin, N. W. Ockwig, M. O'Keeffe, A. J. Matzger and O. M. Yaghi, *Science*, 2005, **310**, 1166.
58. H. Furukawa and O. M. Yaghi, *J. Am. Chem. Soc.*, 2009, **131**, 8875.
59. S. Wan, J. Guo, J. Kim, H. Ihee and D. Jiang, *Angew. Chem., Int. Ed.*, 2008, **47**, 8826.
60. A. Li, R. Lu, Y. Wang, X. Wang, K. Han and W. Deng, *Angew. Chem., Int. Ed.*, 2010, **49**, 3330.
61. N. B. McKeown and P. M. Budd, *Chem. Soc. Rev.*, 2006, **35**, 675.
62. J. Schmidt, M. Werner and A. Thomas, *Macromolecules*, 2009, **42**, 4426.
63. C. D. Wood, B. Tan, A. Trewin, H. Niu, D. Bradshaw, M. J. Rosseinsky, Y. Z. Khimyak, N. L. Campbell, R. Kirk, E. Stöckel and A. I. Cooper, *Chem. Mater.*, 2007, **19**, 2034.
64. C. F. Martin, E. Stöckel, R. Clowes, D. J. Adams, A. I. Cooper, J. J. Pis, F. Rubiera and C. Pevida, *J. Mater. Chem.*, 2011, **21**, 5475.

65. T. Ben, H. Ren, S. Ma, D. Cao, J. Lan, X. Jing, W. Wang, J. Xu, F. Deng, J. M. Simmons, S. Qiu and G. Zhu, *Angew. Chem., Int. Ed.*, 2009, **48**, 9457.

66. T. Ben, C. Pei, D. Zhang, J. Xu, F. Deng, X. Jing and S. Qiu, *Energy Environ. Sci.*, 2011, **4**, 3991.

67. Y. Yuan, F. Sun, H. Ren, X. Jing, W. Wang, H. Ma, H. Zhao and G. Zhu, *J. Mater. Chem.*, 2011, **21**, 13498.

68. T. Ben and S. Qiu, *CrystEngComm*, 2013, **15**, 17.

69. L. J. Murray, M. Dinca and J. R. Long, *Chem. Soc. Rev.*, 2009, **38**, 1294.

70. J. R. Li, R. J. Kuppler and H. Zhou, *Chem. Soc. Rev.*, 2009, **38**, 1477.

71. J. Lee, O. K. Farha, J. Roberts, K. A. Scheidt, S. T. Nguyen and J. T. Hupp, *Chem. Soc. Rev.*, 2009, **38**, 1450.

72. L. Ma, C. Abney and W. Lin, *Chem. Soc. Rev.*, 2009, **38**, 1248.

73. M. Yoon, R. Srirambalaji and K. Kim, *Chem. Rev.*, 2012, **112**, 1196.

74. L. E. Kreno, K. Leong, O. K. Farha, M. Allendorf, R. P. Van Duyne and J. T. Hupp, *Chem. Rev.*, 2012, **112**, 1105.

75. C. A. Kent, D. Liu, L. Ma, J. M. Papanikolas, T. J. Meyer and W. Lin, *J. Am. Chem. Soc.*, 2011, **133**, 12940.

76. C. A. Kent, B. P. Mehl, L. Ma, J. M. Papanikolas, T. J. Meyer and W. Lin, *J. Am. Chem. Soc.*, 2010, **132**, 12767.

77. S. Jin, H. Son, O. K. Farha, G. P. Wiederrecht and J. T. Hupp, *J. Am. Chem. Soc.*, 2013, **135**, 955.

78. H. Son, S. Jin, S. Patwardhan, S. J. Wezenberg, N. C. Jeong, M. So, C. E. Wilmer, A. A. Sarjeant, G. C. Schatz, R. Q. Snurr, O. K. Farha, G. P. Wiederrecht and J. T. Hupp, *J. Am. Chem. Soc.*, 2013, **135**, 862.

79. M. So, S. Jin, H. Son, G. P. Wiederrecht, O. K. Farha and J. T. Hupp, *J. Am. Chem. Soc.*, 2013, **135**, 15698.

80. S. Horike, D. Umeyama and S. Kitagawa, *Acc. Chem. Res.*, 2013, **46**, 2376.

81. A. Shigematsu, T. Yamada and H. Kitagawa, *J. Am. Chem. Soc.*, 2011, **133**, 2034.

82. M. L. Aubrey, R. Ameloot, B. M. Wiers and J. R. Long, *Energy Environ. Sci.*, 2014, 7, 667.

83. G. K. H. Shimizu, J. M. Taylor and S. Kim, *Science*, 2013, **341**, 354.

84. C. Wang, T. Zhang and W. Lin, *Chem. Rev.*, 2012, **112**, 1084.

85. R. Dawson, D. J. Adams and A. I. Cooper, *Chem. Sci.*, 2011, **2**, 1173.

86. H. Wang, Z. Du and Q. Duan, *Chem. Ind. Eng. Prog.*, 2002, **21**, 836.

87. G. Ferry, C. M. Draznieks, C. Serre, F. Millange, J. Dutour, S. Surble and I. Margiolaki, *Science*, 2005, **309**, 2040.

88. M. Eddaoudi, J. Kim, N. L. Rosi, D. T. Vodak, J. Wachter, M. O'Keeffe and O. M. Yaghi, *Science*, 2002, **295**, 469.

89. C. Pei and F. Qu, *Chem. J. Chinese U.*, 2011, **32**, 1234.

90. Z. Wang and S. M. Cohen, *Chem. Soc. Rev.*, 2009, **38**, 1315.

91. K. K. Tanabe and S. M. Cohen, *Chem. Soc. Rev.*, 2011, **40**, 498.

92. S. M. Cohen, *Chem. Rev.*, 2012, **112**, 970.

93. B. F. Hoskins and R. Robson, *J. Am. Chem. Soc.*, 1990, **112**, 1546.

94. A. D. Burrows, F. L. I. Xamena and J. Gascon, *Metal Organic Frameworks as Heterogeneous Catalysts*, The Royal Society of Chemistry, 2013, p. 31.

95. W. Lu, D. Yuan, J. Sculley, D. Zhao, R. Krishna and H. Zhou, *J. Am. Chem. Soc.*, 2011, **133**, 18126.
96. V. Presser, J. McDonough, S. Yeon and Y. Gogotsi, *Energy Environ. Sci.*, 2011, **4**, 3059.
97. A. F. Oliveira, G. Seifert, T. Heine and H. A. Duarte, *J. Brazil. Chem. Soc.*, 2009, **20**, 1193.
98. D. Britt, C. Lee, F. J. Uribe-Romo, H. Furukawa and O. M. Yaghi, *Inorg. Chem.*, 2010, **49**, 6387.
99. C. Volkringer and S. M. Cohen, *Angew. Chem., Int. Ed.*, 2010, **49**, 4644.
100. D. N. Dybtsev, H. Chun and K. Kim, *Angew. Chem., Int. Ed.*, 2004, **43**, 5033.
101. M. J. Ingleson, J. P. Barrio, J.-B. Guilbaud, Y. Z. Khimyak and M. J. Rosseinsky, *Chem. Commun.*, 2008, 2680.
102. K. L. Mulfort, O. K. Farha, C. L. Stern, A. A. Sarjeant and J. T. Hupp, *J. Am. Chem. Soc.*, 2009, **131**, 3866.
103. M. H. Weston, O. K. Farha, B. G. Hauser, J. T. Hupp and S. T. Nguyen, *Chem. Mater.*, 2012, **24**, 1292.
104. K. K. Tanabe, N. A. Siladke, E. M. Broderick, T. Kobayashi, J. F. Goldston, M. H. Weston, O. K. Farha, J. T. Hupp, M. Pruski, E. A. Mader, M. J. A. Johnson and S. T. Nguyen, *Chem. Sci.*, 2013, **4**, 2483.
105. S. J. Kraft, R. H. Sánchez and A. S. Hock, *ACS Catal.*, 2013, **3**, 826.
106. R. K. Totten, M. H. Weston, J. K. Park, O. K. Farha, J. T. Hupp and S. T. Nguyen, *ACS Catal.*, 2013, **3**, 1454.
107. G. Sartori, R. Ballini, F. Bigi, G. Bosica, R. Maggi and P. Righi, *Chem. Rev.*, 2004, **104**, 199.
108. T. Gadzikwa, G. Lu, C. L. Stern, S. R. Wilson, J. T. Hupp and S. T. Nguyen, *Chem. Commun.*, 2008, 5493.
109. T. Gadzikwa, O. K. Farha, C. D. Malliakas, M. G. Kanatzidis, J. T. Hupp and S. T. Nguyen, *J. Am. Chem. Soc.*, 2009, **131**, 13613.
110. R. K. Deshpande, J. L. Minnaar and S. G. Telfer, *Angew. Chem., Int. Ed.*, 2010, **49**, 4598.
111. K. Oisaki, Q. Li, H. Furukawa, A. U. Czaja and O. M. Yaghi, *J. Am. Chem. Soc.*, 2010, **132**, 9262.
112. Y. Li, T. Ben, B. Zhang, Y. Fu and S. Qiu, *Sci. Rep.*, 2013, **3**, 2420.
113. J. Lan, D. Cao, W. Wang, T. Ben and G. Zhu, *J. Phys. Chem. Lett.*, 2010, **1**, 978.
114. Y. Sun, T. Ben, L. Wang, S. Qiu and H. Sun, *J. Phys. Chem. Lett.*, 2010, **1**, 2753.
115. L. Wang, Y. Sun and H. Sun, *Faraday Discuss.*, 2011, **151**, 143.
116. Y. Yuan, F. Sun, F. Zhang, H. Ren, M. Guo, K. Cai, X. Jing, X. Gao and G. Zhu, *Adv. Mater.*, 2013, **25**, 6619.
117. A. P. Nelson, O. K. Farha, K. L. Mulfort and J. T. Hupp, *J. Am. Chem. Soc.*, 2009, **131**, 458.
118. J. An, C. M. Shade, D. A. Chengelis-Czegan, S. Petoud and N. L. Rosi, *J. Am. Chem. Soc.*, 2011, **133**, 1220.

119. J. An and N. L. Rosi, *J. Am. Chem. Soc.*, 2010, **132**, 5578.

120. G. Calleja, J. A. Botas, M. Sanchez-Sanchez and M. G. Orcajo, *Int. J. Hydrogen Energy*, 2010, **35**, 9916.

121. S. Han, Y. Wei, C. Valente, R. S. Forgan, J. J. Gassensmith, R. A. Smaldone, H. Nakanishi, A. Coskun, J. F. Stoddart and B. A. Grzybowski, *Angew. Chem., Int. Ed.*, 2011, **50**, 276.

122. Z. Xiang, Z. Hu, D. Cao, W. Yang, J. Lu, B. Han and W. Wang, *Angew. Chem., Int. Ed.*, 2011, **50**, 491.

123. D. Dybstev, C. Serre, B. Schmitz, B. Panella, M. Hirscher, M. Latroche, P. L. Llewellyn, S. Cordier, Y. Molard, M. Haouas, F. Taulelle and G. Férey, *Langmuir*, 2010, **26**, 11283.

124. M. Meilikhov, K. Yusenko, D. Esken, S. Turner, G. Van Tendeloo and R. A. Fischer, *Eur. J. Inorg. Chem.*, 2010, 3701.

125. C. Zlotea, R. Campesi, F. Cuevas, E. Leroy, P. Dibandjo, C. Volkringer, T. Loiseau, G. Férey and M. Latroche, *J. Am. Chem. Soc.*, 2010, **132**, 2991.

126. O. Karagiaridi, W. Bury, A. A. Sarjeant, C. L. Stern, O. K. Farha and J. T. Hupp, *Chem. Sci.*, 2012, **3**, 3256.

127. O. Karagiaridi, W. Bury, E. Tylianakis, A. A. Sarjeant, J. T. Hupp and O. K. Farha, *Chem. Mater.*, 2013, **25**, 3499.

128. O. Karagiaridi, M. B. Lalonde, W. Bury, A. A. Sarjeant, O. K. Farha and J. T. Hupp, *J. Am. Chem. Soc.*, 2012, **134**, 18790.

129. M. Lalonde, W. Bury, O. Karagiaridi, Z. Brown, J. T. Hupp and O. K. Farha, *J. Mater. Chem. A*, 2013, **1**, 5453.

130. M. Kim, J. F. Cahill, Y. Su, K. A. Prather and S. M. Cohen, *Chem. Sci.*, 2012, **3**, 126.

131. Y. E. Cheon and M. P. Suh, *Angew. Chem., Int. Ed.*, 2009, **48**, 2899.

132. X. Zhang, F. X. Llabrés i Xamena and A. Corma, *J. Catal.*, 2009, **265**, 155.

133. M. Savonnet, S. Aguado, U. Ravon, D. Bazer-Bachi, V. Lecocq, N. Bats, C. Pinel and D. Farrusseng, *Green Chem.*, 2009, **11**, 1729.

134. M. Banerjee, S. Das, M. Yoon, H. J. Choi, M. H. Hyun, S. M. Park, G. Seo and K. Kim, *J. Am. Chem. Soc.*, 2009, **131**, 7524.

135. K. L. Taylor-Pashow, J. D. Rocca, Z. Zie, S. Tran and W. Lin, *J. Am. Chem. Soc.*, 2009, **131**, 14261.

136. S. C. Jones and C. A. Bauer, *J. Am. Chem. Soc.*, 2009, **131**, 12516.

137. D. Cao, J. Lan, W. Wang and B. Smit, *Angew. Chem., Int. Ed.*, 2009, **48**, 4730.

138. S. S. Han and W. A. Goddard III, *J. Am. Chem. Soc.*, 2007, **129**, 8422.

139. X. Xu and W. A. Goddard, *Proc. Natl. Acad. Sci. U. S. A.*, 2004, **101**, 2673.

140. N. L. Rosi, A. R. Millward, K. S. Park and O. M. Yaghi, *J. Am. Chem. Soc.*, 2004, **126**, 5666.

141. A. G. Wong-Foy, A. J. Matzger and O. M. Yaghi, *J. Am. Chem. Soc.*, 2006, **128**, 3494.

142. A. Dailly, J. J. Vajo and C. C. Ahn, *J. Phys. Chem. B*, 2006, **110**, 1099.

143. K. L. Mulfort and J. T. Hupp, *J. Am. Chem. Soc.*, 2007, **129**, 9604.

144. M. Inagaki and O. Tanaike, *Synth. Met.*, 1995, **73**, 77.

145. C. D. Stevenson, C. V. Rice, P. M. Garland and B. K. Clark, *J. Org. Chem.*, 1997, **62**, 2193.
146. K. Konstas, J. W. Taylor, A. W. Thornton, C. M. Doherty, W. X. Lim, T. J. Bastow, D. F. Kennedy, C. D. Wood, B. J. Cox, J. M. Hill, A. J. Hill and M. R. Hill, *Angew. Chem., Int. Ed.*, 2012, **51**, 6639.
147. K. K. Tanabe, Z. Wang and S. M. Cohen, *J. Am. Chem. Soc.*, 2008, **130**, 8508.
148. E. Dugan, Z. Wang, M. Okamura, A. Medina and S. M. Cohen, *Chem. Commun.*, 2008, **29**, 3366.
149. Z. Wang and S. M. Cohen, *Angew. Chem., Int. Ed.*, 2008, **47**, 4699.
150. S. J. Garibay, Z. Wang, K. K. Tanabe and S. M. Cohen, *Inorg. Chem.*, 2009, **48**, 7341.
151. S. J. Garibay, Z. Wang and S. M. Cohen, *Inorg. Chem.*, 2010, **49**, 8086.
152. S. Bhattacharjee, D. A. Yang and W. S. Ahn, *Chem. Commun.*, 2011, **47**, 3637.
153. M. Ma, A. Gross, D. Zacher, A. Pinto, H. Noei, Y. Wang, R. A. Fischer and N. Metzler-Nolte, *CrystEngComm*, 2011, **13**, 2828.
154. P. Serra-Crespo, E. V. Ramos-Fernandez, J. Gascon and F. Kapteijn, *Chem. Mater.*, 2011, **23**, 2565.
155. T. Ahnfeldt, N. Guillou, D. Gunzelmann, I. Margiolaki, G. Férey, J. Senker and N. Stock, *Angew. Chem., Int. Ed.*, 2009, **48**, 5163.
156. S. J. Garibay and S. M. Cohen, *Chem. Commun.*, 2010, **46**, 7700.
157. M. Kandiah, M. H. Nilsen, S. Usseglio, S. Jakobsen, U. Olsbye, M. Tilset, C. Larabi, E. A. Quadrelli, F. Bonino and K. P. Lillerud, *Chem. Mater.*, 2010, **22**, 6632.
158. M. Kandiah, S. Usseglio, S. Svelle, U. Olsbye, K. P. Lillerud and M. Tilset, *J. Mater. Chem.*, 2010, **20**, 9848.
159. A. D. Burrows, C. G. Frost, M. F. Mahon and C. Richardson, *Angew. Chem., Int. Ed.*, 2008, **47**, 8482.
160. A. Phan, C. J. Doonan, F. J. Uribe-Romo, C. B. Knobler, M. O'Keeffe and O. M. Yaghi, *Acc. Chem. Res.*, 2010, **43**, 58.
161. A. S. Huang, W. Dou and J. Caro, *J. Am. Chem. Soc.*, 2010, **132**, 15562.
162. M. H. Mir, L. L. Koh, G. K. Tan and J. J. Vittal, *Angew. Chem., Int. Ed.*, 2010, **49**, 390.
163. S. Bernt, V. Guillerm, C. Serre and N. Stock, *Chem. Commun.*, 2011, **47**, 2838.
164. W. Lu, J. P. Sculey, D. Yuan, R. Krishna, Z. Wei and H. Zhou, *Angew. Chem., Int. Ed.*, 2012, **51**, 7480.
165. W. Lu, J. P. Sculley, D. Yuan, R. Krishna and H. Zhou, *J. Phys. Chem. C*, 2013, **117**, 4057.
166. A. Sayari, A. Heydari-Gorji and Y. Yang, *J. Am. Chem. Soc.*, 2012, **134**, 13834.
167. E. Merino, E. Verde-Sesto, E. M. Maya, M. Iglesias, F. Sánchez and A. Corma, *Chem. Mater.*, 2013, **25**, 981.

168. A. D. Burrows, C. G. Frost, M. F. Mahon and C. Richardson, *Chem. Commun.*, 2009, 4218.
169. T. Ben, Y. Li, L. Zhu, D. Zhang, D. Cao, Z. Xiang, X. Yao and S. Qiu, *Energy Envrion. Sci.*, 2012, **5**, 8370.
170. Z. Xiang, D. Cao, L. Huang, J. Shui, M. Wang and L. Dai, *Adv. Mater.*, 2014, **26**, 3315.
171. S. Tanaka, N. Nishiyama, Y. Egashira and K. Ueyama, *Chem. Commun.*, 2005, 2125.
172. H. B. Wu, S. Wei, L. Zhang, R. Xu, H. H. Hng and X. W. Lou, *Chem. - Eur. J.*, 2013, **19**, 10804.
173. A. Banerjee, R. Gokhale, S. Bhatnagar, J. Jog, M. Bhardwaj, B. Lefez, B. Hannoyer and S. Ogale, *J. Mater. Chem.*, 2012, **22**, 19694.
174. Y. Wang, Y. Shao, D. W. Matson, J. Li and Y. Lin, *ACS Nano*, 2010, **4**, 1790.
175. H. M. Jeong, J. W. Lee, W. H. Shin, Y. J. Choi, H. J. Shin, J. K. Kang and J. W. Choi, *Nano Lett.*, 2011, **11**, 2472.
176. B. D. Guo, Q. A. Liu, E. D. Chen, H. W. Zhu, L. A. Fang and J. R. Gong, *Nano Lett.*, 2010, **10**, 4975.
177. P. Kuhn, A. Thomas and M. Antonietti, *Macromolecules*, 2009, **42**, 319.
178. M. J. Bojdys, J. Jeromenok, A. Thomas and M. Antonietti, *Adv. Mater.*, 2010, **22**, 2202.
179. P. Katekomol, Je. Roeser, M. Bojdys, J. Weber and A. Thomas, *Chem. Mater.*, 2013, **25**, 1542.
180. P. Kuhn, A. Forget, D. Su, A. Thomas and M. Antonietti, *J. Am. Chem. Soc.*, 2008, **130**, 13333.
181. C. E. Chan-Thaw, A. Villa, P. Katekomol, D. Su, A. Thomas and L. Prati, *Nano Lett.*, 2010, **10**, 537.
182. R. Babarao, S. Dai and D. Jiang, *Langmuir*, 2011, **27**, 3451.

Applications of Porous Polymers

10.1 Gas Storage

Environmentally friendly characteristics with high chemical abundance and high energy density have made hydrogen gas interesting as an energy resource. The US Department of Energy (DOE) has set targets of 5.5 wt% and 81 kg H_2 m^{-3} at 253–323 K with a pressure of 100 atm for hydrogen storage by the year 2015. Porous organic polymers have potential advantages for H_2 storage due to their light weight and the possibility of introducing a variety of functionalities to bind gas molecules.

Methane, CH_4, is a transitional fuel source because of its high ratio of hydrogen to carbon. The US DOE target for CH_4 storage is 180 (v/v) at 35 bar and 298 K.[1] The optimal heat of adsorption for CH_4 storage at room temperature has been determined to be 18.8 kJ mol^{-1}.[2]

The capture and storage of CO_2 is an important goal due to its role in global warming. Under pre-combustion conditions, CO_2 needs to be separated from either hydrogen or methane at high pressure; post-combustion conditions require atmospheric pressures and temperatures, with a low concentration of CO_2 mixed with N_2.[3] Thus, carbon capture and storage requires separation of CO_2 from the other gases present, along with the high capture ability of the material used.[4]

A convenient method for H_2 storage is physisorption within a highly porous material. That's where porous polymers come into play. What is crucial for efficient storage of gas molecules under suitable and practicable conditions is the presence of a large number of small micropores, as multiple sorbate–sorbent interactions increase the enthalpy of adsorption to enhance the storage capability. For reversible H_2 storage at room

Monographs in Supramolecular Chemistry No. 17
Porous Polymers: Design, Synthesis and Applications
By Shilun Qiu and Teng Ben
© The Royal Society of Chemistry 2016
Published by the Royal Society of Chemistry, www.rsc.org

temperature, the optimum enthalpy is calculated to be 15.1 kJ mol^{-1}. Other aspects that have made porous frameworks potential gas storage materials are their low densities and extremely high surface areas. Incorporation of metal ions and metal alkoxide groups, and naked halide functionalities into some porous frameworks enhances the interaction of hydrogen molecules with the metal or halide in the networks, enhancing the uptake by the framework to a significant level.

Isosteric heats of adsorption for methane sorption in carbon-based compounds fall within the exact storage range close to ambient temperatures, which makes organic porous compounds important candidates for methane storage.

High physicochemical stability is an important requirement for porous polymers if they are to be successfully used as gas storage materials. Covalent organic frameworks (COFs) can adsorb significant gas molecules, especially CO_2, but they lack the physicochemical stability, thus demolishing their potential for gas storage. Porous aromatic framework-1 (PAF-1), being very stable towards harsh conditions, is reusable and has a good lifetime, which makes it an excellent storage material. The well-defined pore structure of porous networks also allows effective gas storage inside them.

10.1.1 Hyper-crosslinked Polymers (HCPs)

HCPs represent a class of mainly microporous organic compounds with high surface areas.[5] A hyper-crosslinked polystyrene with a Brunauer–Emmett–Teller (BET) surface area of 1466 m^2 g^{-1} was shown to adsorb up to 3.04 wt% H_2 at 15 bar and 77 K.[6] Germain *et al.* described the properties of similar low-pressure sorption for a hyper-crosslinked polystyrene material,[7] comparing it with a wide range of commercially available macroporous polymer sorbents that showed a decreased H_2 storage capacity. HCPs with higher surface areas ($\sim$ 1900 m^2 g^{-1} BET; $\sim$ 3000 m^2 g^{-1} Langmuir) may be obtained from the polycondensation of dichloroxylene (DCX).[8] However, other bischloromethyl monomers were shown to adsorb up to 3.7 wt% H_2 at 15 bar and 77 K. Hyper-crosslinked polystyrene synthesized by the Cooper group with a BET surface area of 1466 m^2 g^{-1} was shown to adsorb 1.27% H_2 by mass at 1 bar/77 K and 3.04% at 15 bar/77 K, which is much larger than a previously reported group of polymers of intrinsic microporosity (PIMs).[6] The reported hyper-crosslinked polymer adsorbed almost 1 wt% more H_2 than CTC-PIM at 10 bar. However, this difference is much smaller at a pressure of 1 bar, when the adsorption of the hyper-crosslinked polymer was 1.28 wt% H_2 and the adsorptions of the PIMs were in the range of 1.04–1.43 wt%. Another report has demonstrated the comparison of H_2 adsorption between a group of commercially available HCPs and hyper-crosslinked polystyrene. The HCP displayed an H_2 uptake of 1.4% by mass at 1 bar/77 K and 3.9% H_2 uptake at 40 bar/77 K.[9] The HCPs were prepared by the Friedel–Crafts alkylation of chloromethylaromatics with densities of 0.78 g cm^{-3} and uptakes 5.5 mmol g^{-1} CH$_4$. It is seen that porous polymers having lower densities adsorb higher amounts of methane; thus, HCP networks are quite useful[10] (Figure 10.1).

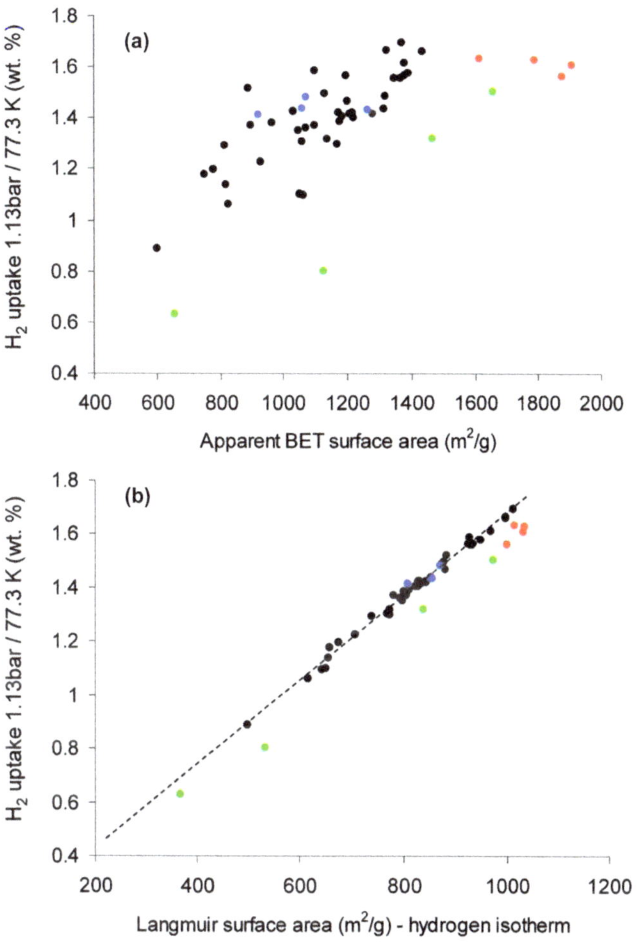

Figure 10.1   (a) Hydrogen uptake at 1.13 bar/77.3 K as a function of apparent BET surface area for a series of hyper-crosslinked polymers: dichloro-(xylene) (DCX) networks (black symbols); bis(chloromethyl)biphenyl (BCMBP) networks (red symbols); bis(chloromethyl)anthracene (BCMA) networks (blue symbols); hypercrosslinked poly(vinyl-benzyl chloride) (HCPVBC) "Davankov resins" (green symbols). (b) Hydrogen uptake at 1.13 bar/77.3 K as a function of the Langmuir surface area as calculated from the H$_2$ isotherm for the same series of hyper-crosslinked polymer (sample labeled as in part a). Dashed line shows linear fit ($r_2 = 0.9927$) for the combined DCX dataset, dichloro-(xylene) (DCX) networks.[5]
Reprinted with permission from C. D. Wood, B. Tan, A. Trewin, H. Niu, D. Bradshaw, M. J. Rosseinsky, Y. Z. Khimyak, N. L. Campbell, R. Kirk, E. Stöckel and A. I. Cooper, *Chem. Mater.*, 2007, **19**, 2034. Copyright 2007, American Chemical Society.

Table 10.1 Apparent surface areas from N_2 and H_2 sorption, and gravimetric hydrogen uptake at 77 K.[9]

Sample	N_2 BET surface area ($m^2\ g^{-1}$)	H_2 Langmuir surface area ($m^2\ g^{-1}$)	H_2 uptake 1 bar (% mass)	H_2 uptake 10 bar (% mass)
PIM-1	750	540	0.95	1.45
PIM-7	680	530	1.00	1.35
HATN-PIM[a]	680	590	1.12	1.56
CTC-PIM	770	630	1.35	1.70
Porph-PIM	960	760	1.20	1.95
Trip-PIM	1050	1100	1.63	2.71
HCP	1466	1250	1.27	2.75

[a]HATN, hexaazatrinaphthylene.

10.1.2 Polymers of Intrinsic Microporosity (PIMs)

A special kind of PIM network has been demonstrated to adsorb around 1.5–1.7 wt% H_2 at 77.3 K and 1 bar, and up to 2.71 wt% at 10 bar. Hydrogen uptake for these materials is observed to be 1.4% by mass adsorbed at 1 bar and 1.7% at 10 bar for the cyclotricatechylene-PIM (CTC-PIM). The triptycene-based polymer (Trip-PIM) shows a better result of 1.6% by mass at 1 bar and 2.7% at 10 bar.[11] Porphyrin-PIM (Porph-PIM) and PIM-7 have also been reported to absorb significant amounts of H_2. Table 10.1 represents the hydrogen uptake for all of these PIMs at 77 K. Trip-(R)-PIMs with shorter alkyl chains at moderate pressure show excellent H_2 adsorption; for Trip-(Me)-PIMs, the adsorption of 3.4% by mass at 18 bar is comparable to different types of microporous compounds with similar BET surface areas[12] (Table 10.1).

10.1.3 Covalent Organic Frameworks (COFs)

Grand canonical Monte Carlo (GCMC) simulations of COFs predict COF-105 and COF-108 to be efficient at H_2 uptake up to 10.0 wt% at 77 K. COF-102 normally stores 40.4 g L^{-1} of H_2, which is comparable with that of several metal–organic frameworks (MOFs).[13] Theoretical calculations predict that the uptake of hydrogen is determined by the free volume of the COFs and by the surface area. COFs show better performance with regards to hydrogen storage at low temperatures compared to ambient temperatures and pressure. Adsorption simulations of the COFs show that doping with charged species could affect their capacity. It has already been proposed that doping with electropositive elements, like Li, Na, K *etc.*, may increase H_2 adsorption. Lithium ions especially enhance storage capacity. Simulation results suggest that pillared COFs, spill-over COFs and novel three-dimensional (3D) COF architectures are most promising for hydrogen storage to meet the DOE target.[14]

For CH_4 storage, the highest values were observed for COF-103 (175 mg g^{-1}) and COF-102 (187 mg g^{-1}), and were comparable to MOFs (MOF-210; 220 mg g^{-1}). Theoretical simulation studies predict the strong interaction of methane molecules with the faces of the aromatic rings or boroxine rings. Doping the COFs with Li^+ significantly enhances their storage capacity at relatively low pressures. It is also predicted that COF-5 and COF-6 are more selective than zeolites and other similar frameworks, including MOFs.[14]

Yaghi and co-workers reported that the carbon dioxide uptake of COF-102 (27 mmol g^{-1} at 298 K and 35 bar) is higher than the CO_2 uptake of MOF-5 (22 mmol g^{-1}) and zeolites (5–8 mmol g^{-1}). Simulations show that CO_2 uptake is directly related to the total pore volume. The dense lattice packing of COF-102 and COF-103 made them better with regards to adsorption at low pressures compared to COF-105 and COF-108, and the trend is reversed at higher pressures due to the larger pore volumes for COF-105 and COF-108, which help their higher storage capacities. Lithium ion and other alkali metal ion doping is again most effective at enhancing the CO_2 storage capacity.[14]

COFs with boronate-ester linkages show effective ammonia sorption due to the Lewis acid–base interaction. Among them, COF-10 exhibits an uptake capacity of 15 mol kg^{-1} at 298 K and 1 bar. COFs can be recycled several times without much loss of activity.[14]

Conjugated Microporous Polymers (CMPs)

CMPs exhibit enthalpies of adsorption for hydrogen up to 18 kJ mol^{-1}. Such a high value is indicative of reversible hydrogen storage at room temperature.[7] CMP-0 demonstrated an H_2 capacity of 1.4 wt% at 77 K and 1 bar. The introduction of Pd nanoparticles into CMP-0 enhances its capacity.[4] Incorporation of acid functionalities into the CMP material decreases the CO_2 adsorption capacity more than for the non-functionalized CMP analogues.[4]

Porous Aromatic Frameworks (PAFs)

The H_2 uptakes at 77 K/760 mm Hg have been reported as 111 and 142 cm^3 g^{-1} for JUC-Z5 and JUC-Z4, respectively. PAF-1 exhibits a better result with regards to uptake (186 cm^3 g^{-1} at 77 K and 760 mm Hg). In the low-pressure region, JUC-Z4 showed a significant hike to a value of 120 cm^3 g^{-1} at a pressure lower than 370 mm Hg.

JUC-Z5 and JUC-Z4 showed reversible methane adsorption isotherms. CH_4 uptake of JUC-Z5 was 16 cm^3 g^{-1} while that of JUC-Z4 was 20 cm^3 g^{-1} and that of PAF-1 was 18 cm^3 g^{-1} at 273 K/760 mm Hg. In spite of the much lower specific surface areas of JUC-Z5 and JUC-Z4 compared to PAF-1, the strong interactions between CH_4 and the phosphorus (P) moieties make JUC-Z5 and JUC-Z4 comparable with PAF-1 in the case of methane uptake.

Key features that make JUC-Z5 and JUC-Z4 very effective for CO_2 capture are the excellent physico-chemical stability with well-defined micropore sizes (pore size <1 nm). For JUC-Z5, JUC-Z4 and PAF-1, the CO_2 storage capacities at 273 K and 760 mmHg are 56, 59 and 46 cm^3 g^{-1}, respectively, while the values are 38, 34 and 25 cm^3 g^{-1} at 298 K and 1 bar.

The small size of the pore and the strong interactions between the P moieties in JUC-Z5 and the CO_2 molecules are the key factors for high CO_2 uptake.[15]

PAF-3 shows increased gas adsorption at low pressure due to the increased heat of adsorption between the gas molecules and PAF-3 by introduction of Si moieties. The low pressure sorption of PAF-3 shows a high CO_2 sorption capacity of 78 cm^3 g^{-1} (corresponding to 15.3 wt%), while this is only 9.1 wt% (46 cm^3 g^{-1}) for PAF-1 at 273 K and 1 atm. For H_2 sorption, PAF-3 exhibited a high uptake of 232 cm^3 g^{-1} (2.07 wt%) at 1 atm and 77 K, much higher than that of PAF-1 (232 cm^3 g^{-1}; 1.66 wt%) under the same conditions. An increased CH_4 adsorption was also observed for PAF-3 compared to PAF-1.[16]

The derivatives of PAF-1 also show high gas sorption. The almost fully carbonized PAF-1-450 and the partially carbonized PAF-1-350, PAF-1-380 and PAF-1-400 exhibit very strong affinity toward CO_2 due to their high quadrupole moments. Carbonized PAF-1s exhibit exceptionally greater CO_2 adsorption capacities, which show complete reversibility at 273 K and 1 bar. Compared to PAF-1, partially carbonized PAF-1 shows a tendency for an increase in CO_2 uptake. More significantly, completely carbonized PAF-1-450 shows a remarkable increase in CO_2 uptake, with a value of 100 cm^3 g^{-1} (equivalent to 16.5 wt%, 4.5 mmol g^{-1}).[17]

The gravimetric hydrogen uptakes of PAF-301 and PAF-302 are 0.83 and 2.21 wt% at temperature $T = 77$ K and pressure $p = 100$ bar. PAF-303 and PAF-304 reach 4.16 and 6.53 wt% at $T = 77$ K and $p = 100$ bar, which are better than the previous two PAFs. Moreover, PAF-304 exhibits one of the highest uptakes of hydrogen, at about 6.53 wt% at $T = 298$ K and $p = 100$ bar, among all of the current porous compounds without surface modification.[18]

Using quantized liquid density functional theory (QLDFT) within the LIE-0 approximation, it is predicted that the hydrogen uptake capacity of PAF-qtph (also called PAF-304 (ref. 18), qtph represents quaterphenyl) may be 7.32 wt%, which exceeds the 2015 DOE target; however, this is a slightly higher capacity of storage compared with GCMC results in the literature.[19]

PAF-301 shows much higher uptakes of CO_2 at low pressure due to the small pore size. Because of the small size of the pore, 5.2 Å (275 mg g^{-1} at 298 K and 1 bar), the uptakes of CO_2 in PAF-304 and PAF-303 reach 3124 and 3432 mg g^{-1} at 50 bar and at 298 K, respectively, which are larger than MOF-200 (2437 mg g^{-1}) and MOF-210 (2396 mg g^{-1}).[20]

The derivatives of carbonized PAF-1 formed by high-temperature KOH activation exhibit good carbon dioxide, methane and hydrogen sorption abilities in both the high pressure and low pressure regions, while the methane storage of K-PAF-1-750 is one of the greatest at 35 bar. High physicochemical stability make these compounds very promising for applications in industry such as carbon dioxide capture and clean energy storage at a high density.[21]

10.2 Selective Separation

Gas molecule, like hydrogen, that have a small size can permeate through smaller pores, which are inaccessible to larger gas molecules like nitrogen, argon or carbon dioxide. It has been seen that H_2 molecules can diffuse through ultramicropores with a radius of around 0.29 nm, while N_2 and CO_2 can pass through 0.36 and 0.7 nm pores of minimum radii, respectively.[22,23]

CO_2 as a byproduct during the combustion of fuels in industrial plants and automobiles is a major contributor to global warming. Hence, the selective capture of carbon dioxide requires special attention from the scientific community. The principle sources of CO_2 that cause harm to the environment are contaminated natural gas, containing a mixture of methane and CO_2 (known as pre-combustion), and exhaust gas generated in industry or from automobiles (post-combustion).[3] Separate physical conditions are needed for CO_2 capture from pre- and post-combustion mixtures, and a variety of porous polymer networks, both soluble and insoluble, have been involved in CO_2 capture with consideration of their pore dimensions.[24–26]

HCP-1,4-benzenedimethanol (HCP-BDM) and HCP-benzyl alcohol (HCP-BA) networks synthesized by the Friedel–Crafts self-condensation method have shown high selectivity for CO_2 over N_2, measured by nitrogen adsorption isotherm at two different temperatures, 273 and 298 K.[27] Non-local density functional theory (NLDFT) calculations confirmed the pore sizes to be below 2 nm for both of the networks. Such small pore sizes, as well as the high oxygen content in both the polymers, is most probably the reason for the strong interactions with polar CO_2 rather than N_2, making these good candidates for the selective separation of CO_2 from N_2.

Polyaniline and aminobenzene, due to their aromatic frameworks, constitute HCPs with pores small enough to disallow the penetration of N_2 but large enough for H_2 permeation.[22] Trivalent and rigid linking groups with aromatic rings form a dense and highly cross-linked network of comparatively light elements to show high enthalpy of adsorption towards H_2. Such materials exhibit a large enthalpy of H_2 adsorption, having pore sizes similar to the kinetic diameter of the hydrogen molecule, thus establishing the fact that selective permeation here is solely dependent on thermodynamic rather than kinetic parameters. Formation of a large number of well-controlled pores in such polymers allows size discrimination between nitrogen and hydrogen molecules.

In comparison to other polymer membranes, soluble porous networks hold an advantage in gas separations in the form of thin solution-cast films.[28,29] Linear PIMs are soluble and thus good candidates in this regard.[30] The introduction of triazole groups in PIMs (TZPIMs) by post-synthetic modification favors CO_2 uptake and greater selectivity for gas separation.[31] TZPIM membranes exhibit exceptional selective permeation as polymeric membranes[29] for gas mixtures like CO_2 and N_2.

A chiral PIM based on a binaphthalene monomer[32] has been functionalized by incorporation of polyimide with binaphthalene and 4,4-(9-fluorenylidene)dianiline. The polymeric film of this network demonstrates selective gas uptakes to separate CO_2 from H_2.[33]

Proper tuning of the pore size in COFs has been proven to be useful for selective gas separations. Successive reduction of the pore size has been seen in COF-18A with addition of bigger alkyl functionalities such as methyl and propyl. With the reduction of the pore size of the networks from COF-18A to COF-11A, the ability to adsorb bigger gas molecules, like nitrogen, also decreased while adsorption of the smaller gases (H_2) increased. Hence, selective separation can be tuned by controlling the pore dimensions.[34]

COF-6 has a smaller pore diameter and is a tremendous material for the selective capture of CO_2, showing a better performance than other COFs under ambient conditions.[34] In COF materials, the pore volume or intrinsic surface area play an important role in determining the gas storage capacity. Compounds with bigger pore volumes are advantageous for storing a higher number of guest molecules, which has been supported by several simulation calculations.[35,36] Mesoporous materials display some deviation from such a prediction, which indicates that, for micropore networks, the gas uptake varies linearly with the total pore volume.

Methane has a strong interaction with the faces of the boroxine or aromatic rings of COFs.[14] COF-102 and COF-103 can be considered as promising candidates for methane storage. Lithiated COFs significantly enhance their CH_4 uptakes at low pressures. Such COFs show excellent separation capacities with regards to CH_4 from H_2 with mass transport of the adsorbed species.

Another porous material, PAF-30, based on tetra-(4-anilyl)-methane and cyanuric chloride, also shows high selectivity for CO_2 compared to CH_4.[37] Triazines of cyanuric chloride have a high nitrogen content, which is useful for selective CO_2 separation due to the strong interactions of CO_2 with nitrogen (N) moieties, and the resultant high dipole moment of the framework also facilitates strong dipole–quadrupole interactions between CO_2 and the pore surface.[31] Highly polar frameworks and small porosity contribute to the high heat of adsorption (Q_{st}) value and the increased heat of adsorption of PAF-30. Such small porous structures are responsible for the simultaneous strong interaction of CO_2 molecules with the multiple pore surfaces, thus enhancing selectivity towards CO_2.

Porous polymer networks (PPNs), including PPN-1, PPN-2 and PPN-3, have been reported by Zhou *et al.* and exhibit similar affinity towards CO_2 and CH_4.[38] Among them, PPN-1 has been proven to be the best candidate, probably due to its high surface area density in terms of the volume. In addition, the framework exerts the largest difference with regards to the heat of adsorption between CO_2 and CH_4. By ideal adsorption solution theory (IAST), they predicted selectivity for CO_2 over CH_4 using a dual-site Langmuir–Freundlich model to fit the pure isotherms of CO_2 and CH_4 for the three PPNs. CO_2 has the higher saturation capacity, which in turn favors its selectivity under higher pressure conditions.

JUC-Z2[15] has good physicochemical stability and is capable of high and reversible gas uptake with high gas selectivity. It can efficiently separate CO_2 and CH_4 from H_2, Ar and N_2. JUC-Z2 shows a strong interaction with CO_2 by virtue of the aromatic and nitrogenous, electron rich building blocks, exhibiting excellent selectivity for CO_2 over N_2. Such high electron density also leads to a strong $H-\pi$ interaction between CH_4 and JUC-Z2, which allows it to show high CH_4 uptake and selectivity.

PAF-1, PAF-3 and PAF-4 can recognize greenhouse gases to separate them even from dry air containing Ar, H_2, N_2 and O_2.[16] At atmospheric pressure, the uptake of CO_2 by PAF-1 is 38 times higher than that of N_2 and the adsorption amount of CH_4 is almost 15 times higher than that of N_2, while PAF-3 exhibits an extraordinarily high selectivity of 87 times for the adsorption of CO_2 over N_2. This exceptional selectivity of PAF-3 possibly comes from the strong interactions between CO_2 and the Si center of the polymer.

PAF-1 (also known as PPN-6) networks have pore dimensions in the range of 5.0 to 10.0 Å, which is suitable for CO_2 separation from gases with larger diameters like N_2 and CH_4.[39] Polar groups, like SO_3H and SO_3Li, can be attached to PPN-6 to obtain open and interconnected pores formed by the carbon scaffold of the network and the tetrahedral monomers. Polar group incorporation in the PPN network leads to the strong interaction of electric fields created on the surface of the networks with that of the quadrupole moment of CO_2, enhancing the uptake capacity. PPN-6-SO_3Li shows even more promising CO_2 adsorption, having a stronger interaction of CO_2 with three open coordination sites of Li^+. The pore volume of the frameworks plays a major role in the selective uptake of greenhouse gases as polymers with the smallest pore volume lead to the best selectivity. The calculated pore volumes for PPN-6, PPN-6-SO_3H and PPN-6-SO_3Li are 2.44, 0.58 and 0.52 cm^3 g^{-1}, respectively, where the lithiated moiety exhibits the largest adsorption selectivity towards CO_2 as it has the smallest pore volume.

Carbonization seems to be an effective method to adjust the pore size of PAF-1 to increase the gas selectivity. PAF-1-450 (PAF-1 carbonized at 450 °C), with a narrow micropore distribution of 0.8 nm, shows obvious increased CO_2 sorption. Besides, on the basis of single component isotherm data, the dual-site Langmuir–Freundlich adsorption model-based IAST prediction indicates that the CO_2/N_2 adsorption selectivity may be as high as 209 at a $15:85$ $CO_2:N_2$ ratio. Also, the CO_2/CH_4 adsorption selectivity should be in the range of 7.8–9.8 at a $15:85$ $CO_2:CH_4$ ratio at $0 < p < 40$ bar, which is highly desirable for landfill gas separation. The calculated CO_2/H_2 adsorption selectivity could be about 392 at 273 K and 1 bar for the $20:80$ $CO_2:H_2$ mixture (Figure 10.2).[17]

Figure 10.3 compares the uptakes of CO_2, CH_4, N_2, H_2 and Ar adsorbed on JUC-Z4 at 273 K. JUC-Z4 exhibits high adsorption selectivity for CO_2 over N_2 specifically at low pressure, while at higher pressure selectivity decreases. This selectivity is predicted to be higher than several porous networks like MOFs. CO_2 separation from H_2 by JUC-Z4 is also found to be pretty good at ambient pressure and temperature. Such exceptional separation performance makes JUC-Z4 a potential candidate as a selective separator and absorber for CO_2 capture.[40]

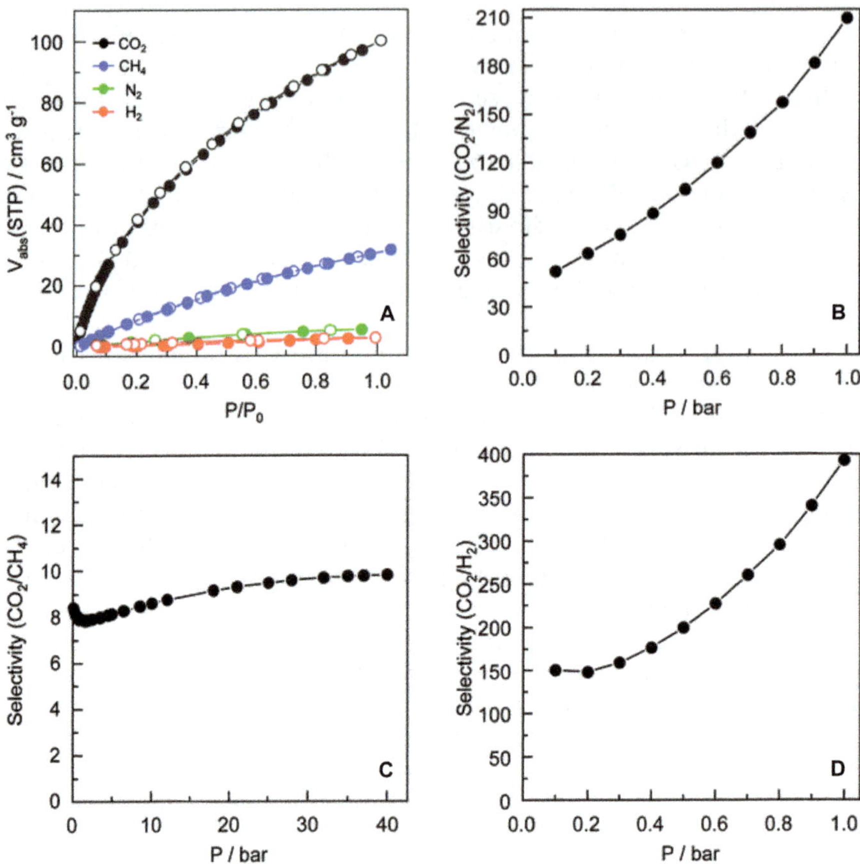

Figure 10.2 (A) CO_2, CH_4, N_2 and H_2 sorption of PAF-1-450 at 273 K; (B) IAST-predicted adsorption selectivity of CO_2/N_2 of PAF-1-450 using a 15 : 85 $CO_2 : N_2$ ratio; (C) IAST-predicted adsorption selectivity of CO_2/CH_4 of PAF-1-450 using a 15 : 85 $CO_2 : CH_4$ ratio; (D) IAST-predicted adsorption selectivity of CO_2/H_2 of PAF-1-450 using a 20 : 80 $CO_2 : H_2$ ratio. Reproduced from Ref. 17 with permission from The Royal Society of Chemistry.

PAF-30X ($X = 1$–4) is a series of PAFs with a diamond-like structure and it has been predicted to be useful in molecular gas separation using GCMC simulations. With the increase of the pore size in such PAFs, the isosteric heats of adsorption decrease from PAF-301 to PAF-304 along with the selectivity. PAFs with smaller pores and high isosteric heats of adsorption, *e.g.*, PAF-301, exhibit a high CO_2 adsorption capacity at low pressure and high selectivity for CO_2/H_2, CO_2/N_2, CO_2/CH_4 and CH_4/H_2 mixtures, whereas frameworks with larger pores and lower isosteric heats of adsorption, such as PAF-303 and PAF-304, show high CO_2 adsorption at higher pressure.[20]

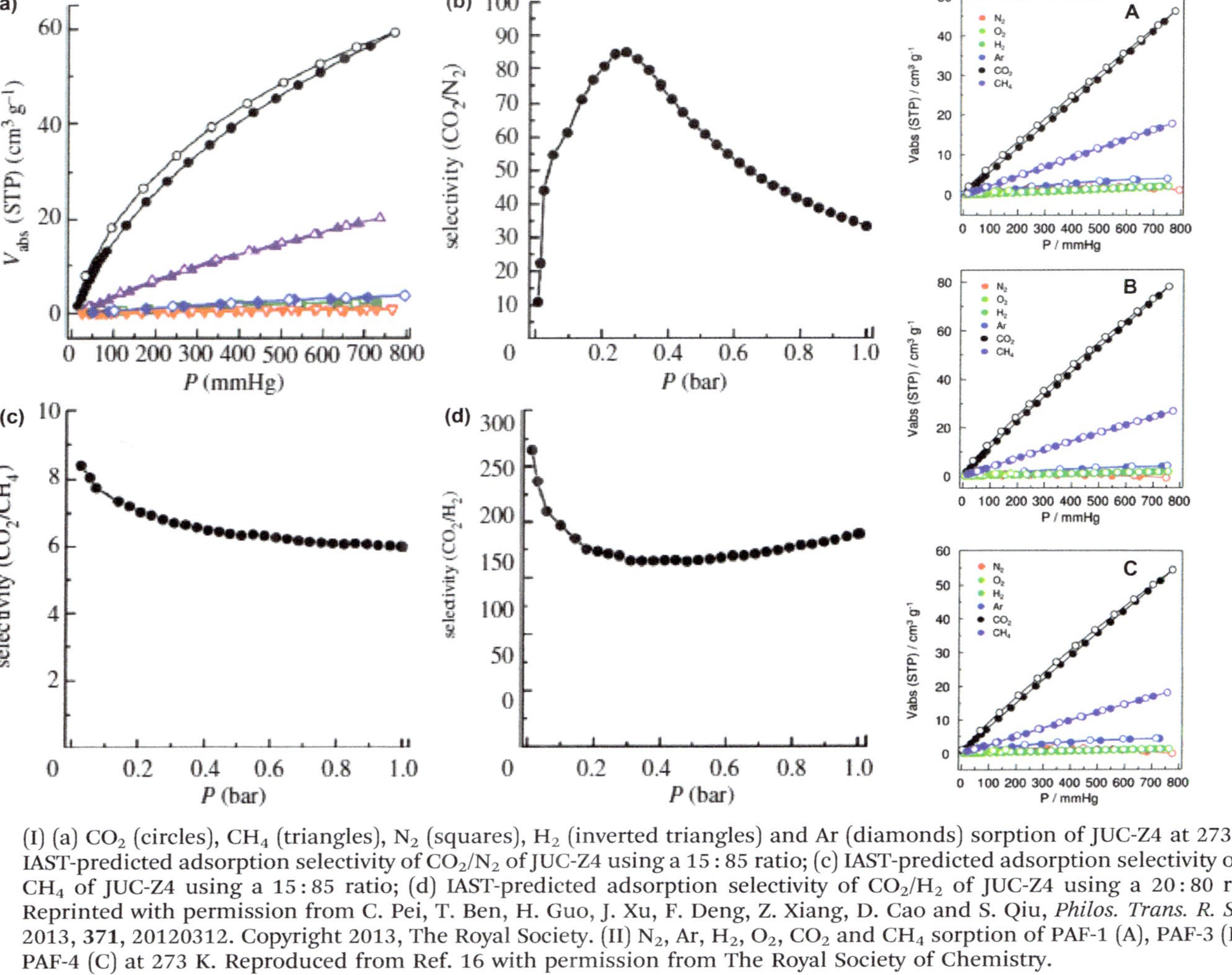

Figure 10.3 (I) (a) CO_2 (circles), CH_4 (triangles), N_2 (squares), H_2 (inverted triangles) and Ar (diamonds) sorption of JUC-Z4 at 273 K; (b) IAST-predicted adsorption selectivity of CO_2/N_2 of JUC-Z4 using a 15 : 85 ratio; (c) IAST-predicted adsorption selectivity of CO_2/CH_4 of JUC-Z4 using a 15 : 85 ratio; (d) IAST-predicted adsorption selectivity of CO_2/H_2 of JUC-Z4 using a 20 : 80 ratio.[40] Reprinted with permission from C. Pei, T. Ben, H. Guo, J. Xu, F. Deng, Z. Xiang, D. Cao and S. Qiu, *Philos. Trans. R. Soc., A*, 2013, **371**, 20120312. Copyright 2013, The Royal Society. (II) N_2, Ar, H_2, O_2, CO_2 and CH_4 sorption of PAF-1 (A), PAF-3 (B) and PAF-4 (C) at 273 K. Reproduced from Ref. 16 with permission from The Royal Society of Chemistry.

10.3   Catalysis by Porous Polymers

The applicability of porous polymers lies in their structural framework, porosity and surface area. Thus, by modification of the framework, controlling the porosity and changing the surface area, it is very much possible to tune their applicability over a huge range.

The surface area of these porous networks plays a significant role in catalysis. Interconnected pores and high surface area also increase catalytic activity by enhancing the accessibility of the catalytic sites and facilitating the mass transport process. This also provides very fast catalysis by the frameworks, enhancing their usage as effective catalysts. Frameworks containing rigid and hard networks, like the spiro-linked network, create a lot of space around the active catalytic centers, which enhances the accessibility for the reagents. Rigid functionalities inside the framework provide significant steric restriction to prevent structural relaxation, which in turn helps obtain high catalytic properties. In the case of conjugated porous networks, due to the conjugated structure, they harvest visible photons for photocatalytic reactions, thus providing excellent catalytic activities. Visible photon absorption allows them to generate the triplet excited state, which triggers the activation of molecular oxygen quite efficiently. A well-defined, ordered structure in the porous architecture is also useful for controlling the photo-generated excited states.

Catalytic activity depends on the number of catalytically active centers per molecule. Porous frameworks containing multiple metal or non-metallic catalytic active centers in their pores provide excellent catalysis due to the more open accessibility of those centers for reagents to be catalyzed.

Transition metal ions can be introduced in phthalocyanine-, porphyrin- or hexaazatrinaphthylene- (HATN-) containing PIM networks to obtain catalytic activity towards a wide variety of reactions like the degradation of H_2O_2 or oxidation of cyclohexene to 2-cyclohexene-1-one. Pd-containing PIMs are such a catalytic material. Pd^{2+} ions here act as a bridge among PIMs to form an extended network, thus providing catalytic activity[30] (Figure 10.4).

Incorporation of metallo-phthalocyanines and metallo-porphyrins into PIM networks helps to allow access to the catalytic metal centers.[37,41] It was shown that such network-PIMs demonstrate effective heterogeneous catalysis in reactions such as the oxidation of hydroquinone. Desulfurization of salt water has been shown to be catalyzed by a cobalt phthalocyanine network-PIM. Suzuki carbon–carbon coupling reactions can be catalyzed effectively by a hexaazatrinaphthylene-based Pd-loaded PIM network.[42]

Rigid spirocyclic linking groups can be introduced between porphyrin subunits to provide significant steric restriction to prevent structural relaxation, which in turn helps promote fruitful catalytic properties in PIMs. Phthalocyanine network PIMs are important catalysts; for example, iron–porphyrin derivatives can permit the catalysis of hydrocarbon hydroxylations and alkene epoxidations.[43]

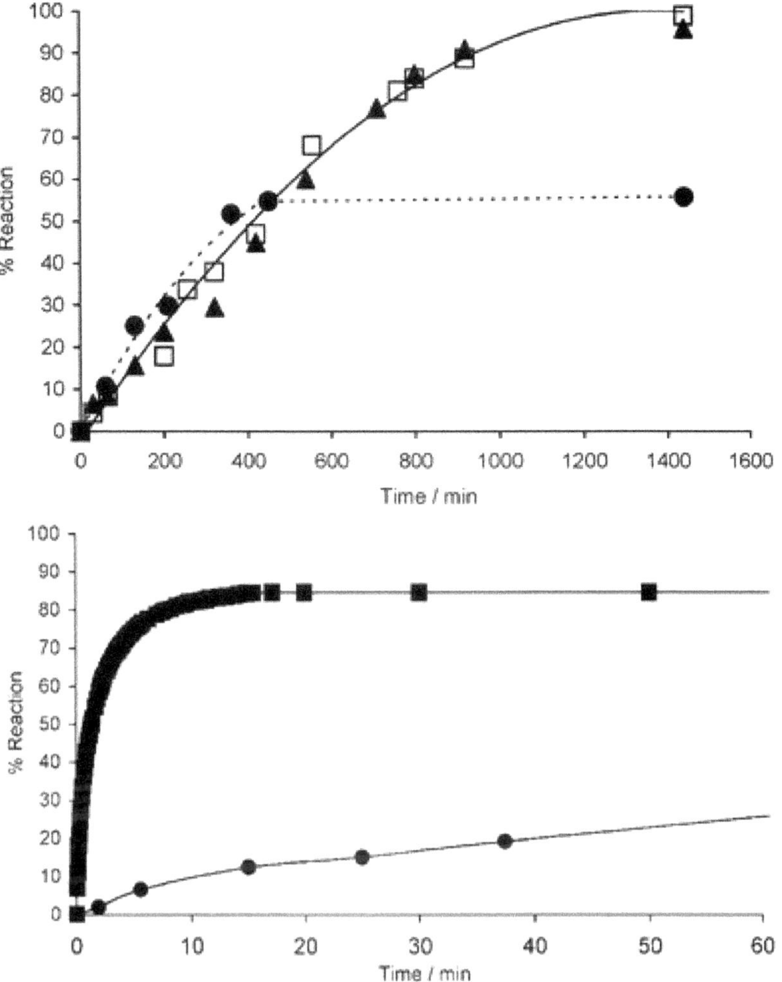

Figure 10.4 (Top) Dependence of the extent of reaction on time for the Suzuki aryl–aryl coupling reaction between 4-fluorobromobenzene and 4-methoxyphenylboronic acid catalyzed homogeneously by (●) Pd(PPh$_3$)$_4$ and heterogeneously by the palladium-containing HATN-network-PIM (▲) recycled twice and (□) recycled three times. (Bottom) Dependence of the extent of reaction on time for the degradation of H$_2$O$_2$ with (●) low molar mass cobalt phthalocyanine (CoPc) and (■) the CoPc-network-PIM as a catalyst. Reproduced from Ref. 30 with permission from The Royal Society of Chemistry.

Rigid and nonlinear linking units in phthalocyanines provide spirocentres and nonlinearity to inhibit facial association and prevent solidification through π–π and other non-covalent interactions. This also improves the catalytic properties of phthalocyanines as they can catalyze H_2O_2 degradation more effectively than solidified, low molar mass phthalocyanines.[44]

Functionalization of COFs with various species in their pores empowers one to exploit their catalytic properties. Two strategies have been employed for the construction of catalytic COFs. The first attaches groups possessing catalytic sites under harsh synthetic conditions. The second uses the post-synthetic integration of catalytic sites into a crystalline COF to minimize the effect of the bulky catalytic sites on COFs and eliminates harsh conditions. An example of a COF-based catalyst is the Pd-loaded COF catalyzing the Suzuki cross-coupling reaction.[45] The pore surface engineering approach has been utilized for the construction of covalently linked and highly active organocatalytic COFs using boronate-linked COFs based on thiol–ene and alkyne–azide reactions. Controlled integration of organocatalytic sites into the pore walls allows the synthesis of organocatalytic COFs that display greater activity in asymmetric Michael addition reactions, while retaining stereoselectivity. A mesoporous imine-linked porphyrin COF was used as a scaffold, where the porphyrin units were situated at the vertices and the phenyl groups were placed at the edges of the frameworks. Such a pore surface engineering strategy offers a common strategy to build the molecular design of COF skeletons, with functional groups on the pore walls, to obtain effective catalytic properties. Incorporation of pyrrolidine groups onto the pore walls creates aqueous organocatalytic COFs with improved catalytic features and recyclability, while retaining stereoselectivity[46] (Figure 10.5).

An imine-based COF can uptake Pd^{2+} ions into its pores by coordinating them *via* nitrogen atoms. These Pd^{2+} ions on the COF walls show catalytic activity in heterogeneous catalysis as they are accessible for both the substrate and the reactant. One effective catalytic reaction by Pd/COF is the Suzuki–Miyaura coupling reaction, in which the catalyst shows tremendous activity with great recyclability.[14]

High stability and crystallinity, even in common organic solvents and water, enhances the possibility of CuP-SQ (squaraine) COF to be used as a heterogeneous catalyst. CuP-SQ COF is a strong photocatalyst for the activation of molecular oxygen, while monomeric CuP exhibits pure catalytic activity for the same. This can be explained in terms of the light harvesting ability of the CuP-SQ COF for photocatalysis. CuP-SQ COF can effectively generate the triplet excited state that allows the activation of molecular O_2 upon absorption in the visible region. The organized, ordered architecture of the COF is also beneficial for stabilizing the photo-generated excited states, and this architecture is absent in monomeric CuP, which differentiates the catalytic abilities of CuP and CuP-SQ COF. Noble metals, for example, palladium and platinum, may be incorporated in to metalloporphyrins and are known for the activation of molecular O_2, with high abilities to generate the triplet state. CuP-SQ COF exhibits better photocatalysis and it contains no noble metals. The SQ linkage extends the π-conjugation over the 2D COF skeleton, providing the charge-carrier

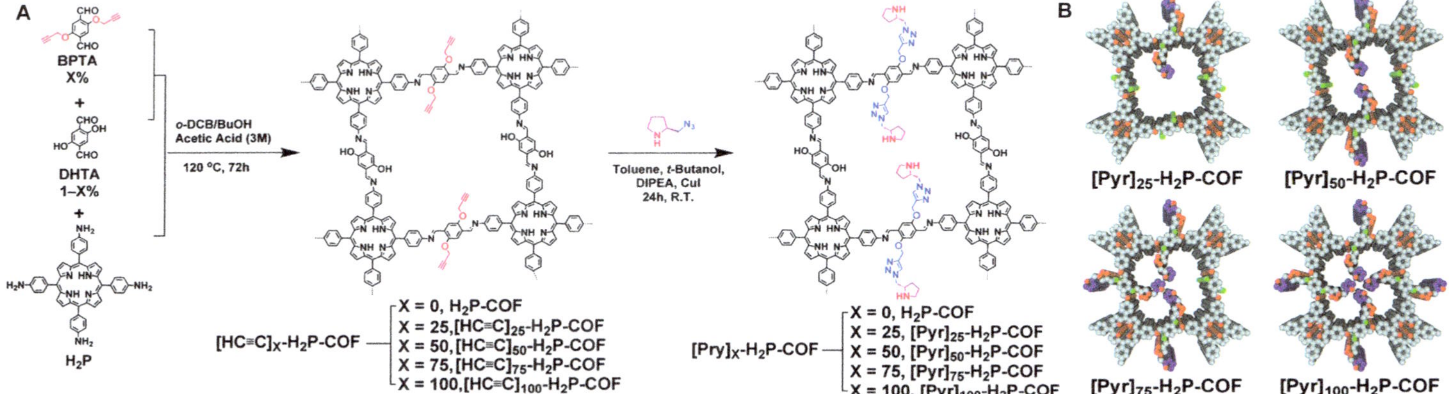

Figure 10.5 (A) The general strategy for the pore surface engineering of imine-linked COFs *via* a condensation reaction and click chemistry (BPTA, 2,5-bis(2-propynyloxy)terephthaladehyde; DHTA, 2,5-dihydroxyterephthaladehyde; *o*-DCB, *ortho*-dichlorobenzene; DIPEA, *N,N*-diisopropylethylamine). (B) A graphical representation of [Pyr]X-H₂P-COF with different densities of catalytic sites on the pore walls.
Reproduced from Ref. 46 with permission from The Royal Society of Chemistry.

capability. Their enhanced light-harvesting capacity, lowered band gap, layered π-stacking porphyrin arrays and open mesopores are useful features for photocatalytic systems.[47]

Spiro-linked CMPs functionalized with metal phthalocyanine units show enhanced catalytic activity towards different reactions.[48] The Co-phthalocyanine-incorporated CMP acts as a catalyst with improved activity for cyclohexene oxidation, hydroquinone oxidation and H_2O_2 decomposition, whereas the spiro-linked Fe-porphyrin network shows increased catalytic activity for hydroquinone oxidation. The spiro linkages in these networks open up a lot of free space around the catalytic sites to enhance the accessibility of substrates to reach more catalytic sites. More functionalization in this way of conjugated networks by various metals improves the scope of these networks in heterogeneous catalysis. Oxidation of sulfides, reductive aminations and photocatalyzed aza-Henry reactions are reactions effectively catalyzed by different metal-incorporated CMPs[49–51] (Figure 10.6).

Other functional moieties, such as phenolphthalein and benzodifuran, when incorporated in to CMP networks showed effective photocatalytic activity towards heterogeneous photosensitization for methyl methacrylate photopolymerization and the photocoupling of primary amines,[53] respectively. The

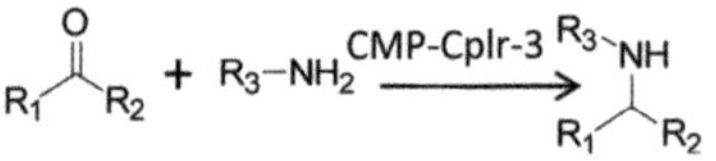

entry	ketone	amine	yield[%]
1	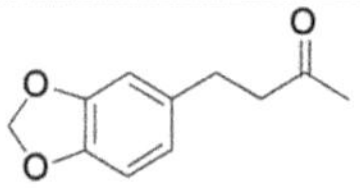	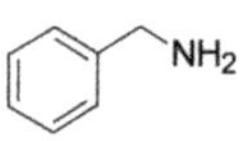	95
2	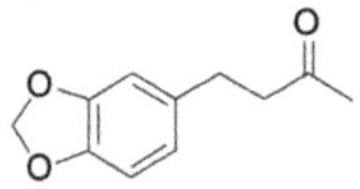		91
3	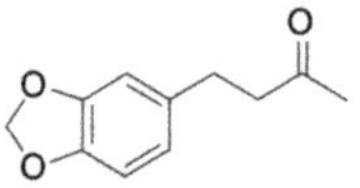	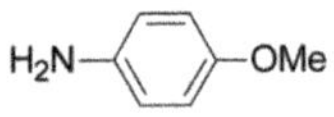	90
4	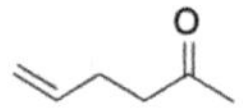		93
5	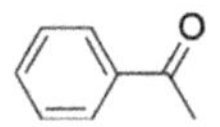	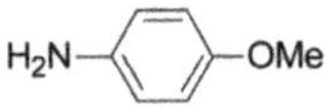	86

Figure 10.6 Reductive amination catalyzed by CMP-CpIr-3.[52]
Reprinted with permission from J.-X. Jiang, C. Wang, A. Laybourn, T. Hasell, R. Clowes, Y. Z. Khimyak, J. Xiao, S. J. Higgins, D. J. Adams and A. I. Cooper, *Angew. Chem., Int. Ed.*, 2011, **50**, 1072. Copyright 2011 WILEY-VCH Verlag GmbH & Co. KGaA, Weinheim.

alkyne units in poly(aryleneethynylene)- (PAE-) derived CMPs may further be modified by organometallic groups to get enhance heterogeneous catalysis.[54]

Stitching of binding sites in the pore walls of CMPs might be effective in attaching catalytically active nanoparticles.[55] Conjugated microporous poly(thienylene arylene) networks containing sulfur-based or thiol ligands in the pore walls are highly accessible for guest molecules or ions. Pd-catalyzed hydrogenation is an important application of such networks, as Pd clusters are accessible in the porous network. Such accessibility of the pores, along with the strong ligation effect of the multiple atoms attached, allows these networks to act as catalyst supports, as is also used in electrocatalysis.

Rose Bengal dye-incorporated CMPs (RB-CMPs) are used as highly active, recyclable and reusable heterogeneous organo-photocatalysts in aza-Henry reactions. The interconnected pores in RB-CMPs provide easy access of the catalytic centres for the incoming substrate molecules, resulting in excellent catalytic activity. The catalyst is highly reusable without significant loss of conversion. The RB-CMP network may also play a dominant role in heterogeneous noble-metal-free photocatalysis, including chiral porous metal-free photocatalysis[56] (Figure 10.7).

The Pd-incorporated covalent triazine framework (CTF) acts as a catalyst in the oxidation of glycerol. The longer lifetime of the catalyst is due to the large number of nitrogenous moieties coordinated in the Pd/CTF frameworks, which offer stabilization of the metal nanoparticles, as well as better confinement. It exhibits longer activity and selectivity than activated carbon in the oxidation of glycerol[57] (Figure 10.8).

As-synthesized CTF-1 delivers excellent catalytic activity in the conversion of CO_2 to cyclic carbonates. The catalyst can be activated through the incorporation of basic sites of nitrogen within it. The catalytic activity of CTF-1 can be

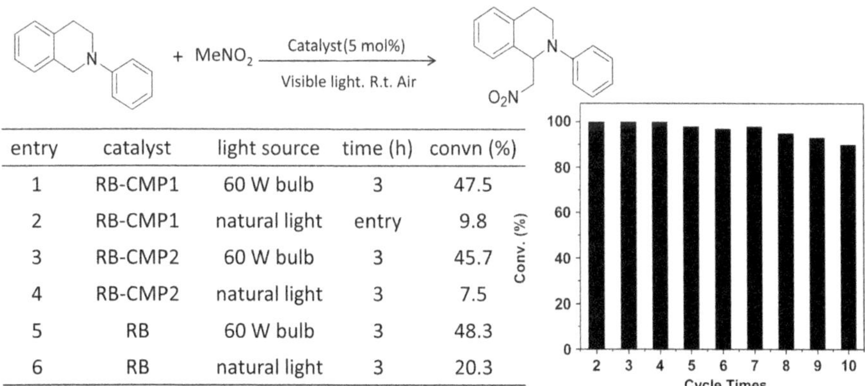

entry	catalyst	light source	time (h)	convn (%)
1	RB-CMP1	60 W bulb	3	47.5
2	RB-CMP1	natural light	entry	9.8
3	RB-CMP2	60 W bulb	3	45.7
4	RB-CMP2	natural light	3	7.5
5	RB	60 W bulb	3	48.3
6	RB	natural light	3	20.3

Figure 10.7   (Left) Screening conditions for the aza-Henry reaction. (Right) Recycling experiments of RB-CMP1 for the aza-Henry reaction.[56]
Reprinted with permission from J.-X. Jiang, Y. Li, X. Wu, J. Xiao, D. J. Adams and A. I. Cooper, *Macromolecules*, 2013, **46**, 8779. Copyright 2013, American Chemical Society.

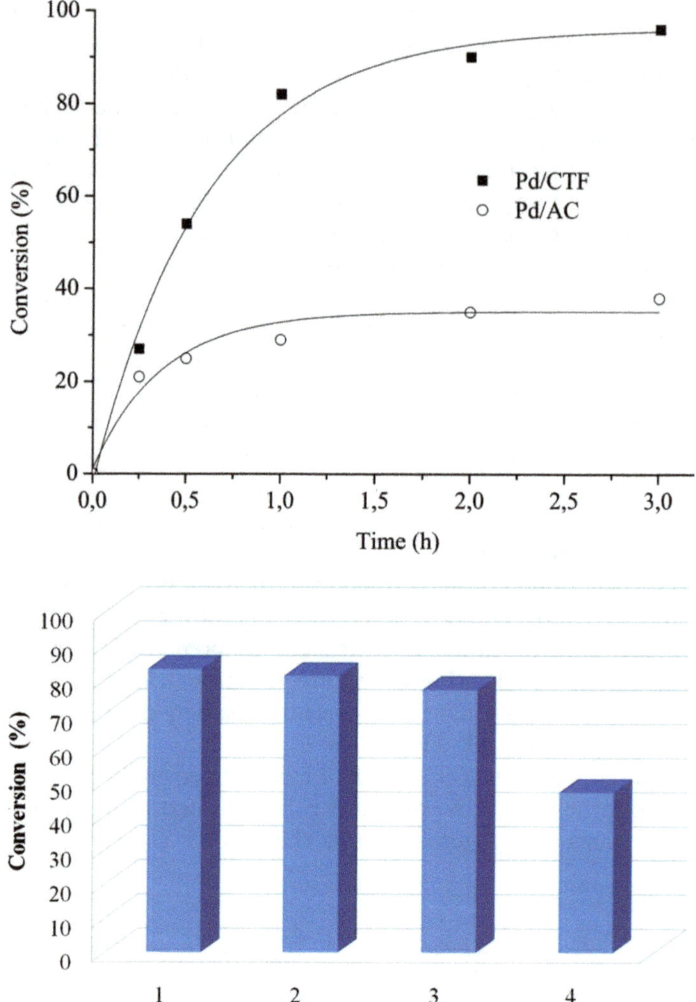

Figure 10.8   (Top) Supported Pd-based catalysts for glycerol oxidation (AC, activated carbon). (Bottom) Conversion on recycling of 1% Pd/CTF for glycerol oxidation.[57]
Reprinted with permission from C. E. Chan-Thaw, A. Villa, P. Katekomol, D. Su, A. Thomas and L. Prati, *Nano Lett.*, 2010, **10**, 537. Copyright 2010, American Chemical Society.

enhanced by structural modifications of the synthesized network to build a hierarchical porous system. The diffusion rate to the catalytic sites can be increased by incorporation of excess mesopores and higher nitrogen contents.[58] A high surface area and increased pore sizes enhance the catalytic activity.

The JUC-Z4 composite with heptamolybdate soft-oxometalate has been used as a catalyst for benzaldehyde oxidation. This composite has also been patterned on a substrate using laser power to obtain catalytic trails to perform site-specific catalysis of benzaldehyde oxidation.[59]

10.4 Other Applications

Porous polymers have been proven to be very useful in semi-conduction, photoconduction, light harvesting, sensing and in antimicrobial applications, along with their usual gas confinement properties, catalytic activity and their selective separation of gases. This section shall describe these applications of polymers.

10.4.1 Conjugated Microporous Polymers

10.4.1.1 *Light Harvesting*

Recently, light harvesting properties have been seen in a CMP formed with polyphenylene.[60] Blue luminescence with a wavelength of 443 nm has been observed for a network polymer. When coumarin 6 is incorporated into the vacant spaces of the polymer, on excitation, emission at 512 nm predicts energy transfer from the polymer to coumarin 6, while no emission has been observed for the polymer itself on doping.[4]

10.4.1.2 *Photovoltaic and Photocatalytic Applications*

There are other potential applications of CMPs. Owing to the massive surface area of CMPs, they can be fused with other types of conjugated polymers, which otherwise are not able to form a composite network. The pores of CMPs can be used for doping with various functional materials, *e.g.*, TiO_2[61] or quantum dots, which could be utilized in applications of photovoltaics and photocatalysis. A brand new microporous network can be obtained when the CMP network is deposed out.[62]

10.4.1.3 *Semi-conduction*

There are a few examples of CMPs that possess inherent conducting properties. Semi-conductivity can also be introduced in the CMP through doping. Such materials can be very useful in battery applications, the capacitor industries[62] and as sensors such as PAE-derived CMPs (Figure 10.9). The conjugation in these CMP networks makes them suitable for semi-conduction purposes.

10.4.2 Porous Aromatic Frameworks

10.4.2.1 *Luminescence Quenching*

Among PAFs, PAF-15 may be used as a quencher and sensor. Insertion of elements like Ge into the network of crystalline PAFs decreases the reduction potential and offers enhancement of the delocalization of electrons.[63]

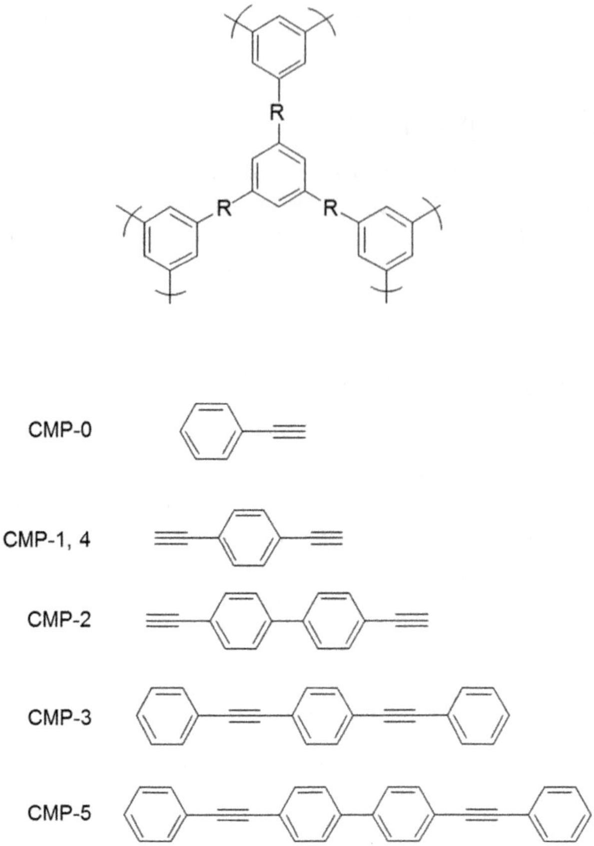

CMP-0

CMP-1, 4

CMP-2

CMP-3

CMP-5

Figure 10.9 PAE-derived CMPs.[62]
Reprinted with permission from A. I. Cooper, *Adv. Mater.*, 2009, **21**, 1291. Copyright 2009 WILEY-VCH Verlag GmbH & Co. KGaA, Weinheim.

10.4.2.2 Antibacterial Polymer Coatings

PAF-50 and its derivatives with silver chloride within the pores, *i.e.*, AgCl-PAF-50, are known to exhibit[64] antibacterial activity. This activity is comparable to commonly used antimicrobials and sometimes even shows better results. These PAFs are highly soluble in various organic solvents and are also compatible with common polymers. This makes them quite useful in this regard (Figure 10.10).

PAF-50 and AgCl-PAF-50 polymer mixtures can be fabricated and coated as antimicrobials using casting or spraying methods. Such easy and flexible operations are inaccessible for presently used antimicrobials. Thus, the effectiveness, versatility and durability make such a kind of PAF immensely important in various applications as antimicrobial coatings.[64]

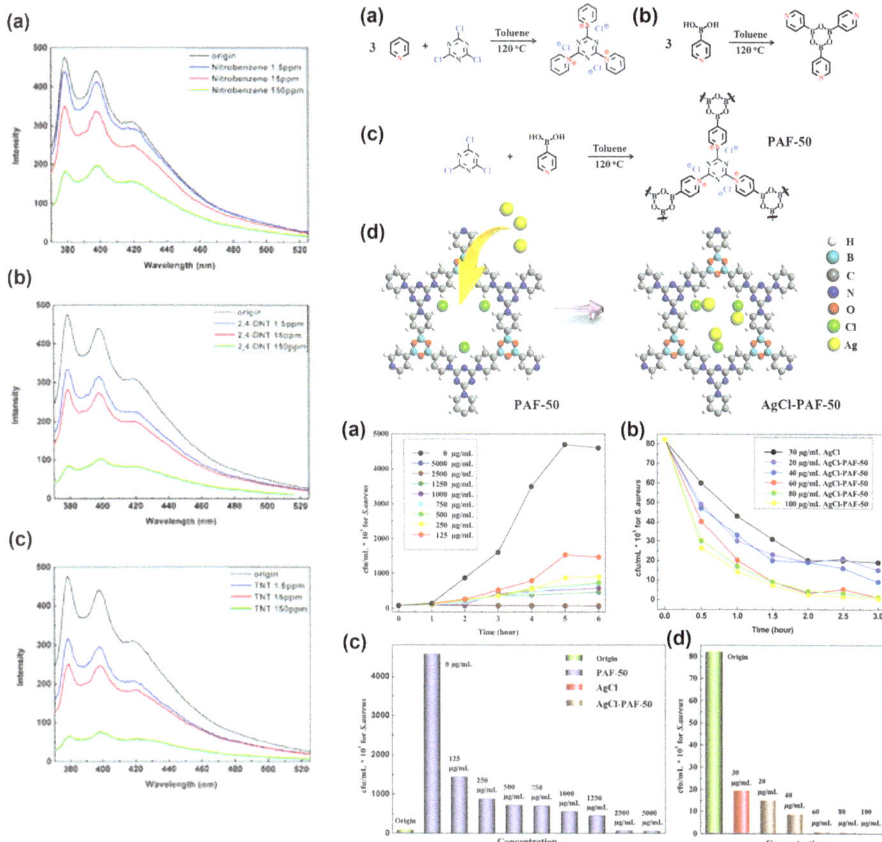

Figure 10.10 Applications of PAFs in luminescence quenching and as antibacterial coatings.[63,64] In the top right corner of the image: (a and b) Condensation reaction used to produce discrete molecules and extended PAFs; (c) scheme of synthesis of PAF-50 and (d) AgCl-PAF-50. In the bottom right corner of the image: plot of the viability of *S. aureus*, recovered from LB plates after 24 h of incubation at 37 °C *versus* the time of incubation with the different concentration of PAF-50 (a) and AgCl-PAF-50 (b). Chart of the viability of *S. aureus* (cfu mL^{-1}) versus the concentration of PAF-50 after 6 h incubation (c) and AgCl-PAF-50 after 3 h incubation (d). In the left side of the image, PL spectra of the CHCl$_3$ solution of PAF-15 (a, b, c) with different analytes: nitrobenzene, 2,4-dinitrotoluene(2,4-DNT) and 2,4,6-trinitrotoluene (TNT) concentration (excited at 346 nm). (left) Reprinted with permission from Y. Yuan, H. Ren, F. Sun, X. Jing, K. Cai, X. Zhao, Y. Wang, Y. Wei and G. Zhu, *J. Phys. Chem. C*, 2012, **116**, 26431. Copyright 2012, American Chemical Society. (right) Reprinted with permission from Y. Yuan, F. Sun, F. Zhang, H. Ren, M. Guo, K. Cai, X. Jing, X. Gao and G. Zhu, *Adv. Mater.*, 2013, **25**, 6619. 2013 WILEY-VCH Verlag GmbH & Co. KGaA, Weinheim.

10.4.2.3   *Sensing and Optoelectronics*

Another useful PAF with applications in sensing and optoelectronic devices is JUC-Z2, which shows electroactivity as the first porous organic framework (Figure 10.11). Such electrochemically active PAFs offer chemical and electric energy conversion inside the framework in a size-specific manner with guest ions.[65] Such size specificity towards ion guests allows usage of JUC-Z2 as a sensor. The p-type semi-conducting and optoelectronic properties of JUC-Z2 are due to the easy formation of an interpenetrating

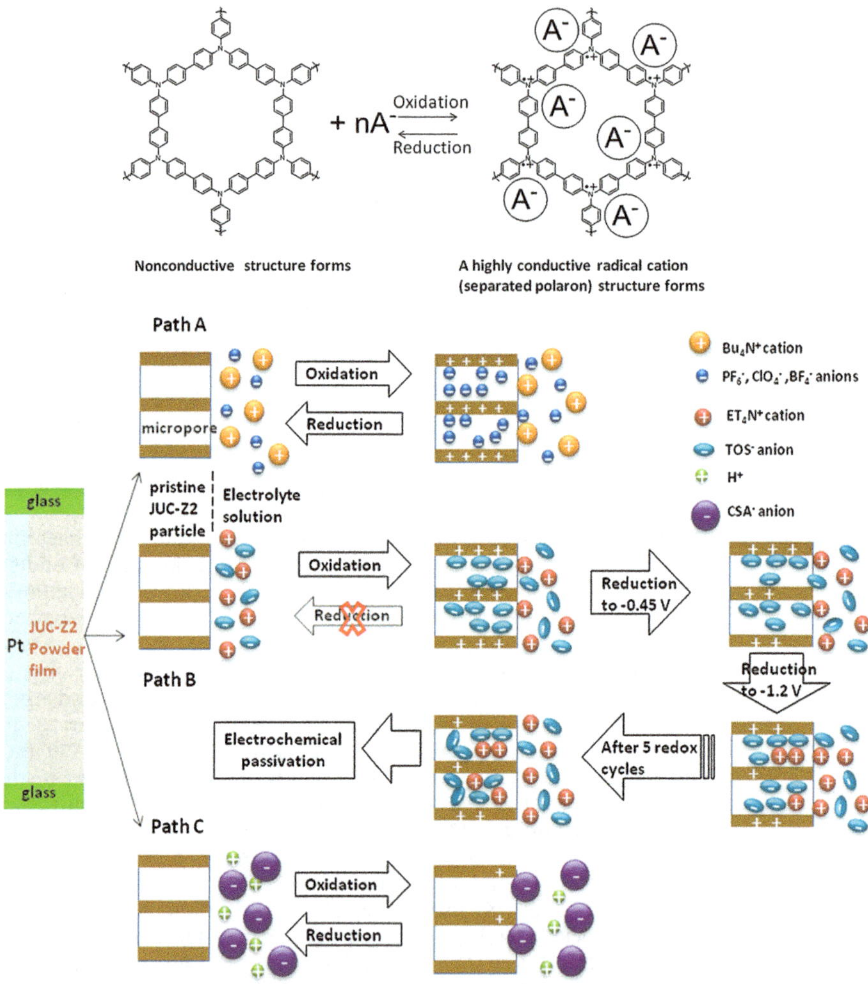

Figure 10.11   Sensing applications of JUC-Z2.[65]
Reproduced from Ref. 65 with permission from The Royal Society of Chemistry.

polymeric network with electron deficient materials owing to the uniform porosity.[66]

10.4.3 Polymers of Intrinsic Microporosity

PIM-1 thin films exhibit excellent optical sensing properties by changing their color on adsorbing organic vapor, even at very low concentration (Figure 10.12). Although the performance of a fiber-optic spectrometer is better than this,[67] for PIM-1 the hindrance effect due to humidity is less due to the hydrophobicity. The microporous nature, as well as the behavior towards solvent, is the key to fabricate sensor-based devices with PIMs.[68]

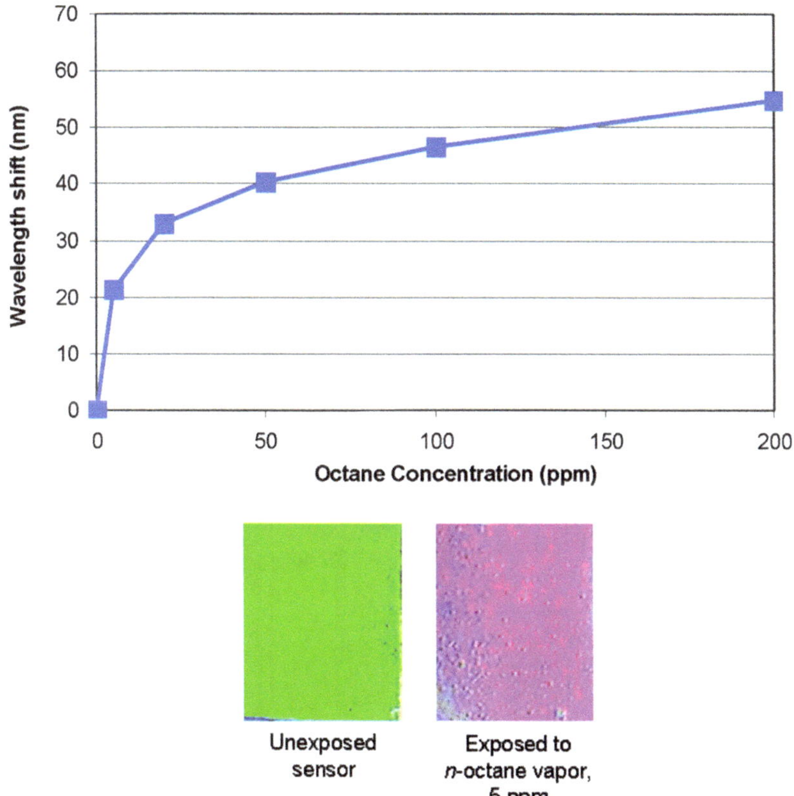

Figure 10.12 (Top) Response of the indicator to various concentrations of *n*-octane vapor in dry air. (Bottom) A distinctive shift from green to magenta is observed at 5 ppm.[67]
Reprinted with permission from N. A. Rakow, M. S. Wendland, J. E. Trend, R. J. Poirier, D. M. Paolucci, S. P. Maki, C. S. Lyons and M. J. Swierczek, *Langmuir*, 2010, **26**, 3767. Copyright 2010, American Chemical Society.

10.4.4   Covalent Organic Frameworks

10.4.4.1   Chemosensing

Azine-linked COFs with pyrene columns inside show high luminescence properties. The unique stacking of pyrene and supramolecular bonding sites provided by azine groups is responsible for the good sensitivity towards chemosensing and the selective detection of explosives like 2,4,6-trinitrophenol[69] (Figure 10.13).

10.4.4.2   Semi-conduction, Photoconduction and Luminescence

The physics of the donor–acceptor junction is the principle tool to monitor charge regulation in recent-day devices like LED, transistors, photovoltaics *etc.*[70–72] Crystalline networks in COFs provide structural rigidity to form a repetitive array of networks.[73–81] Several COFs with porphyrin, arene and phthalocyanine have been shown to offer excellent semi-conductivity, as well as photoconductivity.[75] The covalent organic framework with a donor–acceptor conformation is advantageous over bulk-material junction systems as the former provides a more arranged and ordered network. An aggregated, continuous donor–acceptor arrangement is the key configuration offered by COFs to control charge separation and transfer[82] (Figure 10.14).

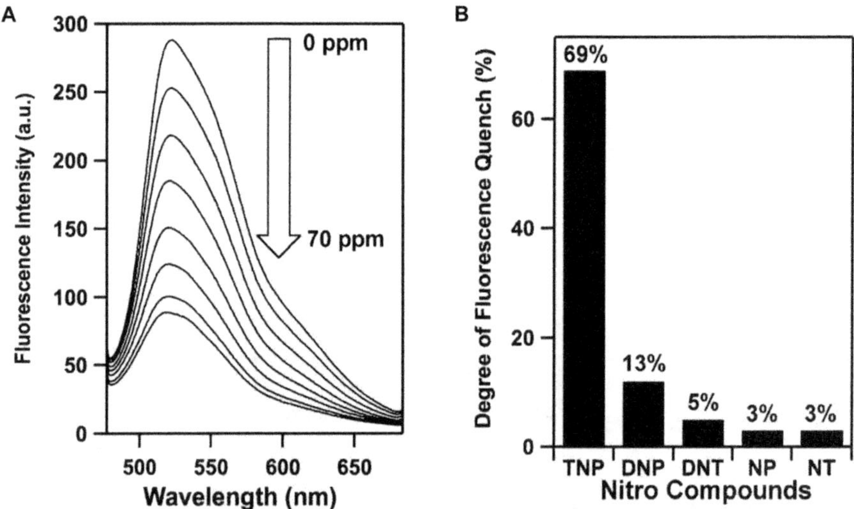

Figure 10.13   (A) Fluorescence quenching of the Py-Azine COF upon addition of 2,4,6-trinitrophenol (TNP) (0–70 ppm) in acetonitrile. (B) Degree of fluorescence quenching upon addition of the nitro compounds (70 ppm).
Reprinted with permission from S. Dalapati, S. Jin, J. Gao, Y. Xu, A. Nagai and D. Jiang, *J. Am. Chem. Soc.*, 2013, **135**, 17310. Copyright 2013, American Chemical Society.

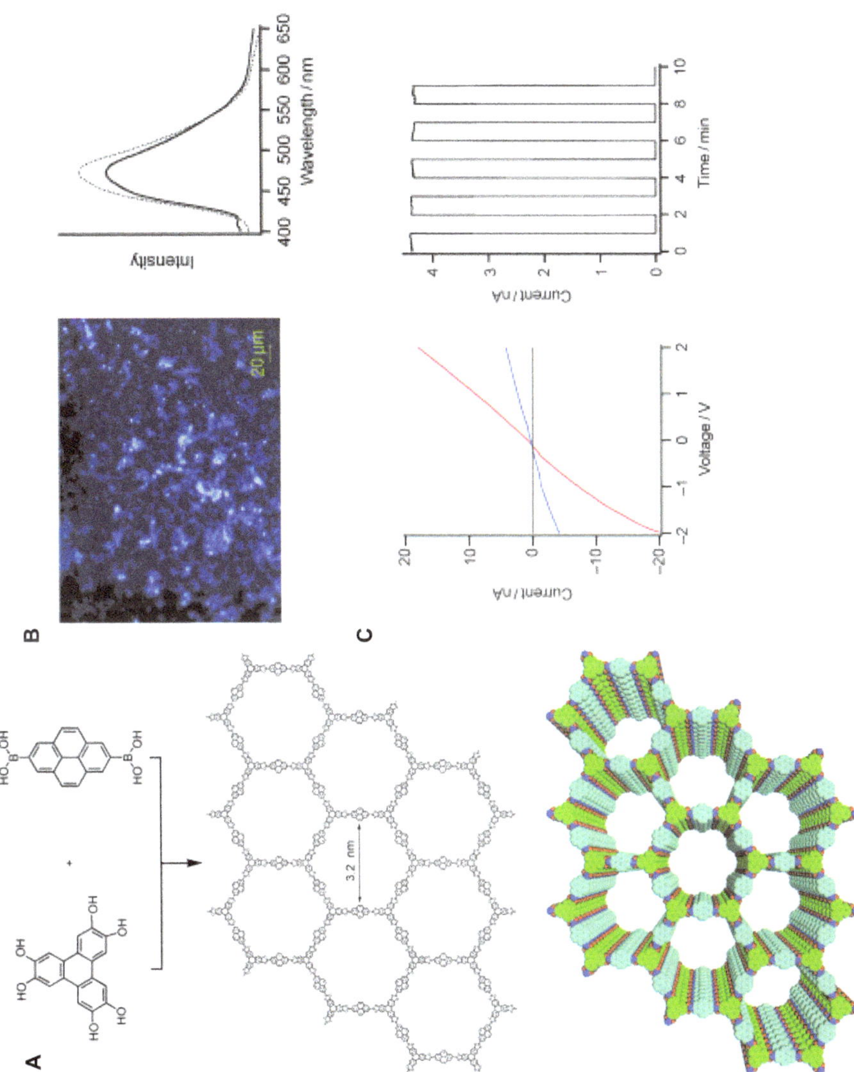

Figure 10.14 (A) Synthesis of TP-COF (upper), Schematic representation of TP-COF (structure is based on quantum calculation and crystal lattice parameters; B purple, O red, triphenylene green, pyrene blue, H atoms are omitted for clarity) (lower). (B) Fluorescence image of TP-COF (left) and fluorescence spectra of TP-COF upon excitation at 340 nm (black curve) and 376 nm (dotted curve) at 25 °C (right). (C) I–V profile of a 10 μm width Pt gap (black curve: without TP-COF; blue curve: with TP-COF; red curve: with iodine-doped TP-COF) (left). Electric current when 2 V bias voltage is turned on or off (right).[75b] Reprinted with permission from S. Wan, J. Guo, J. Kim, H. Ihee and D. Jiang, *Angew. Chem.*, 2008, **120**, 8958; Copyright 2008 WILEY-VCH Verlag GmbH & Co. KGaA, Weinheim.

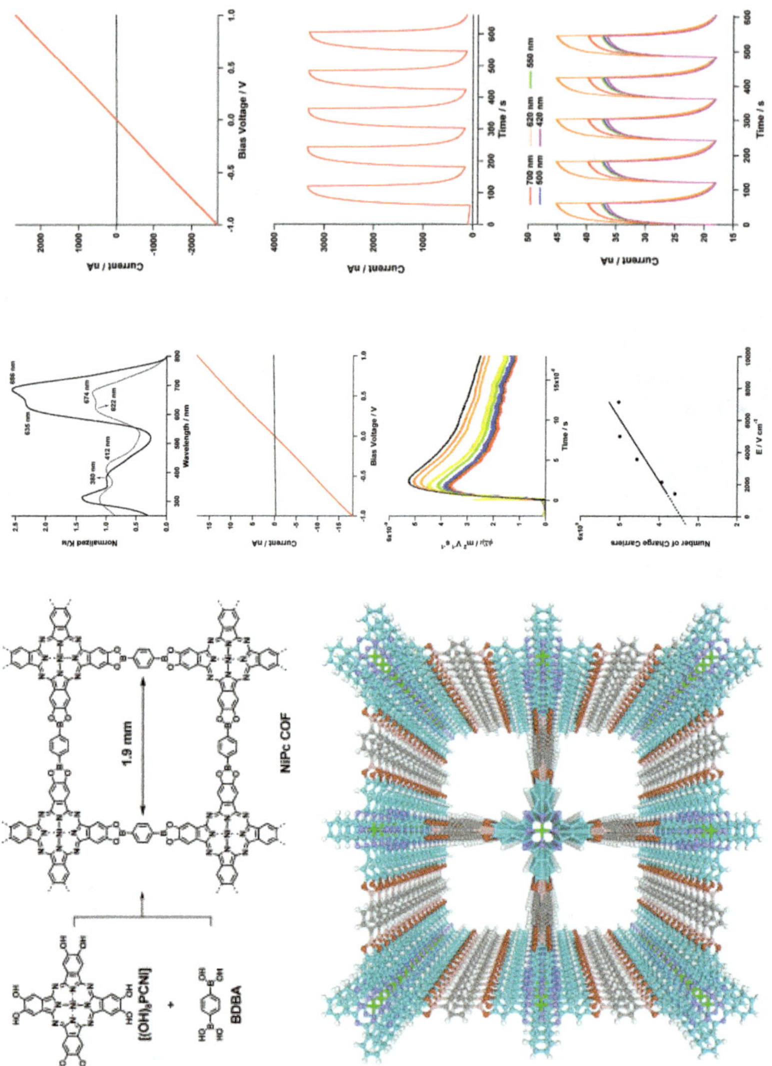

The incorporation of metallo-phthalocyanine blocks within COF networks provides the generation of p-type electronic COF frameworks[83–86] as the phthalocyanine unit is photochemically active and shows redox properties. A two-dimensional NiPc-COF has been achieved from 1,4-benzene diboronic acid (BDBA) holding phthalocyanine derivatives at the vertex of each of the tetragonal-structured lattices.[84] NiPc-COF exists in a flake-like shape with a two-dimensional sheet-like structure. Thus, this COF acts with hole transport properties, which vary with the linkages attached to the COFs. The Ni-phthalocyanine-benzothiadiazole COF is formed with benzothiadiazole, which possesses electron deficiency.[86] This COF shows n-type semi-conducting properties unlike Ni-Pc-COF, which is a p-type semi-conductor. So, on tuning the type of linker molecule, porous networks can be controlled to exhibit both n- and p-type semi-conducting properties. Such a reversal of the charge transport properties make porous networks excellent candidates for applications in industries like electronics[14] (Figure 10.15).

In two-dimensional porphyrin MP-COFs, electron carrier properties come from the stacking of polyporphyrin sheets forming a regular columnar structure. Depending on the basic framework and the metal ion in the framework, MP-COFs can be tuned to vary the type of charge it carries. The free-base porphyrin H_2P-COFs conduct positive charge; on the other hand, CuP-COFs provide the conduction of negative charge. Ion-conducting channels generated by porphyrin rings inside COFs offer such ion transport. Ion conduction is also affected by metal atoms present centrally. Other ion conducting COFs reported are COF-66 (boronate ester linkage) and COF-366 (imine linkage).[79] The ion carrying capacity of COF-66 is half of that of COF-366, though COF-66 possesses smaller interlayer spacing (3.81 Å) than COF-366 (5.64 Å).[87] Thus, it

---

Figure 10.15   (Upper) The synthesis of the nickel phthalocyanine covalent organic framework (NiPc COF) by a boronate esterification reaction, eclipsed stack of phthalocyanine 2D sheets and microporous channels in NiPc COF (a 2×2 grid is shown). Colors used for identification: phthalocyanine unit: sky blue; Ni green; N violet; C gray; O red; B orange; H white. (Middle, from left to right) (First) Absorption profiles of NiPc COF (solid line) and [(MeO)$_8$PcNi] (dotted line) (Second) I–V curves of NiPc COF (red curve) and [(MeO)$_8$PcNi] (black curve) sandwiched between Al/Au electrodes. (Third) Transient conductivity profiles upon irradiation with a 355 nm pulse laser at different photon densities: 4.51014 (red curve) to 1.51016 photon cm^{-2} (black curve). (Forth) Number of charge carriers measured by the time of flight transient current integration at different bias voltages, irradiated with a 355 nm pulse laser at 6.51014 photon cm^{-2}. (Lower, from left to right) (First) I–V curves in the dark (black curve) and upon irradiation (red curve) with a xenon light source. (Second) The photocurrent for NiPc COF (red curve) and [(MeO)$_8$PcNi] (black curve) at a bias voltage of 1.0 V, generated by repeatedly switching the light on and off. (Third) Wave-length-dependent on-off switching of photocurrent at a bias voltage of 1.0 V. Reprinted with permission from X. S. Ding, J. Guo, X. Feng, Y. Honsho, J. D. Guo, S. Seki, P. Maitarad, A. Saeki, S. Nagase and D. Jiang, *Angew. Chem., Int. Ed.*, 2011, **50**, 1289. Copyright 2011 WILEY-VCH Verlag GmbH & Co. KGaA, Weinheim.

can be concluded from this information that ion transport can occur in between the layers and also through the stacked layers. Pyrene-based PPy-COF is known as the first of the COFs to conduct photons and may be synthesized from pyrene-2,7-diboronic acid (PDBA) through a self-condensation method.[88] PPy-COF possesses a cubic shape of micron size, which exhibits blue

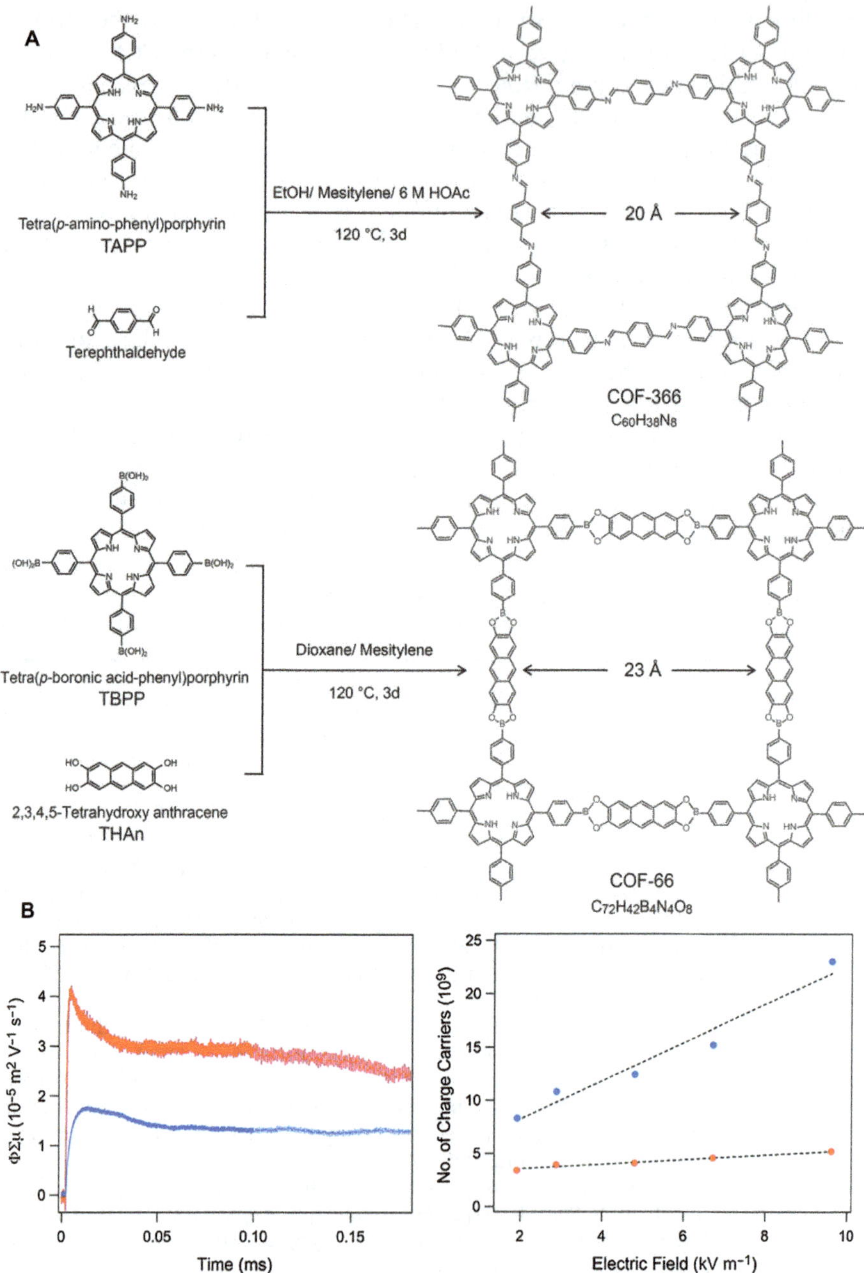

luminescence in the stacking state of pyrene on excitation. PPy-COF shows depolarized fluorescence with a very small anisotropy value, while for COFs with triphenylene (TP) show much higher values. This is due to the fact that PPy-COF has a component with a single band gap and the excitons are delocalized throughout all the layers. Unlike porphyrin-COFs and phthalocyanine-COFs, which absorb throughout a wide region beginning from the ultraviolet (UV) region and ending at near-infrared (NIR), the absorption band of PPy-COF is quite narrow in the visible range.[84] Metal-dependent photoconductivity has been seen in metallo-porphyrin COFs[14] (Figure 10.16).

In very recent times, COFs with both donor and acceptor aggregated networks have been studied. This is quite important in producing organic electronic materials by arranging one-by-one arrays to form less ordered segregation, which can hold charge carriers.[89] Such COFs with hierarchical donor–acceptor covalent bonded aggregations form interlocked planar structures at the molecular level Macroscopically, the D–A heterojunction structure forms a periodical, organized and continuous linkage for both donor and acceptor species. These networks with donor–acceptor junctions form arrays vertically, and are responsive to photoconductivity over a large range and allow fast charge transportation[14] (Figure 10.17).

COFs possessing comparatively high pore volumes have been reported with in-built donor and acceptor moieties. COFs with nano-channels have been synthesized by taking di-imide as an acceptor and triphenylene as a donor. Such large pore COFs exhibiting large surface areas and high crystallinity may be used as benchmark networks to show the effect of lattice structure and the role of D–A pairs in charge transportation. This empowers one to provide a general theory for the design and synthesis of optoelectronic and photovoltaic COFs[90] (Figure 10.18).

TP-COF with PDBA possesses a belt-like network, which is 300 nm wide and 100 nm thick. On excitation of the pyrene and triphenylene moieties, TP-COF exhibits blue luminescence. This is due to the fact that triphenylene has a non-localized excitation, but it can channelize the energy *via* the pyrene framework. That is why TP-COF may be utilized to generate energy in a wide region, from UV to visible, with the help of blue luminescence[14] (Figure 10.19).

Figure 10.16 (A) Condensation Reaction between TAPP and tetraphthaldehyde, TBPP, and THAn produce extended COF-366 and COF-66. (B) Image on the left side is the FP TRMC profile of COF-366 (red) and COF-66 (blue) at 25 °C upon irradiation with 355 nm pulse laser at a power of 1.4×10^{16} and 2.1×10^{16} photons cm^{-2}, respectively. Another in the right side is the accumulated number of photoinduced carriers upon 355 nm pulse exposure to COF-366 (red) and COF-66 (blue) sandwiched by ITO and A1 electrodes. Excitation was carried out at the photo density of 9.[88]

Reprinted with permission from S. Wan, F. Gandara, A. Asano, H. Furukawa, A. Saeki, S. K. Dey, L. Liao, M. W. Ambrogio, Y. Y. Botros, X. Duan, S. Seki, F. Stoddart and O. M. Yaghi, *Chem. Mater.*, 2011, **23**, 4094. Copyright 2011, American Chemical Society.

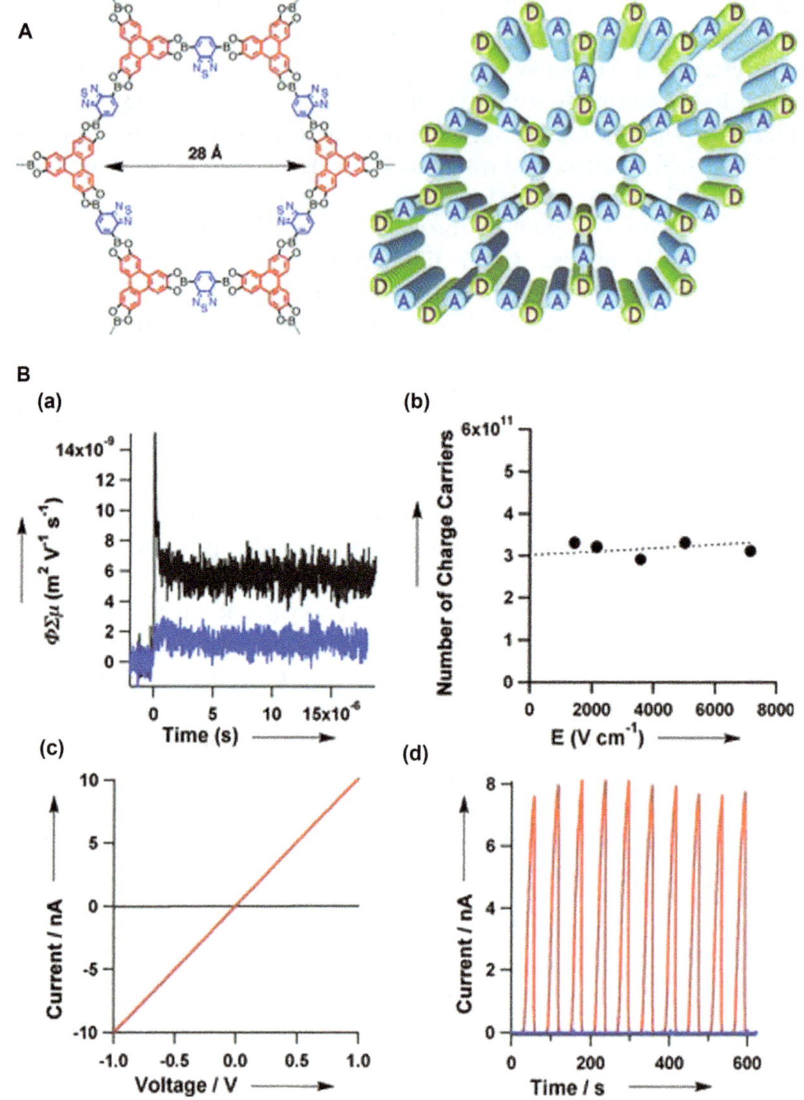

Figure 10.17 (A) Schematic representation of 2D D-A COF with self-sorted and periodic electron donor–accepter ordering and bicontinuous conducting channels (right: structure of one hexagon; left: a 3×3 grid), (B) (a) FP-TRMC profile of the 2D D-A COF in Ar (black) and SF6 atmospheres (blue), (b) Accumulated number of photo-induced charge carriers in the 2D D-A COF upon 355 nm pulse exposure, (c) I–V curve of the 2D D–A COF in the dark (black) and upon irradiation (red) with visible light from a Xenon light source, (d) Photocurrents of the 2D D–A COF (red curve), COF-5 (black), and a simple mixture of D and A at a molar ratio of $3:2$ (blue) upon repeated switching of the light on and off.
Reprinted with permission from X. Feng, L. Chen, Y. Honsho, O. Saengsawang, L. Liu, L. Wang, A. Saeki, S. Irle, S. Seki, Y. Dong and D. Jiang, *Adv. Mater.*, 2012, **24**, 3026. Copyright 2012 WILEY-VCH Verlag GmbH. & Co. KGaA, Weinheim.

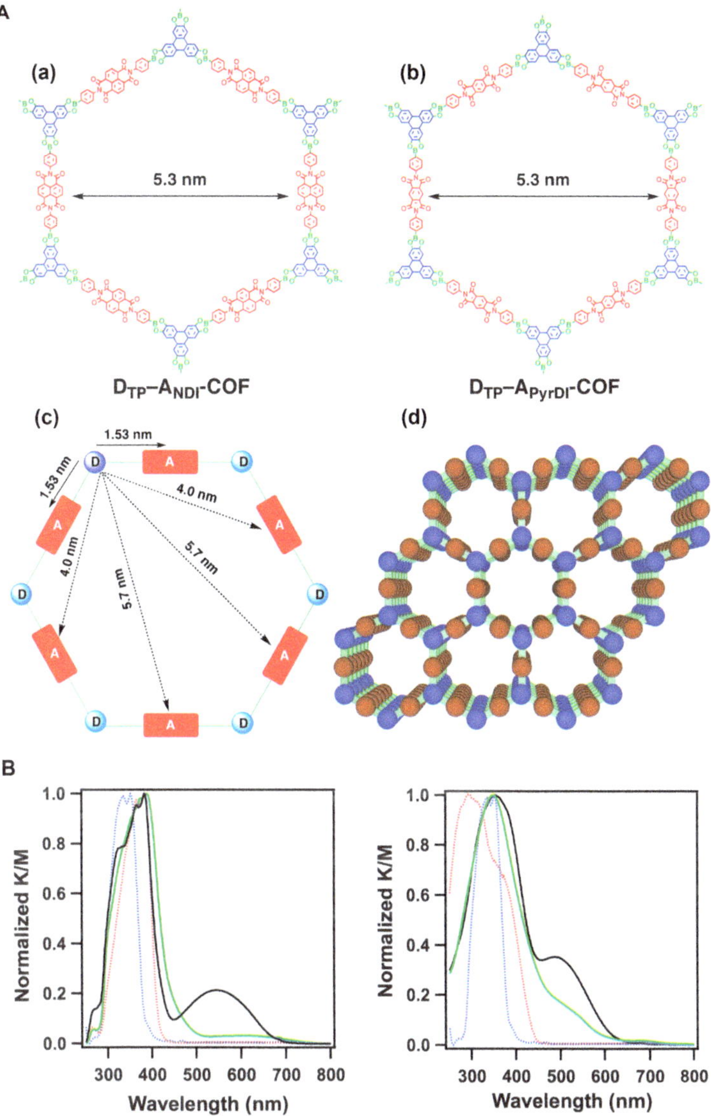

Figure 10.18 (A) Schematic representation of structure of (a) DTP-ANDI-COF and (b) DTP-APyrDI-COF. Dotted lines at the periphery indicate extended structure. (c) Center-to-center distance from a donor to acceptors in the COF. (d) A graphic view of a 3×3 porous framework. (B) Electronic absorption spectra of (left) DTP-ANDI-COF (green) and (right) DTPA-PyrDI-COF, simple mixture of their monomers (black), $TP(OMe)_6$ (dotted blue), and NDI boronate aster (dotted red), PyrDI boronate ester (dotted red). For structure of $TP(OMe)_6$, NDI boronate ester and PyrDI boronate ester.
Reproduced from Ref. 90 with permission from The Royal Society of Chemistry.

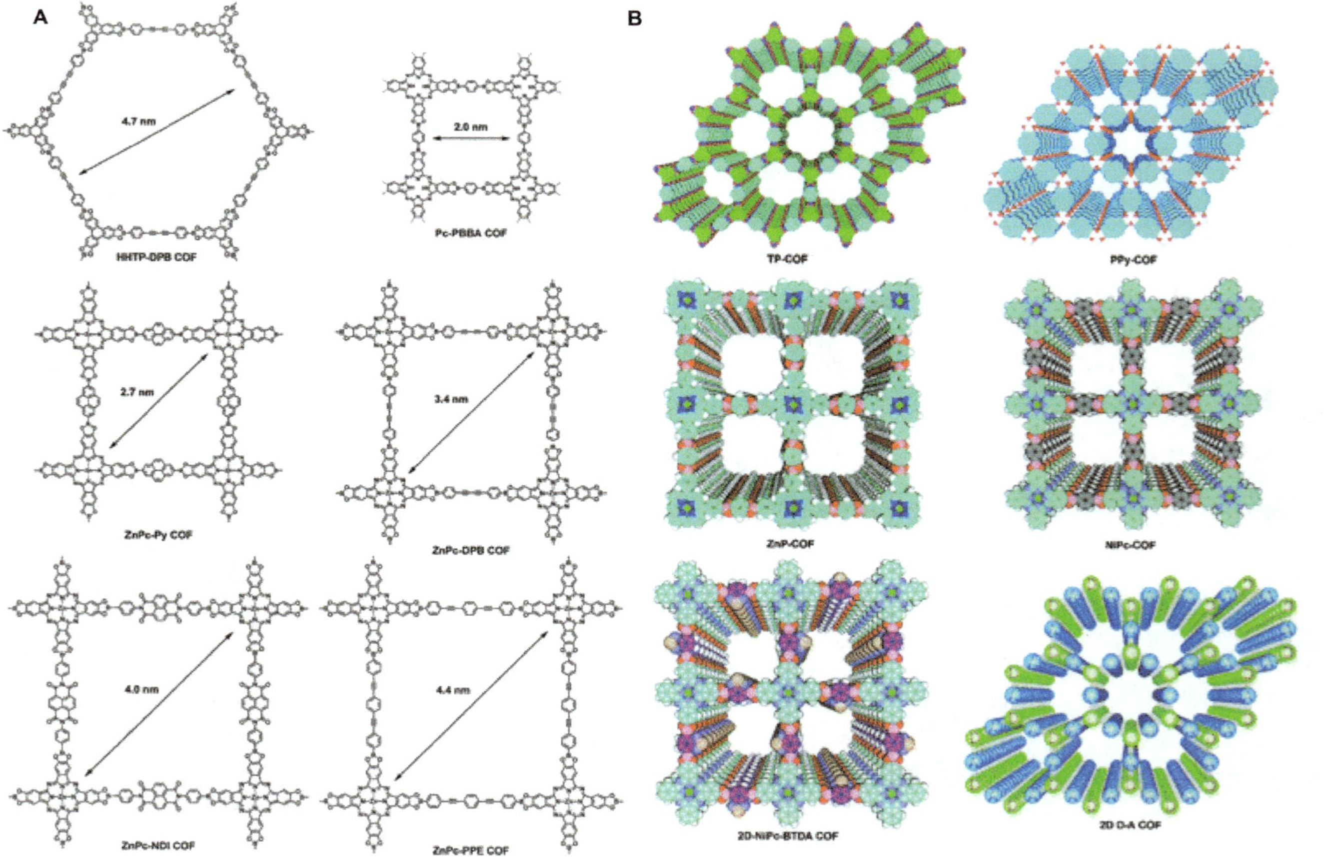

Figure 10.19 (A) Schematic representation of the COFs with the largest pore size (HHTP-DPB COF) and teteagonal phthalocyanine COFs with different pore sizes prepared on graphene. (B) Schematic representation of the AA stacked structure of COFs with preorganized and built-in Π columns.
Reproduced from Ref. 14 with permission from The Royal Society of Chemistry.

References

1. W. Zhou, *Chem. Rec.*, 2010, **10**, 200.
2. S. K. Bhatia and A. L. Myers, *Langmuir*, 2006, **22**, 1688.
3. D. D'Alessandro, B. Smit and J. Long, *Angew. Chem., Int. Ed.*, 2010, **49**, 6058.
4. R. Dawson, A. I. Cooper and D. J. Adams, *Prog. Polym. Sci.*, 2012, **37**, 530.
5. C. D. Wood, B. Tan, A. Trewin, H. Niu, D. Bradshaw, M. J. Rosseinsky, Y. Z. Khimyak, N. L. Campbell, R. Kirk, E. Stöckel and A. I. Cooper, *Chem. Mater.*, 2007, **19**, 2034.
6. J. Y. Lee, C. D. Wood, D. Bradshaw, M. J. Rosseinsky and A. I. Cooper, *Chem. Commun.*, 2006, 2670.
7. J. Germain, J. Hradil, J. M. J. Fréchet and F. Svec, *Chem. Mater.*, 2006, **18**, 4430.
8. M. P. Tsyurupa and V. A. Davankov, *React. Funct. Polym.*, 2002, **53**, 193.
9. N. B. McKeown, P. M. Budd and D. Book, *Macromol. Rapid Commun.*, 2007, **28**, 995.
10. C. D. Wood, B. Tan, A. Trewin, F. Su, M. J. Rosseinsky, D. Bradshaw, Y. Sun, L. Zhou and A. I. Cooper, *Adv. Mater.*, 2008, **20**, 1916.
11. B. S. Ghanem, K. J. Msayib, N. B. McKeown, K. D. M. Harris, Z. Pan, P. M. Budd, A. Butler, J. Selbie, D. Book and A. Walton, *Chem. Commun.*, 2007, 67.
12. B. S. Ghanem, M. Hashem, K. D. M. Harris, K. J. Msayib, M. Xu, P. M. Budd, N. Chaukura, D. Book, S. Tedds, A. Walton and N. B. McKeown, *Macromolecules*, 2010, **43**, 5287.
13. S. S. Han, H. Furukawa, O. M. Yaghi and W. A. Goddard III, *J. Am. Chem. Soc.*, 2008, **130**, 11580.
14. X. Feng, X. Ding and D. Jiang, *Chem. Soc. Rev.*, 2012, **41**, 6010.
15. C. Pei, T. Ben, Y. Cui and S. Qiu, *Adsorption*, 2012, **18**, 375.
16. T. Ben, C. Pei, D. Zhang, J. Xu, F. Deng, X. Jing and S. Qiu, *Energy Environ. Sci.*, 2011, **4**, 3991.
17. T. Ben, Y. Li, L. Zhu, D. Zhang, D. Cao, Z. Xiang, X. Yao and S. Qiu, *Energy Environ. Sci.*, 2012, **5**, 8370.
18. J. Lan, D. Cao, W. Wang, T. Ben and G. Zhu, *J. Phys. Chem. Lett.*, 2010, **1**, 978.
19. B. Lukose, M. Wahiduzzaman, A. Kuc and T. Heine, *J. Phys. Chem. C*, 2012, **116**, 22878.
20. Z. Yang, X. Peng and D. Cao, *J. Phys. Chem. C*, 2013, **117**, 8353.
21. Y. Li, T. Ben, B. Zhang, Y. Fu and S. Qiu, *Sci. Rep.*, 2013, **3**, 2420.
22. J. Germain, F. Svec and J. M. J. Fréchet, *Chem. Mater.*, 2008, **20**, 7069.
23. D. Cazorla-Amorós, J. Alcañiz-Monge and J. Linares-Solano, *Langmuir*, 1996, **12**, 2820.
24. R. Dawson, E. Stöckel, J. R. Holst, D. J. Adams and A. I. Cooper, *Energy Environ. Sci.*, 2011, **4**, 4239.
25. Y. Jin, B. A. Voss, R. D. Noble and W. Zhang, *Angew. Chem. Int. Ed.*, 2010, **49**, 6348.

26. J. A. Mason, K. Sumida, Z. R. Herm, R. Krishna and J. R. Long, *Energy Environ. Sci.*, 2011, **4**, 3030.
27. Y. Luo, S. Zhang, Y. Ma, W. Wang and B. Tan, *Polym. Chem.*, 2013, **4**, 1126.
28. L. M. Robeson, *J. Membr. Sci.*, 1991, **62**, 165.
29. L. M. Robeson, *J. Membr. Sci.*, 2008, **320**, 390.
30. N. B. McKeown and P. M. Budd, *Chem. Soc. Rev.*, 2006, **35**, 675.
31. N. Du, H. B. Park, G. P. Robertson, M. M. Dal-Cin, T. Visser, L. Scoles and M. D. Guiver, *Nat. Mater.*, 2011, **10**, 372.
32. N. Ritter, I. Senkovska, S. Kaskel and J. Weber, *Macromol. Rapid Commun.*, 2011, **32**, 438.
33. N. Ritter, I. Senkovska, S. Kaskel and J. Weber, *Macromolecules*, 2011, **44**, 2025.
34. H. Furukawa and O. M. Yaghi, *J. Am. Chem. Soc.*, 2009, **131**, 8875.
35. H. Frost, T. Düren and R. Q. Snurr, *J. Phys. Chem. B*, 2006, **110**, 9565.
36. R. Rao and J. Jiang, *Energy Environ. Sci.*, 2008, **1**, 139.
37. H. Zhao, Z. Jin, H. Su, J. Zhang, X. Yao, H. Zhao and G. Zhu, *Chem. Commun.*, 2013, **49**, 2780.
38. We. Lu, D. Yuan, D. Zhao, C. I. Schilling, O. Plietzsch, T. Muller, S. Bräse, J. Guenther, J. Blümel, R. Krishna, Z. Li and H.-C. Zhou, *Chem. Mater.*, 2010, **22**, 5964.
39. W. Lu, D. Yuan, J. Sculley, D. Zhao, R. Krishna and H.-C. Zhou, *J. Am. Chem. Soc.*, 2011, **133**, 18126.
40. C. Pei, T. Ben, H. Guo, J. Xu, F. Deng, Z. Xiang, D. Cao and S. Qiu, *Philos. Trans. R. Soc., A*, 2013, **371**, 20120312.
41. N. B. McKeown, S. Hanif, K. Msayib, C. E. Tattershall and P. M. Budd, *Chem. Commun.*, 2002, 2782.
42. P. M. Budd, B. Ghanem, K. Msayib, N. B. McKeown and C. Tattershall, *J. Mater. Chem.*, 2003, **13**, 2721.
43. B. Meunier, *Chem. Rev.*, 1992, **92**, 1411.
44. P. M. Budd, N. B. McKeown and D. Fritsch, *J. Mater. Chem.*, 2005, **15**, 1977.
45. S. Y. Ding, J. Gao, Q. Wang, Y. Zhang, W. G. Song, C. Y. Su and W. Wang, *J. Am. Chem. Soc.*, 2011, **133**, 19816.
46. H. Xu, X. Chen, J. Gao, J. Lin, M. Addicoat, S. Irle and D. Jiang, *Chem. Commun.*, 2014, **50**, 1292.
47. A. Nagai, X. Chen, X. Feng, X. Ding, Z. Guo and D. Jiang, *Angew. Chem., Int. Ed.*, 2013, **52**, 3770.
48. H. J. Mackintosh, P. M. Budd and N. B. McKeown, *J. Mater. Chem.*, 2008, **18**, 573.
49. C. Wang, A. Pettman, J. Basca and J. Xiao, *Angew. Chem., Int. Ed.*, 2010, **49**, 7548.
50. H. Y. Wang, W. G. Wang, G. Si, F. Wang, C. H. Tung and L. Z. Wu, *Langmuir*, 2010, **26**, 9766.
51. Y. Yamamoto, Y. Tamaki, T. Yui, K. Koike and O. Ishitani, *J. Am. Chem. Soc.*, 2010, **132**, 11743.

52. J.-X. Jiang, C. Wang, A. Laybourn, T. Hasell, R. Clowes, Y. Z. Khimyak, J. Xiao, S. J. Higgins, D. J. Adams and A. I. Cooper, *Angew. Chem., Int. Ed.*, 2011, **50**, 1072.

53. H. Furukawa, N. Ko, Y. B. Go, N. Aratani, S. B. Choi, E. Choi, A. O. Yazaydin, R. Q. Snurr, M. O. Keeffe, J. Kim and O. M. Yaghi, *Science*, 2010, **329**, 424.

54. C. Weder, *Chem. Commun.*, 2005, 5378.

55. J. Schmidt, J. Weber, J. D. Epping, M. Antonietti and A. Thomas, *Adv. Mater.*, 2009, **21**, 702.

56. J.-X. Jiang, Y. Li, X. Wu, J. Xiao, D. J. Adams and A. I. Cooper, *Macromolecules*, 2013, **46**, 8779.

57. C. E. Chan-Thaw, A. Villa, P. Katekomol, D. Su, A. Thomas and L. Prati, *Nano Lett.*, 2010, **10**, 537.

58. J. Roeser, K. Kailasam and A. Thomas, *ChemSusChem*, 2012, **5**, 1793.

59. P. Thomas, C. Pei, B. Roy, S. Ghosh, S. Das, A. Banerjee, T. Ben, S. Qiu and S. Roy, *J. Mater. Chem. A*, 2015, **3**, 1431.

60. L. Chen, Y. Honsho, S. Seki and D. L. Jiang, *J. Am. Chem. Soc.*, 2010, **132**, 6742.

61. J. Boucle, P. Ravirajan and J. Nelson, *J. Mater. Chem.*, 2007, **17**, 3141.

62. A. I. Cooper, *Adv. Mater.*, 2009, **21**, 1291.

63. Y. Yuan, H. Ren, F. Sun, X. Jing, K. Cai, X. Zhao, Y. Wang, Y. Wei and G. Zhu, *J. Phys. Chem. C*, 2012, **116**, 26431.

64. Y. Yuan, F. Sun, F. Zhang, H. Ren, M. Guo, K. Cai, X. Jing, X. Gao and G. Zhu, *Adv. Mater.*, 2013, **25**, 6619.

65. T. Ben, K. Shi, Y. Cui, C. Pei, Y. Zuo, H. Guo, D. Zhang, J. Xu, F. Deng, Z. Tian and S. Qiu, *J. Mater. Chem.*, 2011, **21**, 18208.

66. C. J. Brabec, N. S. Sariciftci and J. C. Hummelen, *Adv. Funct. Mater.*, 2001, **11**, 15.

67. N. A. Rakow, M. S Wendland, J. E. Trend, R. J. Poirier, D. M. Paolucci, S. P. Maki, C. S. Lyons and M. J. Swierczek, *Langmuir*, 2010, **26**, 3767.

68. N. B. McKeown and P. M. Budd, *Macromolecules*, 2010, **43**, 5163.

69. S. Dalapati, S. Jin, J. Gao, Y. Xu, A. Nagai and D. Jiang, *J. Am. Chem. Soc.*, 2013, **135**, 17310.

70. *An Introduction to Molecular Electronics*, ed. M. Petty, M. Bryce, D. Bloor, Oxford University Press, New York, 1995.

71. *Photochemical Conversion and Storage of Solar Energy*, ed. J. S. Connolly, Academic Press, New York, 1981.

72. P. Heremans, D. Cheyns and B. P. Rand, *Acc. Chem. Res.*, 2009, **42**, 1740.

73. (a) A. P. Côté, A. Benin, N. Ockwig, M. O'Keeffe, A. Matzger and O. M. Yaghi, *Science*, 2005, **310**, 1166; (b) A. P. Côte, H. M. El-Kaderi, H. Furukawa, J. R. Hunt and O. M. Yaghi, *J. Am. Chem. Soc.*, 2007, **129**, 12914.

74. R. W. Tilford, W. R. Gemmill, H. C. zur Loye and J. J. Lavigne, *Chem. Mater.*, 2006, **18**, 5296.

75. (a) A. Nagai, Z. Guo, X. Feng, S. Jin, X. Chen, X. Ding and D. Jiang, *Nat. Commun.*, 2011, **2**, 536; (b) S. Wan, J. Guo, J. Kim, H. Ihee and D. Jiang, *Angew. Chem.*, 2008, **120**, 8958; (c) S. Wan, J. Guo, J. Kim, H. Ihee and D. Jiang, *Angew. Chem.*, 2009, **121**, 5547; (d) X. Ding, J. Guo, X. Feng, Y. Honsho, J. D. Guo, S. Seki, P. Maitarad, A. Saeki, S. Nagase and D. Jiang, *Angew. Chem.*, 2011, **123**, 1325; (e) X. Ding, L. Chen, Y. Honso, X. Feng, O. Saengsawang, J. D. Guo, A. Saeki, S. Seki, S. Irle, S. Nagase, V. Parasuk and D. Jiang, *J. Am. Chem. Soc.*, 2011, **133**, 14510; (f) X. Feng, L. Chen, Y. P. Dong and D. Jiang, *Chem. Commun.*, 2011, **47**, 1979; (g) X. Feng, L. Liu, Y. Honsho, A. Saeki, S. Seki, S. Irle, Y. P. Dong, A. Nagai and D. Jiang, *Angew. Chem.*, 2012, **124**, 2672; (h) X. Feng, L. Chen, Y. Honsho, O. Saengsawang, L. Liu, L. Wang, A. Saeki, S. Irle, S. Seki, Y. P. Dong and D. Jiang, *Adv. Mater.*, 2012, **24**, 3026; (i) X. Ding, X. Feng, A. Saeki, S. Seki, A. Nagai and D. Jiang, *Chem. Commun.*, 2012, **48**, 8952; (k) X. Feng, Y. P. Dong and D. Jiang, *CrystEngComm*, 2013, **15**, 1508; (l) X. Chen, M. Addicoat, S. Irle, A. Nagai and D. Jiang, *J. Am. Chem. Soc.*, 2013, **135**, 546.
76. S. Patwardhan, A. A. Kocherzhenko, F. C. Grozema and L. D. A. Siebbeles, *J. Phys. Chem. C*, 2011, **115**, 11768.
77. (a) J. W. Colson, A. R. Woll, A. Mukherjee, M. P. Levendorf, E. L. Spitler, V. B. Shields, M. G. Spencer, J. Park and W. R. Dichtel, *Science*, 2011, **332**, 228; (b) E. L. Spitler, J. W. Colson, F. J. Uribe-Romo, A. R. Woll, M. R. Glovino, A. Saldivar and W. R. Dichtel, *Angew. Chem.*, 2012, **124**, 2677.
78. B. Lukose, A. Kuc and T. Heine, *Chem. – Eur. J.*, 2011, **17**, 2388.
79. P. Kuhn, M. Antonietti and A. Thomas, *Angew. Chem.*, 2008, **120**, 3499.
80. N. L. Campbell, R. Clowes, L. K. Ritchie and A. I. Cooper, *Chem. Mater.*, 2009, **21**, 204.
81. I. Berlanga, M. L. Ruiz-Gonzàlez, J. M. Gonzàlez-Calbet, J. L. G. Fierro, R. Mas-Ballest and F. Zamora, *Small*, 2011, **7**, 1207.
82. S. Jin, X. Ding, X. Feng, M. Supur, K. Furukawa, S. Takahashi, M. Addicoat, Md. E. El-Khouly, T. Nakamura, S. Irle, S. Fukuzumi, A. Nagai and D. Jiang, *Angew. Chem., Int. Ed.*, 2013, **52**, 2017.
83. E. L. Spitler and W. R. Dichtel, *Nat. Chem.*, 2010, **2**, 672.
84. X. S. Ding, J. Guo, X. Feng, Y. Honsho, J. D. Guo, S. Seki, P. Maitarad, A. Saeki, S. Nagase and D. Jiang, *Angew. Chem., Int. Ed.*, 2011, **50**, 1289.
85. E. L. Spitler, J. W. Colson, F. J. Uribe-Romo, A. R. Woll, M. R. Giovino, A. Saldivar and W. R. Dichtel, *Angew. Chem., Int. Ed.*, 2012, **51**, 2623.
86. X. Ding, L. Chen, Y. Honsho, X. Feng, O. Saengsawang, J. Guo, A. Saeki, S. Seki, S. Irle, S. Nagase, V. Parasuk and D. Jiang, *J. Am. Chem. Soc.*, 2011, **133**, 14510.
87. S. Wan, F. Gandara, A. Asano, H. Furukawa, A. Saeki, S. K. Dey, L. Liao, M. W. Ambrogio, Y. Y. Botros, X. F. Duan, S. Seki, J. F. Stoddart and O. M. Yaghi, *Chem. Mater.*, 2011, **23**, 4094.

88. S. Wan, F. Gandara, A. Asano, H. Furukawa, A. Saeki, S. K. Dey, L. Liao, M. W. Ambrogio, Y. Y. Botros, X. Duan, S. Seki, F. Stoddart and O. M. Yaghi, *Chem. Mater.*, 2011, **23**, 4094.
89. X. Feng, L. Chen, Y. Honsho, O. Saengsawang, L. Liu, L. Wang, A. Saeki, S. Irle, S. Seki, Y. Dong and D. Jiang, *Adv. Mater.*, 2012, **24**, 3026.
90. S. Jin, K. Furukawa, M. Addicoat, L. Chen, S. Takahashi, S. Irle, T. Nakamura and D. Jiang, *Chem. Sci.*, 2013, **4**, 4505.

Conclusion

11.1 Conclusion

Inclusion of porosity in polymers has opened up a giant arena in research to explore owing to the resulting versatility and fertility. Riding on the porous nature and high surface areas, these polymers have developed tremendously and have been demonstrated to have advanced applicability.

This book has emphasized the importance of the structure–function relationship in the design of the synthetic strategy, covering all the principle sub-families of porous polymers. An emphasis has been placed on assembling the synthetic routes for porous polymers and on building a common thread among them to demonstrate the achievements of the pre-determined targeted porous networks with proper design of the synthetic procedure. The employed synthetic routes themselves pave the way for the formation of a targeted framework with desired properties owing to its individual components. Theoretical studies conform to the experimental data and provide a view of how to plan the design of these network structures. Pre- or post-synthetic modification further allows pre-designed properties of the modified networks to be obtained. With controlled functionalization, attributes such as high surface area, conjugation of the π electrons, catalytically active moieties *etc.* can be exploited in various fields of gas sorption, optoelectronics and catalysis. By tuning the pore size and pore constitution it is possible to control properties like selective permeation and molecular separation.

Understanding of the structural motifs with targeted functional features at the molecular scale and the correlation with their functions can enable one to predict more sophisticated builds and manipulate useful attributes, as well as allowing researchers to look into the possibility of designing new materials.

Monographs in Supramolecular Chemistry No. 17
Porous Polymers: Design, Synthesis and Applications
By Shilun Qiu and Teng Ben
© The Royal Society of Chemistry 2016
Published by the Royal Society of Chemistry, www.rsc.org

In this book, we systematically introduced five typical porous polymers: hyper-crosslinked polymers (HCPs), polymers of intrinsic microporosity (PIMs), covalent organic frameworks (COFs), conjugated microporous polymers (CMPs) and porous aromatic frameworks (PAFs). These five porous polymers represent the historical development and milestones with regards to porous polymers (Table 11.1). HCPs are usually cheap and were the first developed porous polymers with an apparent surface area of 1000 $m^2\ g^{-1}$. They represent the first time that polymer materials with permanent porosity were combined. In recent times, HCPs have been widely used as organic sorbents and packing materials in high-performance liquid chromatography (HPLC) and ion size-exclusion chromatography. Unlike other porous polymers, PIMs show interpenetrated pores by cross-linked frameworks and are linear polymers with high intrinsic intermolecular free volume (*e.g.*, surface area of PIM-1 is 760 $m^2\ g^{-1}$). This unique structure endows PIMs with the possibility of solubility in normal solvents and they can be easily processed *via* solution methods. By reversible condensation reactions involving boroxine ring formation, Yaghi-type COFs show high surface areas with crystalline frameworks. This gives researchers the chance to explore the structure–function correlation of porous polymers more clearly. Prior to the development of CMPs, no porous polymers combined extended π-conjugation along the cross-linked framework with porosity. CMPs expanded greatly the potential applications of porous polymer so they may be utilized not only as sorbents or carriers of catalysts but also as photovoltaic materials. PAF-1 exhibits an ultra-high surface area of 5600 $m^2\ g^{-1}$ with high physicochemical stability. It proved that high surface area can be achieved by an amorphous porous polymer for the first time.

Porous polymers are one of the most active cross-subject areas of chemistry, physics and materials, and have developed rapidly in recent times. We anticipate that, in the next decades, the research within this field will focus

Table 11.1 Milestones in the development of porous polymers.

Milestones	Synthesis method	Surface area/$m^2\ g^{-1}$	Pore size/nm	Pore volume/ $cm^3\ g^{-1}$	Crystallinity
HCP-1	Friedel–Crafts reaction	1000	—	—	Disorder
PIM-1	Dioxane ring formation	760	—	—	Disorder
COF-1	Condensation reactions, boroxine ring formation	711	0.8	0.34	Crystalline
CMP-1	Sonogashira–Hagihara cross-coupling	834	1.1	0.47	Disorder
PAF-1	Yamamoto-type Ullmann cross-coupling	5640	1.4	2.43	Disorder

more on the applications of porous polymers. At the same time, deeper understanding of the fundamental theories of porous polymers is also needed, for example:

1. **Nature of the amorphous porous polymer**: Until now, real structures of many important porous polymers have not been very clearly understood, though efforts from the simulation of functions from the macroscopic view have been reported. Crystalline COFs and covalent triazine frameworks (CTFs) are good examples to understand the structure–function correlation, although much research is still required for the amorphous porous polymer. Accurate description of the interpenetration, defects, topology *etc.* of porous polymers is needed.
2. **Precisely targeted synthesis**: Polymerization is sophisticated, thus not all reactions aim at the "target". Fast, effective, consistent and neat synthesis methods and techniques are required. This will benefit the preparation of previously theoretically designed, "undiscovered" porous polymers.
3. **Processing technique**: It is very difficult to process porous polymers because most of the amorphous porous polymers are in the powder form, while crystalline porous polymers are nanocrystals. Only PIMs and soluble CMPs are solution-processable. A breakthrough with regards to the processing technique will expand the applications of porous polymers.
4. **Environmental and energy applications**: Porous polymers may be the new platforms for electrodes and proton exchange membranes in new batteries, and carbonized porous polymers are readily available to prepare novel supercapacitors.

Subject Index